T0136992

Advances in Intelligent Systems and Computing

Volume 873

Series editor

Janusz Kacprzyk, Polish Academy of Sciences, Warsaw, Poland
e-mail: kacprzyk@ibspan.waw.pl

The series "Advances in Intelligent Systems and Computing" contains publications on theory, applications, and design methods of Intelligent Systems and Intelligent Computing. Virtually all disciplines such as engineering, natural sciences, computer and information science, ICT, economics, business, e-commerce, environment, healthcare, life science are covered. The list of topics spans all the areas of modern intelligent systems and computing such as: computational intelligence, soft computing including neural networks, fuzzy systems, evolutionary computing and the fusion of these paradigms, social intelligence, ambient intelligence, computational neuroscience, artificial life, virtual worlds and society, cognitive science and systems, Perception and Vision, DNA and immune based systems, self-organizing and adaptive systems, e-Learning and teaching, human-centered and human-centric computing, recommender systems, intelligent control, robotics and mechatronics including human-machine teaming, knowledge-based paradigms, learning paradigms, machine ethics, intelligent data analysis, knowledge management, intelligent agents, intelligent decision making and support, intelligent network security, trust management, interactive entertainment, Web intelligence and multimedia.

The publications within "Advances in Intelligent Systems and Computing" are primarily proceedings of important conferences, symposia and congresses. They cover significant recent developments in the field, both of a foundational and applicable character. An important characteristic feature of the series is the short publication time and world-wide distribution. This permits a rapid and broad dissemination of research results.

More information about this series at http://www.springer.com/series/11156

Mohammad S. Obaidat · Tuncer Ören
Floriano De Rango
Editors

Simulation and Modeling Methodologies, Technologies and Applications

7th International Conference, SIMULTECH 2017,
Madrid, Spain, July 26–28, 2017,
Revised Selected Papers

 Springer

Editors
Mohammad S. Obaidat
King Abdullah II School of Information
  Technology
The University of Jordan
Amman, Jordan

and

Nazarbayev University
Astana, Kazakhstan

and

University of Science and Technology
  Beijing
Beijing, China

Tuncer Ören
School of Electrical Engineering and
  Computer Science
University of Ottawa
Ottawa, ON, Canada

Floriano De Rango
Department of Informatics, Modeling,
  Electronics and System Engineering
University of Calabria
Rende, Italy

ISSN 2194-5357   ISSN 2194-5365   (electronic)
Advances in Intelligent Systems and Computing
ISBN 978-3-030-01469-8   ISBN 978-3-030-01470-4   (eBook)
https://doi.org/10.1007/978-3-030-01470-4

Library of Congress Control Number: 2018955974

This Springer imprint is published by the registered company Springer Nature Switzerland AG
The registered company address is: Gewerbestrasse 11, 6330 Cham, Switzerland

Preface

The present book includes extended and revised versions of a set of selected papers from the 7th International Conference on Simulation and Modeling Methodologies, Technologies, and Applications (SIMULTECH 2017), held in Madrid, Spain, in the period of July 26–28, 2017.

SIMULTECH 2017 received 88 paper submissions from 31 countries, of which 20% were included in this book. The papers were selected by the event chairs and their selection is based on a number of criteria that includes the reviews and suggested comments provided by the program committee members, the session chairs' assessments, and also the program chairs' global view of all papers included in the technical program. The authors of selected papers were then invited to submit a revised and extended version of their papers having at least 30% new material.

The purpose of the 7th International Conference on Simulation and Modeling Methodologies, Technologies, and Applications (SIMULTECH 2017) was to bring together researchers, engineers, applied mathematicians, and practitioners interested in the advances and applications in the field of system simulation. Four simultaneous tracks were held, covering on one side domain independent methodologies and technologies and on the other side practical work developed in specific application areas. The specific topics listed under each of these tracks highlight the interest of this conference in aspects related to computing, including conceptual modeling, agent-based modeling and simulation, interoperability, ontologies, knowledge-based decision support, Petri nets, business process modeling and simulation, among others.

The papers selected to be included in this book contribute to the understanding of relevant trends of current research on modeling and simulation methodologies and applications. The field of modeling and simulation is advancing rapidly and affecting many fields of knowledge with an accelerating pace. This volume of carefully selected and edited articles from our conference in Madrid includes several state-of-the-art articles representative of the recent advances of Modeling and Simulation, such as cyber-physical systems, big data analytics, artificial intelligence, computer and network systems as well as UAV systems.

We would like to thank all the authors for their contributions as well as to the reviewers who have helped ensuring the quality of this publication. We also thank the staff of INTICCC and Springer for their good efforts and cooperation.

July 2017

Mohammad S. Obaidat
Fellow of IEEE, Fellow of SCS, and Past President of the
Society for Modeling and Simulation International, SCS

Tuncer Ören
Floriano De Rango

Organization

Conference Chair

Mohammad S. Obaidat   King Abdullah II School of Information
Technology, University of Jordan, Jordan,
Nazarbayev University, Kazakhstan, and
University of Science and Technology
Beijing, China

Program Chairs

Floriano De Rango   University of Calabria, Italy
Tuncer Ören (honorary)   University of Ottawa, Canada

Program Committee

Magdiel Ablan   Universidad de Los Andes, Venezuela
Erika Ábrahám   RWTH Aachen University, Germany
Nael Abu-Ghazaleh   University of California, Riverside, USA
Lyuba Alboul   Sheffield Hallam University, UK
Manuel Alfonseca   Universidad Autonoma de Madrid, Spain
Gianfranco Balbo   University of Torino, Italy
Simonetta Balsamo   University of Venezia Ca' Foscari, Italy
M. Gisela Bardossy   University of Baltimore, USA
Isaac Barjis   City University of New York, USA
Jordi Mongay Batalla   National Institute of Telecommunications, Poland
Mohamed Bettaz   Philadelphia University, Jordan
Wolfgang Borutzky   Bonn-Rhein-Sieg University of Applied
Sciences, Germany

Christos Bouras	University of Patras and CTI&P Diophantus, Greece
Robin T. Bye	Norwegian University of Science and Technology, Norway
Christian Callegari	RaSS National Laboratory, CNIT, Italy
Srinivas Chakravarthy	Kettering University, USA
Franco Cicirelli	Università della Calabria, Italy
Tanja Clees	Fraunhofer Institute for Algorithms and Scientific Computing (SCAI), Germany
Andrea D'Ambrogio	Università di Roma Tor Vergata, Italy
Paul Davis	RAND and Pardee RAND Graduate School, USA
Anatoli Djanatliev	Friedrich-Alexander-Universität Erlangen-Nürnberg, Germany
Atakan Dogan	Anadolu University, Turkey
Julie Dugdale	Laboratoire d'Informatique de Grenoble, France
Sabeur Elkosantini	King Saud University, Saudi Arabia
Zuhal Erden	ATILIM University, Turkey
Denis Filatov	Institute of Physics of the Earth, Russian Academy of Sciences, Russian Federation
Jason Friedman	Tel Aviv University, Israel
Marco Furini	Università di Modena e Reggio Emilia, Italy
José Manuel Galán	Universidad de Burgos, Spain
Charlotte Gerritsen	Vrije Universiteit Amsterdam, Netherlands
Alexandra Grancharova	University of Chemical Technology and Metallurgy, Bulgaria
Francisco Grimaldo	Universitat de València, Spain
Mykola Gusti	International Institute for Applied System Analysis, Austria
Sigurdur Hafstein	University of Iceland, Iceland
Maamar El Amine Hamri	LSIS, France
Samer Hassan	Universidad Complutense de Madrid, Spain
Cathal Heavey	University of Limerick, Ireland
Tsan-Sheng Hsu	Institute of Information Science, Academia Sinica, Taiwan
Xiaolin Hu	Georgia State University, USA
Eric Innocenti	IUT DE CORSE—University of Corsica, France
Nobuaki Ishii	Kanagawa University, Japan
Mura Ivan	University EAN, Colombia
Segismundo Samuel Izquierdo	University of Valladolid, Spain
Syed Waqar ul Qounain Jaffry	University of the Punjab, Pakistan
Mats Jägstam	University of Jönköping, Sweden
Anniken Karlsen	NTNU in Ålesund, Norway

Peter Kemper | College of William and Mary, USA
Etienne Kerre | Ghent University, Belgium
William Knottenbelt | Imperial College London, UK
Juš Kocijan | Jozef Stefan Institute, Slovenia
Petia Koprinkova-Hristova | Institute of Information and Communication Technologies, Bulgaria
Vladik Kreinovich | University of Texas at El Paso, USA
Claudia Krull | Otto von Guericke University Magdeburg, Germany
Mike Lees | University of Amsterdam, Netherlands
Alberto Leva | Politecnico di Milano, Italy
Fengyuan Li | Freescale Semiconductor, USA
Richard Lipka | University of West Bohemia, Czech Republic
Antonio Mendes Lopes | University of Porto, Portugal
Maria Celia Santos Lopes | COPPE/UFRJ, Brazil
Johannes Lüthi | FH Kufstein Tirol, Austria
José Tenreiro Machado | Institute of Engineering, Polytechnic of Porto, Portugal
Emilio Jiménez Macías | Universidad de La Rioja, Spain
Maciej Malawski | AGH University of Science and Technology, Poland
Carla Martin-Villalba | UNED, Spain
Moreno Marzolla | University of Bologna, Italy
Geraldo Robson Mateus | DCC/UFMG, Brazil
Radek Matušu | Tomas Bata University in Zlin, Czech Republic
Roger McHaney | Kansas State University, USA
Nuno Melão | Instituto Politécnico de Viseu, Escola Superior de Tecnologia e Gestão de Viseu, Portugal
Adel Mhamdi | RWTH Aachen University, Germany
Bozena Mielczarek | Wroclaw University of Science and Technology, Poland
Michael Möhring | University of Koblenz and Landau, Germany
Roberto Montemanni | IDSIA—Dalle Molle Institute for Artificial Intelligence (USI-SUPSI), Switzerland
Jairo R. Montoya-Torres | Universidad de La Sabana, Colombia
Bertie Müller | University of South Wales, UK
Nazmun Nahar | University of Jyvaskyla and University of Tartu, Finland
Angela Nebot | Universitat Politècnica de Catalunya, Spain
Guyh Dituba Ngoma | Université du Québec en Abitibi-Témiscamingue, Canada
Lialia Nikitina | Fraunhofer Institute for Algorithms and Scientific Computing (SCAI), Germany
Michael J. North | Argonne National Laboratory, USA
Paulo Novais | Universidade do Minho, Portugal

James J. Nutaro   Oak Ridge National Laboratory, USA
Paulo Moura Oliveira   Universidade de Tras-os-Montes e Alto Douro,
                                     Portugal
Feng Pan   Liaoning Normal University, China
Krzysztof Pawlikowski   University of Canterbury, New Zealand
L. Felipe Perrone   Bucknell University, USA
Alexandre Petrenko   Centre de Recherche Informatique de Montreal,
                                     Canada
Régis Plateaux   SUPMECA, France
Tomas Potuzak   University of West Bohemia, Czech Republic
Francesco Quaglia   Sapienza Università di Roma, Italy
Manuel Resinas   Universidad de Sevilla, Spain
Jerzy Respondek   Silesian University of Technology, Poland
M. R. Riazi   Kuwait University, Kuwait
José Risco-Martín   Universidad Complutense de Madrid, Spain
Ella E. Roubtsova   Open University of the Netherlands, Netherlands
Willem Hermanus le Roux   CSIR, South Africa
Jaroslav Rozman   Brno University of Technology, Czech Republic
Katarzyna Rycerz   Institute of Computer Science, AGH, Krakow,
                                     Poland, Poland
Cristina Montañola Sales   Universitat Politècnica de Catalunya, Spain
Janos Sallai   Vanderbilt University, USA
Paulo Salvador   Instituto de Telecomunicações, DETI, University
                                     of Aveiro, Portugal
Antonella Santone   University of Molise, Italy
María Teresa Parra Santos   University of Valladolid, Spain
Jean-François Santucci   SPE UMR CNRS 6134, University of Corsica,
                                     France
Massimo La Scala   Politecnico di Bari, Italy
Philippe Serré   Supméca, France
Clifford A. Shaffer   Virginia Tech, USA
Flavio S. Correa Da Silva   University of Sao Paulo, Brazil
Xiao Song   Beihang University, China
Yuri Sotskov   United Institute of Informatics Problems of the
                                     National Academy of Belarus, UIIP, Minsk,
                                     Belarus
James C. Spall   Johns Hopkins University, USA
Giovanni Stea   University of Pisa, Italy
Mu-Chun Su   National Central University, Taiwan
Nary Subramanian   University of Texas at Tyler, USA
Halina Tarasiuk   Warsaw University of Technology, Poland
Pietro Terna   Università di Torino, Italy
Mamadou K. Traoré   Université Clermont Auvergne, France
Klaus G. Troitzsch   University of Koblenz and Landau, Koblenz
                                     Campus, Germany

Giuseppe A. Trunfio	University of Sassari, Italy
Zhiying Tu	Harbin Institute of Technology, China
Bruno Tuffin	Inria Rennes Bretagne Atlantique, France
Alfonso Urquia	Universidad Nacional de Educación a Distancia, Spain
Svetlana Vasileva-Boyadzhieva	Bulgarian Modeling and Simulation Association (BULSIM), Bulgaria
Maria Joao Viamonte	Instituto Superior de Engenharia do Porto, Portugal
Manuel Villen-Altamirano	Universidad de Malaga, Spain
Friederike Wall	Alpen-Adria-Universität Klagenfurt, Austria
Frank Werner	Otto von Guericke Universität Magdeburg, Germany
Philip A. Wilsey	University of Cincinnati, USA
Kuan Yew Wong	Universiti Teknologi Malaysia, Malaysia
Yiping Yao	National University of Defense Technology, China
František Zboril	Faculty of Information Technology, Czech Republic
Durk Jouke van der Zee	University of Groningen, Netherlands
Lin Zhang	Beihang University, China
Suiping Zhou	Middlesex University, UK

Invited Speakers

Miquel Àngel Piera Eroles	Universitat Autònoma de Barcelona (UAB), Spain
Franco Davoli	University of Genoa, Italy
Jerry Couretas	Booz Allen Hamilton, USA

Contents

Complex Systems Modeling and Simulation

Application Domains

Simulation Tools and Platforms

Simulation of Surface Plasmon Waves Based on Kretschmann Configuration Using the Finite Element Method

Tanaporn Leelawattananon[1(✉)] and Suphamit Chittayasothorn[2]

[1] Department of Physics, Faculty of Sciences, King Mongkut's Institute of Technology Ladkrabang, Bangkok 10520, Thailand
tanaporn.le@kmitl.ac.th
[2] Department of Computer Engineering, Faculty of Engineering,
King Mongkut's Institute of Technology Ladkrabang, Bangkok 10520, Thailand
suphamit.ch@kmitl.ac.th

Abstract. This paper presents the simulation of optically activated surface plasmon waves based on Kretschmann configuration by using prism. Simulated electric fields of the surface plasmon wave which appears at the interface between the metal thin film and dielectric layer are observed. The occurences of surface plasmon wave can be applied to biomolecular sensing and high speed data communications at the THz level. The simulations employ the finite element method (FEM). The light source is the 632.5 nm red laser which is economical and easy to obtained commercially. Two simulation models are conducted. The first simulation model employs copper thin film on the prism and air as the dielectric layer. This one is intended to find the most suitable copper thin film thickness to produce surface plasmon waves. Copper thin film is used because it is a noble metal which is less expensive than gold but has better conductivity than gold. The second simulation model employs silver, another noble metal which is also less expensive than gold. Silver thin film on prism together with magnesium chloride solution as dielectric layer are simulated. Concentrations of the magnesium chloride solution are varied to find the one which produces good surface plasmon wave pattern. Thus suitable to be used as sensors for biomoleculars such as DNAs.

Keywords: Surface plasmon wave · Kretschmann configuration
Finite Element method · Copper thin film · Silver thin film

1 Introduction

During the past decades, the need to verify and process data by using optics have been growing and developing rapidly. Optical sensors which are able to detect bio-molecular objects such as DNA protein are widely available. They have both sensitivity and size advantages over non-optical sensors. These optical sensors use the Surface Plasmon Resonance (SPR) principle which works by the optical excitation at the interface between the metal thin film layer and dielectric layer (or the test sample layer). The most

© Springer Nature Switzerland AG 2019
M. S. Obaidat et al. (Eds.): SIMULTECH 2017, AISC 873, pp. 3–22, 2019.
https://doi.org/10.1007/978-3-030-01470-4_1

widely used noble metals for the metal layer is gold. Gold has less oxidation and is very resistant to atmospheric contaminants. However, gold is very expensive. It has been recently found that copper has better conductivity than gold. It is also cheaper. The main disadvantage of copper is the ease of having oxidation. However, copper is better in term of diffusion. Copper does not diffuse into the silicon substrate when gold does. Copper is therefore used in the standard silicon manufacturing process such as CMOS technology [1]. Furthermore, there is a research result which reports that copper is an excellent plasmonic material [2]. Also, according to a research work [3], the application of Graphene layer over copper layer can significantly reduce the oxidation; thus the copper layer is slowly deteriorated and improves plasmonic characteristic.

Apart from copper, Silver is another noble metal which is popularly deployed since silver yields better characteristics of surface plasmon resonance than copper and less expensive than gold. In the case of silver, there are applications of surface plasmon wave for detecting concentrations of glucose solutions (glucose sensor) [4, 5]. A research work on glucose detections [6] implements Kretschmann-based plasmonic sensor and fabricates metal layer of sensor by using 50 mm silver thin films adhered by silver nanorod of different sizes. The dielectric layer is glucose solution with different concentrations and the light source is 1150 nm. The findings from this work is that the resonance angle increases when the concentration of the glucose is higher. Other SPR sensors applications include drug discovery [7, 8], DNA sensors [9, 10], and researches which are related to surface analytical techniques such as surface enhanced Raman spectroscopy [11, 12].

This research project comprises two optically activated Kretschmann configuration. The light source is a laser with the 632.5 nm wavelength. This is an inexpensive light source and widely available commercially. In our first simulation model, copper thin film is employed and air is used as the dielectric layer. The objective is to find the suitable thickness of the copper thin film for producing applicable surface plasmon waves, in terms of clearness, consistency, and electric field magnitude. The second simulation model is to find proper concentration of magnesium chloride solution dielectric when deploying with silver thin film. This solution helps the metal thin film improves sensitivity of biomolecular detection (such as DNA detection) since the Mg^{2+} ion assists the binding between metal surface and DNA Molecules. There are researches on nanopore sensor experiment which also employs magnesium chloride solution as electrolyte as salt solution to control DNA mobility [13]. Moreover, magnesium is found to have influential roles in enzymology, cell membrane, and muscle cell physiology [14] as well.

2   Theory and Principles

2.1   Surface Plasmon Resonance (SPR) Phenomenon

Surface plasmon waves (SPs waves) or surface plasmon polaritons (SPPs) are generated at the surface between metal-dielectric interfaces when excited by the incoming light with an appropriate frequency. The excitation of surface plasmon waves can be done by having light beams contact prisms and the phase of the light at the metal-dielectric interface is matched to the phase of the surface plasmon waves. In the early years of the research in this area Otto [15] and Kretschmann [16] developed an experimental optical excitation which created SPs waves by using a prism and coated a metal thin film on the surface of the prism. Attenuated total reflection (ATR) is a technique to observe the plasmons. The reflected light intensity are measured by changing the incident angles of the incoming light to various degrees. At a certain angle which is referred as the "resonance angle", the reflected ATRs from the prism signifies the light absorption by the electrons in the metal and their resonance which in turn creates the surface plasmon wave at the metal–dielectric interface. Apart from this, there are also researches which use the grating [17] and optical waveguides [18] for the optical excitations of surface plasmons.

The Kretschmann method employs a detecting microscope which moves to different positions to give different angles as shown in Fig. 1. When the incoming light travels from the medium which has higher refractive index to the medium which has lesser refractive index, and the light impacts the interface between the two medium with the angle greater than the critical angle, the light will be totally reflected. This phenomenon is called Total Internal Reflection (TIR). The TIR creates a kind of electromagnetic wave between the contact surfaces of the two media called the evanescent electromagnetic field. The minimum amount of the total internal reflection is observed when the incoming energy of the incident light is coupled onto the flat metal. This is referred to as "attenuated total reflection" (ATR).

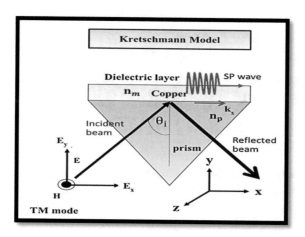

Fig. 1. Surface Plasmon excitation at the surface when activated by light impact to the prism using the Kretschmann's configuration [19].

As shown in Fig. 1, the TM-polarized incident light impacts the prism and activates the excitation of the surface plasmon waves at the interface between the metal thin film and dielectric layer (air layer). The wave vector of the light can be adjusted to be equal to the wave vector of the surface plasmon by launching it from the prism through the metal thin film. The prism is a medium with a higher refractive index than the metal film. Light moving in the prism is reflected at the prism-metal layer interface by means of total internal reflection. The evanescent field of the reflected light at the prism-metal interface penetrates into the metal. With the appropriate thickness of the metal layer, the evanescent wave reaches the metal-dielectric interface (or metal-air interface). In the case that the phase of the incoming light propagating in the prism matches the phase of the surface plasmon waves, the surface plasmon resonance is generated and surface plasmon waves propagate along this metal-dielectric interface. They are generated according to a certain condition which depends on the incident angle and the incident wavelength: [20]

$$k_{sp} = k_x = k_0 n_p \sin\theta \tag{1}$$

k_{sp} is the wave vector of surface plasmon waves
k_x is the wave vector of the incoming light
n_p is the refractive index of the prism
θ is the resonance angle (ATR angle)

According to the equation, the energy and momentum of the incoming light which impact the prism are transfered to the electrons group of the metal thus excites surface plasmon wave. Dispersion relation of surface plasmon wave is shown in the following equation: [20]

$$k_{sp} = k_x = \frac{\omega}{c}\sqrt{\frac{\varepsilon_m \varepsilon_d}{\varepsilon_m + \varepsilon_d}} \tag{2}$$

ε_m is the relative permittivity of the metal
ε_d is the relative permittivity of the dielectric layer

When the dispersion relation is combined with the excited condition, it is found that the minimum incident angle of the incoming light is: [20]

$$\theta = \sin^{-1}\left(\frac{1}{n_p}\sqrt{\frac{\varepsilon_m \varepsilon_d}{\varepsilon_m + \varepsilon_d}}\right) \tag{3}$$

If the wavelength of the incident light and the relative permittivity of the metal layer are known, the suitable incident angle which activates the surface plasmon wave can be calculated.

3   The Simulated Experiments

The excitation of the surface plasmon resonance phenomenon needs to have the k-vector of the activating light which impacts the prism equals to the k-vector of the surface plasmon wave (SP wave) at the interface between the metal thin film layer and the air layer. We therefore simulate the light activation using the Kretschmann's configuration which is a relatively easy to implement activation method. In the first simulation model, a prism with high refractive index and coated with copper thin film is implemented as shown in Fig. 2. Electric fields of the surface plasmon wave at the copper-air interface (metal-air interface) are analyzed using the finite element method (FEM).

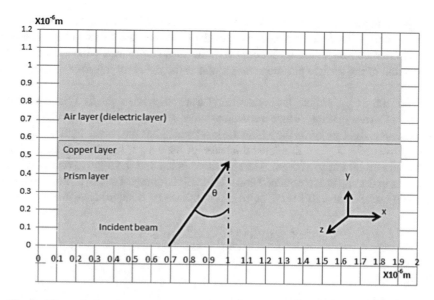

Fig. 2. The optical excitation of Kretschmann's configuration for the first simulation [19].

Parameters as shown in the Table 1 are set according to the dispersion relation. This is to make the k-vector of the p-polarized (TM mode) light which impacts the prism equals to the k-vector of the SP wave. When the light penetrates into the metal thin film, free electrons groups in the metal are coupling with the activating light and vibrate resonantly with the frequency of light. This phenomenon is called the surface plasmon resonance.

Table 1. Important parameters of each medium. [19, 21].

Parameter name	Value
Refractive index of Air	1
Refractive index of Prism (BK7)	1.5151
Relative permittivity of Prism (Real part)	2.2955
Relative permittivity of Prism (Imaginary part)	3.6715e−8
Refractive index of Copper Thin film	0.30730
Relative permittivity of Copper (Real part)	−11.681
Relative permittivity of Copper (Imaginary part)	2.1090
Incident wavelength of Light	632.5 nm
Power of Light	1 W
kx	$k0 * n_{prism} * \sin(\theta)$
ky	$k0 * n_{prism} * \cos(\theta)$

When the incident angle of the light is equal to the resonance angle (also called the Attenuated Total Reflection angle θ_{ATR}), the parallel component of k-vector of the incoming light is matched to the parallel k-vector of the surface plasmon. At this stage, the light transfers its energy to electrons groups in the metal and becomes surface plasmon energy. There is no reflect back of the light from the prism. The evanescent wave at the metal-dielectric layer couples to the surface plasmon which results in the propagation of the surface plasmon wave along to the metal-air interface as previously mentioned.

In this simulation project, the incident angles are changed gradually $1°$ at a time from 30 to $80°$. Corresponding surface resonance waves which are activated by the light are observed. The light source is the TM mode laser which has the wavelength of 632.5 nm with a power of 1 W. The simulated thickness of the copper layer varies from 20 nm, 40 nm, 60 nm, 80 nm to 100 nm. The refractive index and the relative permittivity of the prism and copper are obtained from [20]. Electromagnetic wave propagation as described by the Maxwell's wave equation (in frequency domain) is as follows: [20]

$$\nabla \times \frac{1}{\mu_r}(\nabla \times E) - k_0^2 \left(\varepsilon_r - \frac{j\sigma}{\omega\varepsilon_0} \right) E = 0 \qquad (4)$$

where

ε_r is the relative permittivity of material
E is the electric field equation
k_0 is the wave vector in free space

It is a differential form equation. We use this equation for analyzing the amount of the electromagnetic field at the interface between the thin metal film and the air.

Moreover, we specify Floquet boundary condition both at the left surface and the right surface of every layer in this model to ensure the symmetry of the electric field along and parallel to the interface in the x-y plane.

We also determine the Port boundary condition of this model by using the bottom boundary as Active Port for light impact. The input power of the laser source is set to 1 W. The top boundary is used as the Passive Port which allow light to be transmitted through without reflections.

For the meshing of the 2D geometry in this research work, We partition the subdomain into triangular mesh elements. The resolution of the mesh is set to be extra fine.

For the impact angle from 30 to $80°$, we use the Parametric Sweep to be range(alpha_min, alpha_step, alpha_max) where alpha_min is equal to $30°$, alpha_max is equal to $80°$, and alpha_step is $0.01°$.

After specifying physical quantities such as material properties, constraints, parameter, All related PDE equations automatically compiled by using the finite element analysis. Multiple solvers are used together with adaptive meshing and error control which has been previously specified. Results can be observed from the graphical user interface.

In the second simulation model, we change the incident angle from 30 to 80 °, one degree at a time. The objective is to observe incident angles that causes surface plasmon waves at the interface between silver metal thin film and magnesium chloride solution dielectric with different concentrations. The light source is also the same TM mode light source with 632.5 nm wavelength. The metal thin film is silver with the thickness of 60 nm, adhered by silver nanoparticle in the ellipse shape whose a is 20 nm and b is

50 nm as shown in Fig. 3. Metal parameters which are previously shown in Table 1 are now changed to be silver's: Refractive index of Silver Thin film = 0.15672, Relative permittivity of Silver (Real part) = −14.433 and Relative permittivity of Silver (Imaginary part) = 1.1918. The dielectric layer is magnesium chloride solution in water with concentration range from 5% by weight to 30% by weight by incrementing 5–6% by weight. The refractive index of magnesium chloride solution depends on its concentration. The refractive index and the relative permittivity of the prism and silver are obtained from [21].

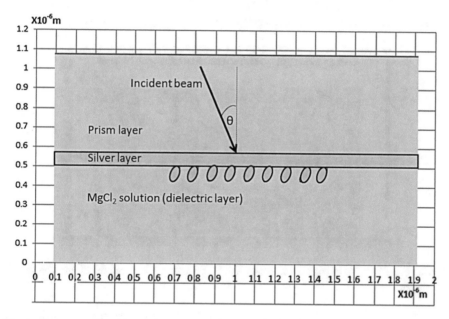

Fig. 3. The optical excitation of the Kretschmann's configuration of the second simulation model.

4 Simulation Results

In the first analysis of the first simulation model using the finite element method, the TM-polarized incident light with the 632.5 nm wavelength impacts the prism with different incident angles. We found that when the surface plasmon resonance phenomenon takes place at the interface between the air layer and the copper layer with the thickness of 20 nm, 40 nm, 60 nm, 80 nm, and 100 nm, there are surface plasmon waves when the incident angle of the incoming light is 44.8°. The incident angle and the light's wavelength correspond with the dispersion relation equation. However, the characteristics of the electric field of surface plasmon wave are different when the thickness of the thin copper film layer is changed.

When the surface plasmon resonance phenomenon is taking place at the contact point between the metal thin film and the air, if the thickness of the metal film is suitable, and the incident angle of the TM mode light is suitable, the simulated electric field at the interface will be clearly seen with high amplitude. This demonstrates that the light reflectance approaches the minimum or even zero. Such an angle is called the resonance angle.

The resonance angle can be shifted if the dielectric layer is changed to other materials such as biomolecular substances. The resonance angle is therefore can be applied to detect the existence of bio-molecular DNA.

When the copper thin film thickness is 20 nm, it is found that the electric field between the interface of the copper film and the air has low amplitudes. The average amplitude is 5.1×10^4 V/m as shown in Fig. 4. The electric field that occurs does not show clear patterns and does not seem to be consistent along the copper-air interface.

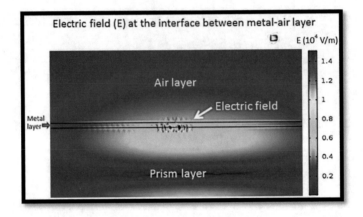

Fig. 4. Electric fields of surface plasmon wave at the interface between the 20 nm copper film and the air [19].

When the copper thin film thickness is 40 nm, it is found that the electric field between the copper-air interface has high amplitudes. The average amplitude is 2.7×10^5 V/m as shown in Fig. 5. The electric field that occurs show clearer patterns and consistent along the copper-air interface.

When the copper thin film thickness is 60 nm, it is found that the electric field between the interface of the copper film and the air has low amplitudes. The average amplitude is 2.2×10^4 V/m as shown in Fig. 6. The electric field that occurs show unclear patterns and not consistent along the copper-air interface.

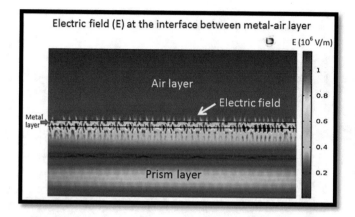

Fig. 5. Electric fields of surface plasmon wave at the interface between the 40 nm copper film and the air [19].

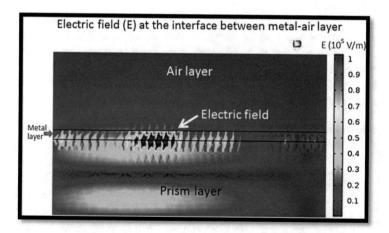

Fig. 6. Electric fields of surface plasmon wave at the interface between the 60 nm copper film and the air [19].

When the copper thin film thickness is 80 nm, it is found that the electric field between the interface of the copper film and the air has low amplitudes. The average amplitude is 2.9×10^4 V/m as shown in Fig. 7. The electric field that occurs show unclear patterns and not consistent along the copper-air interface.

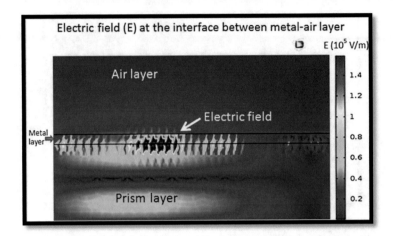

Fig. 7. Electric fields of surface plasmon wave at the interface between the 80 nm copper film and the air [19].

When the copper thin film thickness is 100 nm, it is found that the electric field between the interface of the copper film and the air still has very low amplitudes. The average amplitude is 1.1×10^4 V/m as shown in Fig. 8. However, the electric field that occurs does not show clear patterns and is not consistent along the copper-air interface. The patterns can not be used to identify if they are surface plasmon wave at the interface.

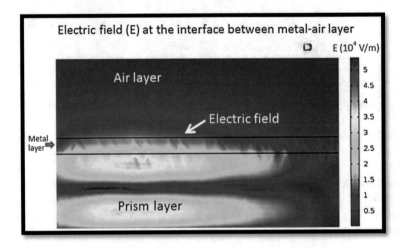

Fig. 8. Electric fields of surface plasmon wave at the interface between the 100 nm copper film and the air [19].

A graph is plotted to show the relationship between the maximum amplitude of the electric field at the copper-air interface and the thickness of the copper film. The graph shows a concave downward pattern in Fig. 9. The highest point of the graph is when the thickness of the copper film is 40 nm. This is the best point when activated by the 632.5 nm light source.

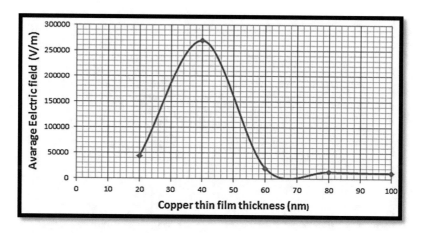

Fig. 9. The graph which shows relationships between the maximum amplitude of the electric field at the copper-air interface and the thickness of the copper thin film [19].

The next step is to find the most suitable impact angle for the given light source. We then change the incident angle parameter to other values around the 44.80° (this is the degree according to the dispersion relation equation) and keep the incident light TM mode to have the wavelength of 632.5 nm to find a suitable impact angle for best surface plasmon resonance phenomenon. The simulated observable surface plasmon wave with highest amplitude at the interface between the copper thin film and the air are considered. Table 2 shows the observation results.

Table 2. The incident angles (Θ_{ATR}) for different copper thin film thickness [19].

Copper thin film thickness (nm)	Optimized incident angle (θ_{ATR}, degrees)	$\Delta\Theta$ (degrees)	E_{avg} (V/m)
20	44.14	0.66	4.4×10^4
40	44.80	0	2.7×10^5
60	44.84	0.04	2.1×10^4
80	44.86	0.06	1.4×10^4
100	45.88	1.08	1.2×10^5

After having the first results of the first simulation model, the second analysis of this model is conducted. The finite element method is still employed. The wavelength of the TM-polarized incident light is changed to be 785 nm with a power of 1 W and the incident angle is now changed to be 44°. The thickness of the copper thin film are 20 nm, 40 nm, 60 nm, and 80 nm. The simulated electric field patterns and amplitudes are observed.

When the copper thin film thickness is 20 nm, it is found that the electric field between the interface of the copper film and the air has low amplitudes. The average amplitude is 2.8×10^4 V/m as shown in Fig. 10. The electric field that occurs show clear patterns and consistent along the copper-air interface.

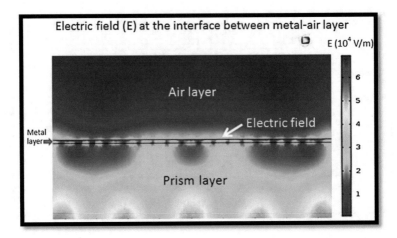

Fig. 10. Electric fields at the interface between the 20 nm copper film and the air when the light has 785 nm wavelength [19].

When the copper thin film thickness is 40 nm, it is found that the electric field between the interface of the copper film and the air has high amplitudes. The average amplitude is 2.5×10^5 V/m as shown in Fig. 11. The electric field that occurs show clear patterns and consistent along the copper-air interface.

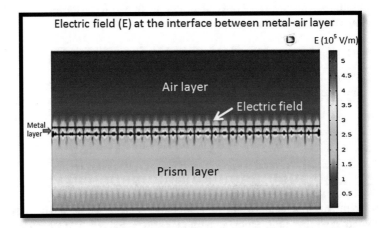

Fig. 11. Electric fields at the interface between the 40 nm copper film and the air when the light has 785 nm wavelength [19].

When the copper thin film thickness is 60 nm, it is found that the electric field between the interface of the copper film and the air has low amplitudes. The average amplitude is 2.5×10^4 V/m as shown in Fig. 12. The electric field that occurs show unclear patterns and not consistent along the copper-air interface.

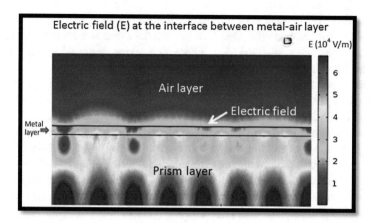

Fig. 12. Electric fields at the interface between the 60 nm copper film and the air when the light has 785 nm wavelength [19].

When the copper thin film thickness is 80 nm, it is found that the electric field between the interface of the copper film and the air has low amplitudes. The maximum amplitude is 2.7×10^4 V/m as shown in Fig. 13. The electric field that occurs show unclear patterns and not consistent along the copper-air interface.

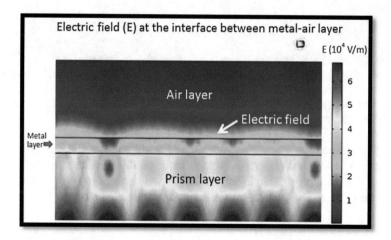

Fig. 13. Electric fields at the interface between the 80 nm copper film and the air when the light has 785 nm wavelength [19].

From the simulation results, it is found that the thickness of the thin copper film has direct effect to the occurrence of the surface plasmon waves. The thickness of 40 nm is considered the best thickness for the activation with the 632.5 nm wavelength. It yields very clear and consistent surface plasmon waves pattern. It also has the highest electric field amplitude at the copper-air interface.

Our analysis results which are obtained by using the finite element method, give similar results to [22] which uses the finite difference method when the incident light source has 785 nm wavelength. That is, when the light source with the 785 nm wavelength is used, the best copper thin film thickness is 40 nm. It gives high and consistent surface plasmon wave amplitudes at the copper-air interface.

For the second simulation model, the thin film is 60 nm silver thin film adhered by elliptical silver nanoparticles (a axis 20 nm, b axis 50 nm). The dielectric layer is magnesium chloride solution and the light source is the 632.5 red laser.

The simulation results show that the activated surface plasmon waves at the interface between silver thin film and the magnesium chloride solution vary as follow:

When the concentration of magnesium chloride solution dielectric is 5% by weight, the result electric field at the interface between the silver thin film and the dielectric layer has the maximum amplitude of 3.4×10^5 V/m as shown in Fig. 14. The electric field is a clearly seen and consistent surface plasmon electric field.

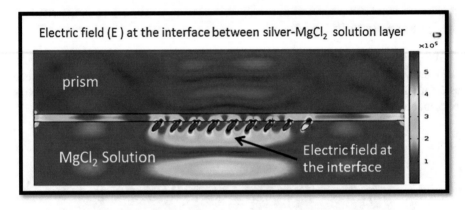

Fig. 14. Electric fields at the interface between the 60 nm silver thin film and the 5% by weight concentration magnesium chloride solution dielectric.

When the concentration of magnesium chloride solution dielectric is 10% by weight, the result electric field at the interface between the silver thin film layer and the dielectric layer has the maximum amplitude of 1.6×10^5 V/m as shown in Fig. 15. The electric field is a clearly seen and consistent surface plasmon electric field.

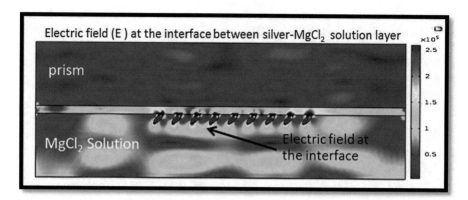

Fig. 15. Electric fields at the interface between the 60 nm silver thin film and the 10% by weight concentration magnesium chloride solution dielectric.

When the concentration of magnesium chloride solution dielectric is 16% by weight, the result electric field at the interface between the silver thin film layer and the dielectric layer has the maximum amplitude of 1.6×10^5 V/m as shown in Fig. 16. The electric field is a clearly seen and consistent surface plasmon electric field.

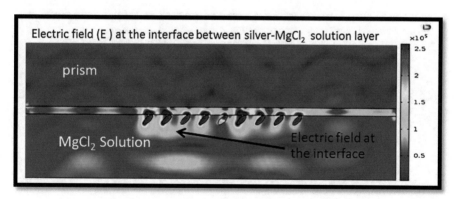

Fig. 16. Electric fields at the interface between the 60 nm silver thin film and the 16% by weight concentration magnesium chloride solution dielectric.

When the concentration of magnesium chloride solution dielectric is 20% by weight, the result electric field at the interface between the silver thin film layer and the dielectric layer has the maximum amplitude of 3.4×10^5 V/m as shown in Fig. 17. The electric field is a clearly seen and consistent surface plasmon electric field.

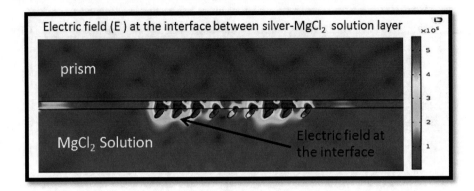

Fig. 17. Electric fields at the interface between the 60 nm silver thin film and the 20% by weight concentration magnesium chloride solution dielectric.

When the concentration of magnesium chloride solution dielectric is 26% by weight, the result electric field at the interface between the silver thin film layer and the dielectric layer has the maximum amplitude of 1.5×10^5 V/m as shown in Fig. 18. The electric field is a clearly seen and consistent surface plasmon electric field.

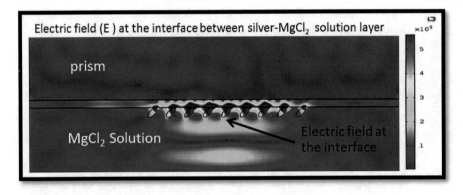

Fig. 18. Electric fields at the interface between the 60 nm silver thin film and the 26% by weight concentration magnesium chloride solution dielectric.

When the concentration of magnesium chloride solution dielectric is 30% by weight, the result electric field at the interface between the silver thin film layer and the dielectric layer has the maximum amplitude of 1.3×10^5 V/m as shown in Fig. 19. The electric field is a clearly seen and consistent surface plasmon electric field.

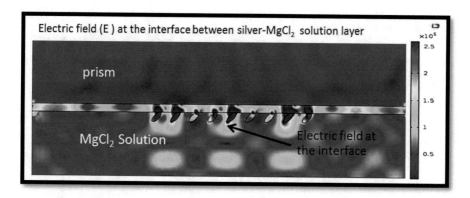

Fig. 19. Electric fields at the interface between the 60 nm silver thin film and the 30% by weight concentration magnesium chloride solution dielectric.

From the simulation results when the dielectric is magnesium chloride solution and the electric field is observed at the interface between the silver thin film layer and the dielectric layer, it is observed that the higher concentration of the dielectric solution, the higher the resonance angle. Table 3 shows the relationships between the dielectric solution and the resonance angle. Figure 20 shows the relationship graph.

Table 3. Magnesium Chloride solution concentration (A% by wt.) and resonance angle (Θ_{SPR}).

Concentration (A % by wt.)	θ_{SPR} (degree) by FEM method
5	68.8
10	70.5
16	72.3
20	74.5
26	77.4
30	79.7

From the graph in Fig. 20, we found that the resonance angle shift is a function of the magnesium chloride solution concentration. The polynomial equation obtained from graph fitting: $0.0095x^2 + 0.0742x + 69.013$ when the magnesium chloride solution has higher concentration, the resonance angle also increases. Moreover, it can also be observed that when the magnesium chloride solution concentration is 5% by weight and 20% by weight, the result electric fields are 3.4×10^5 V/m, which is high. Thus, these levels of concentration can then be used for SPR sensors which detect biomolecular substances such as DNA.

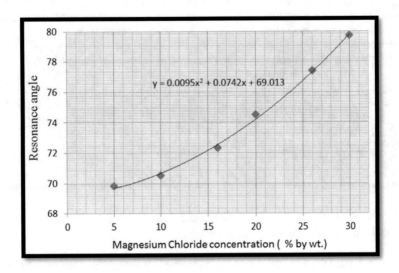

Fig. 20. Graph which shows relationship between magnesium chloride solution concentration and resonance angle SPR.

5 Conclusion

Recently, researchers in the area of plasmonics have been highly active. There are applications in various areas such as biomedical engineering where SPR devices are used as sensors to detect the presence of DNA molecules which are adhered to metal surfaces. Such adhesion layer change the local refractive index which results in the shift of the resonance angle of the incident light. At present, noble metals such as copper and silver are employed as SPR sensors due to the more economical cost.

In this research work, we develop two SPR sensor simulation models. The first one analyses the electric fields at the interface between copper thin film and air dielectric when surface plasmon resonance takes place. The objective is to find the thickness of the copper thin film which is suitable for the surface plasmon resonance excitement using 632.5 nm red laser light source. The second one analyses the effects of surface plasmon resonance using different dielectric solution concentrations when the dielectric layer is magnesium chloride solution and the metal layer is silver thin film. This second simulation model differs from previous other research works in this direction which employ glucose solution dielectric. Instead we employ magnesium chloride solution as the dielectric layer and observe electric fields at the interface between the silver thin film layer and the dielectric layer. Due to the good characteristic of mg^{2+} ion which causes binding between the silver thin film and DNA, the results from our second simulation model show its potential in DNA detection applications.

From the first simulation model, it can be concluded that the thickness of 40 nm copper thin film is the most suitable one when activated by 632.5 nm red laser light source. It gives electric fields with highest amplitudes, clear, and consistent at the interface which means the best surface plasmon waves at the interface. From the second model, it is observed that when the concentration of the magnesium chloride solution in water increases, the resonance angle also increases accordingly in poly-nomial function. Also, when the concentration of the dielectric solution are 5% by weight and 20% by weight, the corresponding electric field is 3.4×10^5 V/m, which is considered high, and likely to be suitable for biomolecular sensor, such as DNA sen-sors development.

References

1. Naik, G.V., Shalaev, V.M., Boltasseva, A.: Alternative plasmonic materials_beyond gold and silver. Adv. Mater. **25**, 3264–3294 (2013)
2. Robusto, P., Braunstein, R.: Optical measurements of the surface plasmon of copper. Phys. Stat. sol. (b) **107**, 443–449 (1981)
3. Kravets, V.G., et al.: Graphene-protected copper and silver plasmonics. Sci. Rep. **4**(5517), 1–7 (2014)
4. Serra, A., Filippo, E., Re, M., Palmisano, M., Vittori-Antisari, M., Buccolieri, A.: Non-functionalized silver nanoparticles for a localized surface plasmon resonance-based glucose sensor. Nanotechnology **20**(16), 165501 (2009)
5. Zhang, X., Wei, M., Bingjing, L., Liu, Y., Liua, X., Wei, W.: Sensitive colorimetric detection of glucose and cholesterol by using Au@Ag core–shell nanoparticles. RSC Adv. **6**, 35001–35007 (2016)
6. Chung, H.Y., et al.: Enhanced sensitivity of surface plasmon resonance phase-interrogation biosensor by using oblique deposited silver nanorods. Nanoscale Res. Lett. **9**(476), 1–5 (2014)
7. Liu, R.J., et al.: Surface plasmon resonance biosensor for sensitive detection of microRNA and cancer cell using multiple signal amplification strategy. Biosens. Bioelectron. **87**, 433–438 (2017)
8. Hsieh, S.C., Chang, C.C., Lu, C.C., Wei, C.F., Lin, C.S., Lai, H.C., Lin, C.W.: Rapid identification of Mycobacterium tuberculosis infection by a new array format-based surface plasmon resonance method. Nanoscale Res. Lett. **7**, 1–6 (2012)
9. Mitchell, J.: Small molecule immunosensing using surface plasmon resonance. Sensors **10**, 7323–7346 (2010)
10. Mariani, S., et al.: Investigating nanoparticle properties in plasmonic nanoarchitectures with DNA by surface plasmon resonance imaging. Chem. Commun. **51**, 6587–6590 (2015)
11. Ahmed, F.E., Wiley, J.E., Weidner, D.A., Bonnerup, C., Mota, H.: Surface plasmon resonance (SPR) spectrometry as a tool to analyze nucleic acid-protein interactions in crude cellular extracts. Cancer Genomics Proteomics **7**(6), 303–310 (2010)
12. Teh, H.F., Peh, W.Y.X., Su, X., Thomsen, J.S.: Characterization of protein-DNA interactions using surface plasmon resonance spectroscopy with various assay schemes. Biochemistry **46**(8), 2127–2135 (2007)
13. Zhang, Y., Liu, L., Sha, J., Ni, Z., Yi, H., Chen, Y.: Nanopore detection of DNA molecules in magnesium chloride solutions. Nanoscale Res. Lett. **8**, 245 (2013)
14. Sternberg, K., et al.: Magnesium used in bioabsorbable stents controls smooth muscle cell proliferation and stimulates endothelial cells in vitro. J. Biomed. Mater. Res. B Appl. Biomater. **100**(1), 41–50 (2012)
15. Suzuki, Y., Shimada, S., Hatta, A., Suëtaka, W.: Enhancement of the IR absorption of a thin film on gold in the otto ATR configuration. Surf. Sci. Lett. (219), L595–L600 (1989)
16. Leskova, T.A., Leyva-Lucero, M., Méndez, E.R., Maradudin, A.A., Novikov, I.V.: The surface enhanced second harmonic generation of light from a randomly rough metal surface in the Kretschmann geometry. Opt. Commun. (183), 529–545 (2000)

17. Iadicicco, A., Cusano, A., Campopiano, S., Cutolo, A., Giordano, M.: Thinned fiber Bragg grating as refractive index sensors. IEEE Sens. J. **5**(6), 1288–1295 (2005)
18. Du, W., Zhao, F.: Surface plasmon resonance based silicon carbide optical waveguide sensor. Mater. Lett. (115), 92–95 (2014)
19. Leelawattananon, T., Lorchalearnrat, K., Chittayasothorn S.: Simulation of copper thin film thickness optimization for surface plasmon using the finite element method. In: the 7th International Conference on Simulation and Modeling Methodologies, Technologies and Applications (SIMULTECH), pp. 188–195 (2017)
20. Raether, H.: Surface Plasmons on Smooth Surfaces. Springer, Heidelberg (1988)
21. Refractive Index Database. https://refractiveindex.info/. Accessed 07 Jan 2018
22. Said, F.A., Menon, P.S., Nawi, M.N., Zain, A.R.M., Jalar, A., Majlis, B.Y.: Copper-graphene SPR-based biosensor for urea detection. In: IEEE International Conference on Semiconductor Electronics (ICSE), pp. 264–267 (2016)

An Object Oriented Approach Towards Simulating Physical Systems with Fluids and Rigid Bodies Based on the Physolator Simulation Framework

Dirk Eisenbiegler[✉] and Waldemar Rose[✉]

University of Furtwangen, Furtwangen Im Schwarzwald, Germany
{eid,waldemar.rose}@hs-furtwangen.de

Abstract. This article is about computer based simulations applied to physical systems that consists both of fluids and rigid bodies. It explains how to build such simulations in a systematic manner applying object oriented programming techniques. This work is based on the Physolator simulation framework. Physolator supports a modular design style. This article shows, how the Physolator framework can be used to build mixed fluid and rigid body simulations in a systematic manner by first building a set of reusable components and then building different kinds of physical systems by assembling these components.

Keywords: Fluid simulation · Particle modeling · Rigid body · Physolator Object oriented modelling

1 Introduction

There are different kinds of domains where scientists and engineers apply physical simulations in order to predict the behavior of certain real world systems. Scientists and engineers are highly specialized. Each professional is focused on a narrow technical area and the physical simulations are to be designed for this particular area. This is why they usually do not just have in mind to build one particular simulation but a variation of simulations. When starting programming, they usually first implement one specific physical system and then start modifying the system. They start altering the parameters of the physical systems and apply different simulation parameters in order to optimize the way the simulation is executed.

Physical systems consist of physical components and subcomponents. Physical systems from the same technical area usually consist of the same physical components. After you have built your first physical system, it makes perfect sense to identify such physical components and then try to assemble them in a different way in order to build other physical systems from the same technical domain. It is useful to provide graphics components that visually represent the actual state of your physical system on the screen. Programming such graphics components can be time consuming. If you are dealing with a set of physical systems from the same technical area, it makes perfect sense not only to share the programming code for the physical components, but also to reuse the program code for the graphics programming.

M. S. Obaidat et al. (Eds.): SIMULTECH 2017, AISC 873, pp. 23–32, 2019.
https://doi.org/10.1007/978-3-030-01470-4_2

Physolator is a simulation framework that supports an object oriented style of developing physical simulations. This framework shall be applied to simulations of physical systems that consist of fluids and rigid bodies. This article explains, how the Physolator is used in order build simulations for this domain in a systematic manner. In the next section, we explain the challenges of fluid simulation and rigid body physics. In the following sections we introduce the Physolator framework and describe the features it provides. Afterwards we explain, how to implement physical components representing fluids and rigid bodies and how to assemble these physical components in order to build physical systems.

2   Particle Based Fluid Simulations and Rigid Body Physics

There are different approaches towards simulating the dynamics of fluids. The approach in this paper is based on particles (see also [1–4, 9]). A computer based simulation computes the actual state of some physical system for a sequence of points in time. For each point in time, a fluid simulation computes the actual location of the fluid, the physical fields with the flow velocity and pressure. Particle simulations are challenging as to computing time (see [10, 11, 13]). It takes a large number of particles to achieve a reasonable accuracy. Therefore, such simulations usually require big amounts of computing time. Programmers have to spend a lot of time for optimizations in order to achieve a reasonable accuracy within a limited computation time.

The challenges that you face when dealing with rigid body systems are completely different. A rigid body is a mechanical object with a fixed shape. During simulation, a rigid body may move or rotate, but its shape remains unchanged. The state of a rigid body is defined by its position and its actuals velocity. Forces applied to a rigid body result in an acceleration. Forces define the rigid body's behavior, i.e. the way the rigid body changes its state. Different kinds of forces can be applied to a rigid body such as gravity, magnetic forces, sliding friction, static friction or rolling friction, etc.. Besides the forces, there are collisions between rigid bodies. Collisions are physical events. Physical events abruptly change the behavior of a physical system. A collision abruptly changes the translational and rotational velocity of the bodies involved.

The aforementioned physical effects from moving physical rigid bodies are described by differential equations. Handling physical events goes beyond differential equations. In order to handle a physical event, you first have to exactly find the point in time, when the event happens. Interval nesting is used to bring the physical system in a state right before the physical event happens. Then some physical rule is applied and the state of the physical system changes abruptly and within the very same point in time. In the case of a collision, the rule to be applied is a partially elastic collision.

The computing power needed for a rigid body simulation very much depends on the number of rigid bodies. Advanced techniques have to be applied to efficiently detect collisions in systems with large number of rigid bodies. With an increasing number of rigid bodies there are also potentially more collisions. Using interval nesting to find the exact points in time where these physical events happen is time consuming. Rigid body simulations are not only used in science and engineering, but also in computer games and in animated films. In computer games and animated films, computing performance matters. It is not necessarily the intention of a computer game or an animated film to produce simulation results that are absolutely accurate. In computer game or an animated movie the accuracy is good enough if the user gets the impression that the movies produced on the screen have a "natural behavior". The mechanical behavior of the objects on the screen should be close enough to the real world, that the viewer can not recognize the difference by watching the movie. In this domain accuracy is not so important, but performance very much matters. Game engines usually contain a component named "physics engine". This kind of physics engine is usually limited to rigid body physics. Physics engines inside computer games are designed for real time execution. For a limited amount of components, they succeed computing frames within milliseconds with an accuracy that is good enough for this purpose.

This paper is about physical systems consisting of both fluids and rigid bodies. Concepts from both worlds, fluid simulation and rigid body simulation, are to be applied. The focus of this article is on high precision simulations that are to be used in science and engineering. With an increasing number of particles and an increasing number of rigid bodies, the time consumption necessarily increases as long as you want to achieve precise results. We will introduce some concepts on how to achieve a reasonable performance. However, it is not in the focus of this paper to achieve high speed real time simulation. Anyhow, in this domain, this could only be achieved with a small number of particles and rigid bodies or by modifying and simplifying the physical laws.

3   The Physolator Simulation Framework

Physolator is a framework based on the programming language Java [5]. Physolator itself has been implemented in pure Java and also the physical systems and graphics components that you can run inside the Physolator are implemented using the Java programming language. Figure 1 shows a snapshot of this framework. As you can see, a physical system has already been loaded. First you have to implement your physical system using Java. Physical systems basically consist of physical variables, derivation relations between the physical variables and formulas for computing the values of dependent variables from the state representing variables. Java annotations are used to describe derivation relationships and they are also used to assign physical units to variables. The tutorials in [7, 8] explain how to implement such physical systems. You may add graphics components to your code. Furthermore your code may also provide some basic settings: simulation parameters describing how the simulation is to be executed, settings for function plotters. Next you have to load this physical system to the Physolator. After loading the physical system to the Physolator, you can see the structure of the physical system with its hierarchy of components and subcomponents (left hand side of Fig. 1). In this hierarchy of components, scalar values are the leaf elements. Every scalar value is represented with its actual value and its physical unit. When loading a physical system, the graphics components attached to the physical system are automatically loaded as well. In the middle of Fig. 1, you can see a graphics showing three balls that are about to fall into a basin filled with water. This graphics represents the actual state of the system.

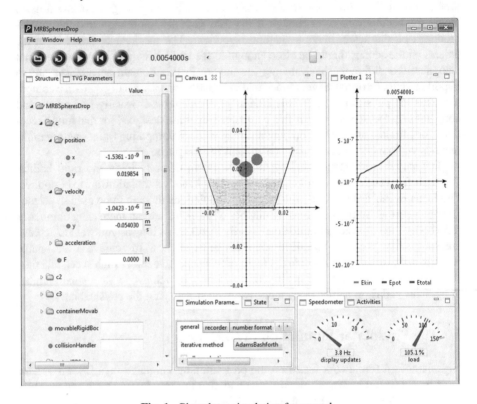

Fig. 1. Physolator simulation framework.

After loading the physical system, you can run it by pressing the start button. This launches the physical simulation. In a sequence of steps, the physical system goes from state to state. For a sequence of points in time, the state of the physical system is computed one by one. The variables start changing their values and also the graphics representing the actual state, starts moving.

There are various simulation parameters. They define, how the simulation is executed. There are several options. First of all, there are different kinds ODE solvers that you can choose from. Some of the ODE solvers work with a fixed step width some use a flexible step width, some use single step methods others use multi step methods. You can choose either to run the simulation in real time mode or you can choose to run the simulation as fast as possible. Also the step width and frequency for updating the screen (frames per second fps) are simulation parameters that you can adjust to your needs. For all these simulation parameters there are reasonable default values. In many cases you do not have to change any of the simulation parameters or at least you only have to change few of them to fulfil your specific requirements.

Physolator provides a built in performance monitor. The performance monitor is used to detect performance bottlenecks during simulation. If the simulation is run in real time mode, then performance monitor tells you about the total load, that the simulations produces. In a situation of an overload the fps value is automatically reduced. The monitor provides the actual fps rate. Furthermore there is also a performance monitoring tool that helps you detecting the reason for the performance bottlenecks. In some cases, the calculations of the ODE solvers are time consuming. In other cases the graphics component is complex and consumes lots of CPU time for drawing. The performance monitoring tool shows what part of the total frame time is spent by the different components inside the Physolator.

4   Particle Modelling

Fluids are to be represented by particles. Each amount of fluid is be represented by a number of particles. This approach is sometimes called "quasi-molecular" modelling. In real world, an amount of fluid consists of a number of molecules. Even for small amounts, this number is very big. Due to the big number, it would be impossible to represent each molecule in a computer simulation – even for very small amounts of liquids. Therefore, the quasi-molecular simulation approach replaces molecules by bigger pieces called particles. For the same reason, a two dimensional simulation will be considered in this paper rather than a three dimensional one. For accurate results, it takes far greater amounts of particles when using three dimensional models.

The behavior of the particles shall be similar to molecules: attraction if two particles are far from one another and repulsion if they get too close. In the molecular world, formulas like the Lennard-Jones-12-6 potential are used to describe the attraction and the repulsion between two molecules. Similar formulas are used for quasi-molecular simulations based on particles. In this approach, the two formulas below are used to describe the potential and forces between two particles, respectively. These formulas have been introduced by Greenspan [1] and have been adapted by many others. The second formula is nothing but the derivation of the first formula with respect to r.

$$\phi(r) = -4\epsilon\frac{\sigma^2}{r^2} + 4\epsilon\frac{\sigma^4}{r^4}$$

$$F(r) = \frac{d\phi(r)}{dr} = 8\,\epsilon\,\frac{\sigma^2}{r^3} - 16\,\epsilon\,\frac{\sigma^4}{r^5}$$

Formula $F(r)$ describes the attraction and repulsion force between two particles. There are two parameters in theses formulas: σ and ϵ. Furthermore, every particle shall have the some mass m. Finally, there are also friction forces whenever there is an inner movement, i.e. some particle moves with respect to the surrounding liquid. Whenever a particle is moving with respect to the surrounding liquid, a friction force will be applied. The strength of this force is defined by parameter c_w.

Parameters σ, ϵ, m and c_w define the behavior of some amount of liquid that is represented by a number particles. Different kinds of fluids are represented by different kinds of values for s σ, $\in$, m and c_w. One has to adjust these parameters appropriately in order to give the particle based liquid the right behavior on a macroscopic level and provide the liquid with the right mass density and viscosity (see [14] for details).

Originally, the formula $F(r)$ is used to compute the force between two particles of the same material. The approach from Greenspan provides a means for also computing the force between two particles A and B, where particle is of a different material than particle B. Given a particle A with parameters σ_A and $\in_A$ and a particle B with parameters σ_B and $\in_B$. The force $F(r)$ between these two particles as well as $\phi(r)$ are computed with the given equations using the following parameters σ_{AB} and $\in_{AB}$.

$$\in_{AB} = \sqrt{\in_A \in_B}$$

$$\sigma_{AB} = \frac{1}{2}\left(\sigma_A + \sigma_B\right)$$

The formulas from Greenspan only describe the forces between particles, i.e. the liquid's inner forces. Besides, there may also be different kinds of external forces applying the fluid and its particles. The particles inside a fluid may be attracted by the gravitational forces from external bodies. Just like a water drop dropping down. The particles inside the water drop are attracted by earth. If a liquid is in touch with rigid bodies, then the liquid is repulsed by the right body. Example: A water drop rests on top a rigid surface. The rigid surface applies an upward force to the liquid hindering it from falling. Many other kinds of external forces may have to be considered.

5 Physical Systems with Fluids and Rigid Bodies

Figure 2 shows an example for a simulations where three balls drop into a basin filled with water. As the balls drop into the water, they dive into the water. As soon as they are inside the water, water breaks down the movement. Since in our physical system the mass density of the balls is smaller than the mass density of the water, the balls are pushed upwards again and finally they swim on the surface of the water. The balls start hitting against one another and they start hitting against side walls of the basin. Due to the inner friction of the water, the water particles and the balls finally come to a standstill. The water surface is even again and the balls are swimming on the water at a fixed position.

In a physical system like in Fig. 2, there are different physical effects that have to be considered. Such physical systems consist of three different kinds of physical components: fluid particles, fixed rigid bodies and movable rigid bodies. Figure 3 originates from our publication [9]. See [9] for more details on this model. The basin with its floor and its side walls on the left and right is a fixed rigid body. The balls are movable rigid body. There are attraction and repulsion forces between the fluid particles and between the fluid particles and the rigid bodies. The formulas from Greenspan are used to describe these forces. In this scenario the rigid bodies also collide with one another. Rigid bodies may collide in an elastic or inelastic manner. Appropriate physical formulas have to be used to describe such physical events.

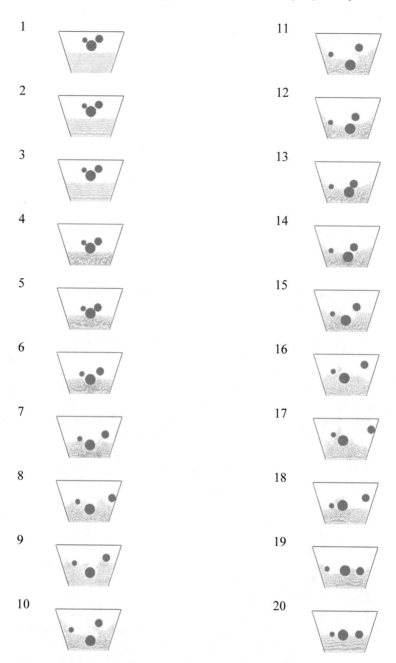

Fig. 2. Physical model with rigid bodies and a fluid.

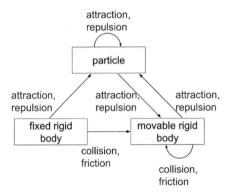

Fig. 3. Physolator simulation framework (from [9]).

6   Using the FRB Library to Implement Physical Systems

FRB stands for "fluid and rigid body". The FRB library provides a set of components designed for simulations containing fluids and liquid bodies. There are physical components for representing amounts of liquids, physical components representing rigid bodies and there is a graphics component for visually representing physical with such fluids and rigid bodies on the screen. This article does describe all the features of the FRB library. Instead, an example is used to explain, how the different components are used in order to build a physical model that can be executed in the Physolator framework.

The following program is the implementation a physical system. The physical system is named *MRBSpheresDrop*. It corresponds to Figs. 1 and 2. As you can see from the program code class *MRBSpheresDrop* inherits form *MRBParticleSystem*. *MRBParticleSystem* is part of the FRB library. It provides some basic infrastructure for building physical systems with fluids and rigid bodies. The entire program code describes, how some rigid bodies and an amount of water are composed in order to build a physical system. Lines 3 through 5 define some constants for the scaling. Modifying them results in resizing the entire physical system and changing the density of the particles. In lines 6 through 8 three circles are defined. They are initialized providing each of them some initial location and speed.

The constructor of this class is used to define the initial state of the system. Lines 12 to 18 are used to define the physical properties of the fluid and of the rigid bodies that are to be placed inside this physical system. Lines 19 to 28 are used to fill a well-defined area with water and to put some rigid bodies to specific positions. Method *fillRectangle* is used to fill the basin with particles. The lambda-term in 20 is a filter. This filter makes sure, that not the entire specified rectangle is filled with particles, but only some well-defined part of it. As a result, a trapezoidal shape is filled with particles. As rigid bodies there are the aforementioned three balls. They are added to the physical system by invoking *addMovableRigidBody*. Furthermore, there are four lines representing the basin's border. They are added by invoking the *addLine* method. The FRB library so far supports only two kinds of rigid bodies: lines and circle segments. Sure, entire circles are supported as well (Fig. 4).

```
1    public class MRBSpheresDrop extends MRBParticleSystem {
2
3       private final double sigma0 = 50e-5;
4       private final double rMax0 = 5 * sigma0;
5       private double sF = 0.01;
6       public MRBCircle c1 = new MRBCircle(0 * sF, 2 * sF, 0.4 * sF, new Vector2D(0, -9.81));
7       public MRBCircle c2 = new MRBCircle(0.6 * sF, 2.5 * sF, 0.3 * sF, new Vector2D(0, -9.81));
8       public MRBCircle c3 = new MRBCircle(-0.5 * sF, 2.4 * sF, 0.2 * sF, new Vector2D(0, -9.81));
9
10      public MRBSpheresDrop() {
11         beginStructure(rMax0);
12         setParticleSchema(MaterialCatalogue.water, sigma0, new Vector2D(0, -9.81));
13         setMRBSchema(MaterialCatalogue.movableRigid, sigma0, new Vector2D(0, -9.81));
14         setRBSchema(MaterialCatalogue.rigid, sigma0);
15         c1.setSchema(actualMRBSchema);
16         c2.setSchema(actualMRBSchema);
17         c3.setSchema(actualMRBSchema);
18         double q = 0.5 * sigma0;
19         fillRectangle(-2.5 * sF + q, q, 2.5 * sF - q, 1.5 * sF,
20            (x,y) -> x>0 ? (y-q)/sF>3*(x+q)/sF-4.5: (y-q)/sF>-3*(x-q)/sF-4.5 );
21         addLine(-1.5 * sF, 0, -2.5 * sF, 3 * sF);
22         addLine(1.5 * sF, 0, -1.5 * sF, 0);
23         addLine(-2.5 * sF, 3 * sF, 2.5 * sF, 3 * sF);
24         addLine(2.5 * sF, 3 * sF, 1.5 * sF, 0);
25         addMovableRigidBody(c1);
26         addMovableRigidBody(c2);
27         addMovableRigidBody(c3);
28         endStructure();
29      }
30   }
```

Fig. 4. Java program code for MRBSpheresDrop.

When loading this system to the Physolator framework, a graphics component is automatically loaded as well. This feature is inherited from *MRBParticleSystem*. The graphics component is generic. It checks, what kinds of components – particles and rigid bodies – have been added and draws them to the screen. The graphics component automatically detects the right range from the canvas to display all components on the canvas.

The code in the FRB library has been optimized for greater amounts of particles. Every particle applies a force to all the other particles. For physical systems with a greater number of particles, this leads to a huge computing effort. With greater distances between particles, however, the forces are very small and can be neglected without significantly having an influence on performance. The algorithms used inside the FRB library efficiently detect neighbourhood relationships and only compute the forces between particles having a distance that is closer than a given limit. This algorithm internally uses a grid. Computations are executed with multiple threads to enhance performance. To synchronize the threads, a concept invented by Bresenham [12] is applied. See [9] for details.

7   Conclusion and Future Work

This article has presented an object oriented approach towards programming mixed fluid and rigid body simulations in a modular manner. The approach is based on the Physolator simulation framework. On top of this framework, a library of components for fluid and rigid body simulations (FRB library) has been built. These components are assembled to build complex physical systems.

This library is currently applied to different kinds of physical systems in this domain. So far, this library still has some limitations. First of all, the library has been designed for two dimensional simulations only. Furthermore, the library is not yet well prepared for a larger number of rigid bodies. With an increasing number of rigid bodies, collision detection gets time consuming. It would take advanced data structures and algorithms to handle a large number of rigid bodies efficiently.

References

1. Greenspan, D.: Particle Modeling. Birkhäuser Boston, Basel, Berlin (1997)
2. Greenspan, D.: N-Body Problems and Models. World Scientific Publishing Co. Pte. Ltd., Singapore (2004)
3. Greenspan, D.: Computer studies in particle modeling of fluid phenomena. In: Mathematical Modeling, vol. 6, pp. 273–294. Pergamon Press Ltd. (1985)
4. Nijmeijer, M., et al.: Molecular dynamics of the surface tension of a drop. J. Chem. Phys. **96**(1), 565–576 (1992)
5. Eisenbiegler, D.: The software architecture of the physolator–a physical simulation framework. In: MSAM 2015, pp. 61–64. Atlantis Press (2015)
6. Eisenbiegler, D.: Objektorientierte Modellierung und Simulation physikalischer Systeme mit dem Physolator. BoD Norderstedt (2015)
7. Eisenbiegler, D.: Physolator – Getting Started. Video Tutorial. http://www.physolator.de/joomla/index.php/en/manual#GettingStarted
8. Eisenbiegler, D.: Physolator Programming. Video Tutorial. http://www.physolator.de/joomla/index.php/en/manual#PhysolatorProgramming
9. Rose, W., Eisenbiegler, D.: Mixed fluid and rigid body simulations - an object oriented component library based on the physolator framework. In: Simultech 2017, ScieTecPress 2017, pp. 36–44 (2017)
10. Korlie, M.: Particle modeling of liquid drop formation on a solid surface in 3-D. Comput. Math Appl. **33**(9), 97–114 (1997)
11. Korlie, M.: Three-dimensional computer simulation of liquid drop evaporation. Comput. Math Appl. **39**, 43–52 (1999)
12. Bresenham, J.E.: Algorithm for computer control of a digital plotter. IBM Syst. J. **4**, 24 (1965)
13. Pozrikidis, C.: Fluid Simulation – Theory, Computation, and Numerical Simulation. Springer, US (2017)
14. Eisenbiegler, D: A generic particle modeling library for fluid simulation. In: AMSM 2016, Conference on Applied Mathematics, Simulation and Modelling. Atlantis Press (2016)

Distributed Co-simulation of Embedded Control Software Using INTO-CPS

Nicolai Pedersen[1,4(✉)], Kenneth Lausdahl[2], Enrique Vidal Sanchez[1],
Casper Thule[3], Peter Gorm Larsen[3], and Jan Madsen[4]

[1] MAN Diesel & Turbo, Teglholmsgade 41, 2450 Copenhagen SV, Denmark
nicolai.pedersen@man.eu
[2] Mjølner Informatics, Finlandsgade 10, 8200 Aarhus N, Denmark
[3] Department of Engineering, Aarhus University,
Finlandsgade 22, Aarhus N, Denmark
[4] Embedded Systems Engineering, Technical University of Denmark,
Anker Engelunds Vej 1, Lyngby, Denmark

Abstract. The systematic engineering of Cyber-Physical Systems is a challenging endeavour. In order to manage the complexity of such multi-disciplinary development collaborative modelling and co-simulation has been proposed. In this setting models are made of different constituent models with different mathematical formalisms using different tools. This paper demonstrates how this can be achieved for a commercial system developed by MAN Diesel & Turbo using such a co-simulation approach. The tool chain is centered around a de-facto standard called Functional Mock-up Interface, and it is open for any tools that can support version 2.0 of this standard for co-simulation. The application support emission reduction control systems for large two-stroke engines which is strategically important. It is demonstrated how this approach can reduce the need for expensive tests on the real system in order to reduce the overall costs of validation.

Keywords: INTO-CPS · Cyber-Physical Systems · Co-simulation
Parallel simulation · Distributed simulation
Functional Mock-up Interface · Embedded control system
Exhaust Gas Recirculation

1 Introduction

With increased complexity in the development of Cyber-Physical Systems (CPS) more advanced modelling and specialised tools are required. Depending upon the CPS in question the main challenges can be on the Discrete Event (DE) side with the controller aspects or due to physical dynamics, control and communication on the Continuous-Time (CT) side. Using a model-based approach it is possible either to make use of a single closed tool for modelling and simulation all parts, or by making use of chains of tools that can combine different

© Springer Nature Switzerland AG 2019
M. S. Obaidat et al. (Eds.): SIMULTECH 2017, AISC 873, pp. 33–54, 2019.
https://doi.org/10.1007/978-3-030-01470-4_3

models of the constituent elements using co-simulation [1, 2]. For differential-equation based CT models, multiple tools are available [3–5], each with their specific specialisation and validity. Discrete event models are often developed within internal software frameworks of companies, or created in one of the many DE tools available. The interconnection between the physical and cyber parts of CPS is becoming increasingly connected and the dynamical influences have to be considered no matter what analysis has been performed of the produced models. The main challenge in connecting these models comes from the fundamental differences in the underlying mathematics, their simulation tools and how they are developed. Many initiatives for connection tools in a so called co-simulation have been published [6, 7]. However, connecting the specific tools making up the holistic simulation is often not the only issue. Deviations in development platforms and performance is as often an issue. A solution for this is a distributed co-simulation, where models can be executed, not only in the tool where they were developed, but also on the correct platform. Furthermore, a distribution of the simulation makes it possible to increase performance by utilising additional hardware, given that the models are prepared for it.

At MAN Diesel & Turbo (MDT) the conventional approach for developing two-stroke combustion engines with a distributed embedded control system is being challenged. In particular for diesel engines pollution is a key element that it is desirable to reduce from a competitive perspective. New emission legislation focuses on the reduction of especially NO_x emission. Widely known emission reduction technologies for reducing NO_x are selective catalytic reduction and Exhaust Gas Recirculation (EGR), both being developed at MDT [8]. These systems require advanced algorithms to control the complexity of the physical dynamics of large engines. MDT is divided into different departments with different responsibilities in the same way as many other large organisations. In the control department at MDT, control algorithms are created directly in the target software framework with the possibility of performing Software In the Loop (SIL) simulation during development. Models of the physical behaviour are created in other departments of MDT using the tools most suitable for the specific constituent system. For the control system development, the physical dynamics models are implemented in an internally developed tool for CT simulation called the Dynamic Simulation Environment (DSE) which is part of the software framework. The primary focus in DSE is SIL/Hardware In the Loop (HIL), and the physics models implemented here are often an abstraction of high-fidelity models. Historically it has been challenging inside MDT to enable heterogeneous collaborations between the different teams producing models in different departments. As a result different models are typically fragmented and solely used within one department for the dedicated purpose each of the models serve. Thus, efforts that goes across these individual insights are only found at the test on the real platform.

At MDT the models used in the control department are based on a software framework and DSE is implemented in C++ and run on a 32-bit Linux platform while the physical modelling tools often require Windows. In this paper it is illustrated how a transition from the current simulation process at MDT to one using co-simulation utilising the Functional Mock-up Interface (FMI) standard can be performed by using the Co-simulation Orchestration Engine (COE) from the Integrated Tool Chain for Modelbased Design of Cyber-Physical Systems (INTO-CPS) project. The aim with the approach suggested in this paper is to reduce redundancy in the development process and reuse and combine models from different departments. One of the main challenges for such a transition is to enable co-simulation across different hardware architectures and Operating System (OS) platforms due to constraints from software frameworks, physical simulation tools and version compatibility. This paper can be considered as an extended version of [9].

In Sect. 2 the overall system is presented. Section 3 describes the previous simulation of a specific subsystem for EGR, and Sect. 4 describes the INTO-CPS tool chain enabling co-simulation. Afterwards, Sect. 5 describes the co-simulation results for the EGR system. Finally, the paper provides concluding remarks and look into future work in Sect. 6.

2   Exhaust Gas Water Handling System

The EGR system presented in this paper recirculates exhaust gas to the intake manifold thereby reducing environmental impact while maintaining efficient combustion. The unclean exhaust gas is potentially damaging to the engine and has to be cleaned before return, which is the purpose of the Water Handling System (WHS). The system is shown in Fig. 1 where the exhaust gas is drawn into the *EGR Unit* using an EGR blower, it is then sprayed with water and cooled so that a Water Mist Catcher (WMC) can collect the damaging particles. Before the gas is returned, the water is collected in the WMC and led to a receiving tank. The water level in this tank is one of the significant parameters that the WHS controls, as discussed in Subsect. 3.4 below. The water is pumped from the receiving tank to an external constituent system called the (Water Treatment System (WTS)) for processing where the water is either cleaned and pumped back to the EGR Unit, pumped overboard or stored for treatment at a harbour.

At the chemical level, EGR is based on exchange of the in-cylinder oxygen (O_2) with carbon dioxide (CO_2) from the exhaust gas, which is re-circulated into the scavenged air. The exchange of O_2 with CO_2 leads to a decrease of combustion speed, resulting in lower peak temperatures during combustion. Furthermore the exchange of O_2 with CO_2 results in a higher in-cylinder heat capacity of the gas which also lowers the combustion temperature. Lower combustion temperatures and especially lower peak temperatures result in lower formation of thermal NO_x during the combustion process. The recirculated exhaust gas is hotter and not as clean as the residual ambient scavenge-air. To prevent Sulphur (SO_2) and other particles from damaging the engine, cleaning and cooling of the recirculated exhaust gas is required. A WHS provides the water used for cleaning the exhaust gas in the EGR unit. To control the flow of exhaust gas to the mixing chamber, an EGR blower is installed. Water from the EGR unit is

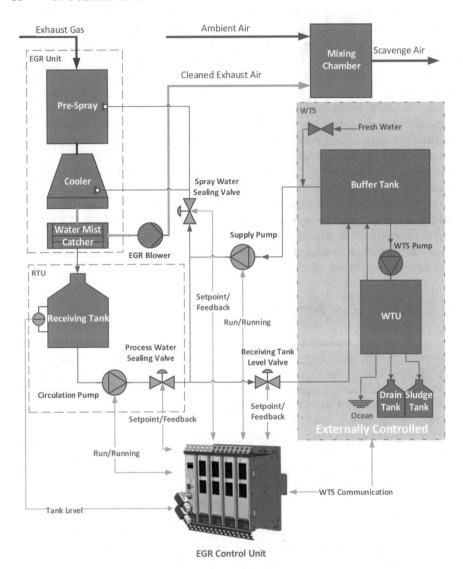

Fig. 1. Water handling system setup (from [9]).

drained to the Receiving Tank Unit (RTU) and recirculated to the EGR unit.
Part of the recirculated water is led to the WTS to be cleaned and returned
to the EGR unit. The surplus of water originating from the combustion process is drained from the WTS as bleed-off water and discharged to the sea. The
residuals from the cleaning process are discharged to the sludge tank. Depending on load of the engine and ambient conditions the combustion process will
accumulate water in the system, which must be discharged as bleed-off water. If
discharged to the sea, the bleed-off water must meet the quality criteria required

by International Maritime Organization (IMO)[1], presently defined in the 2015 Guidelines for Exhaust Gas Cleaning Systems, MEPC 259 (68). Bleed-off water, which does not meet the discharge criteria or cannot be discharged to sea due to local restrictions, is drained to a drain tank for delivery at port.

Vessels operating within an emission control area have to comply with the Tier III emission requirements [10]. This is achieved by activating EGR, at which point the WHS is required to run. The control system for the WHS is divided into two parts, the EGR control which is part of the distributed engine control system and the WTS control which is delegated to the producer of the auxiliary system. The engine control system consists of several multi-purpose controllers. Each controller is composed of a power module, multiple I/O chassis and an Field-Programmable Gate Array (FPGA). All controllers on the engine are identical but the software running on the FPGA determines the specific control objective. The controller controlling the WHS is called the EGR Control Unit (EGRCU) and is seen at the bottom of Fig. 1, with the connections relevant for this simulation. This paper focuses on the control of the WHS. The remaining control of the EGR system will not be covered.

The WHS is controlled and monitored by the EGR control, so that water can be provided to clean and cool the exhaust gas. There are two main water loops that can be distinguished. The recirculation loop where the water from the EGR unit is sent to the RTU and back again by the 'Circulation pump' via the 'Process Water Sealing Valve' and 'Spray Water Sealing Valve'. The other loop is where part of the water from the recirculation loop is sent via the 'Receiving Tank Level Valve' to the externally controlled system, the WTS. The water from the WTS is sent back to the recirculation loop with the 'Supply Pump'. The WTS receives the processed water from the RTU and is collected in the buffer tank. A separate system in the WTS treats the water of the buffer tank. Any excess of water is either sent to the sludge/drain tank or, if the water quality parameters are met, the water can be sent overboard.

The objective of the control loop discussed in this paper is to maintain the water level of the 'Receiving Tank' within specified limits. During start up and shutdown of the WHS the actuation timing of the components has a direct impact on the water level. During running mode, the water level is controlled by the 'Receiving Tank Level Valve' and compensates for deviations in the water flow due to e.g. engine load, exhaust gas and scavenge air pressure changes.

3   WHS Simulation

This section describes the development process of the primary WHS control strategy. The approach and tools used for the first edition of the control system are described and the solution is evaluated.

---

[1] http://www.imo.org.

3.1   The MAN Software Application Framework

As mentioned above the application development at MDT is carried out in a comprehensive in-house C++ software application framework. The framework is developed to enable development of DE control models. The main advantage of the software application framework is the possibility of cross compiling the same application to both SIL, HIL and target platform, see Fig. 2. SIL simulation

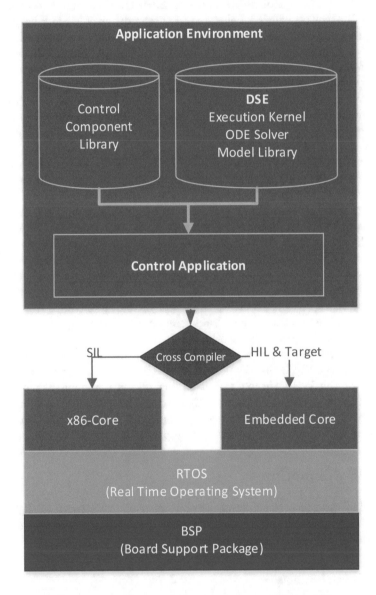

Fig. 2. The software application framework (from [9]).

of target code is made possible by compiling the Board Support Package (BSP) and the Real Time Operating System (RTOS) to an x86 platform. With the SIL simulation, engineers are able to test their application on their own PC. When moving to HIL or target, the same application code, the BSP and the RTOS, are simply cross-compiled to the embedded core of the controller. The primary focus of the framework is control development, where algorithms are directly implemented in C++ with a vast amount of reusable components and macros available, aiding engineers. For CT models, an extension to the framework, called DSE, can be utilised. This includes a kernel for execution, an Ordinary Differential Equation (ODE) solver and a model library of physical components. Models created in DSE are executable on both PC (SIL) and the HIL platform, given that the abstraction of the models allow for real-time execution.

When challenged with a new application, control engineers at MDT often start studying the physical dynamical challenges of the system in MATLAB/Simulink. When a sufficient understanding of the system is achieved, the control strategy is formulated in the software application framework and tested against a DSE model implementation of the MATLAB model. DSE is designed with HIL execution in mind, and while it can simulate complex CT systems, the models implemented are often at a lower abstraction level than e.g. the MATLAB models.

3.2   The WHS Model

The WHS model is divided into a control algorithm created in the software application framework and a model of the physical components in DSE. The control algorithm is created as a component in the controller EGRCU along side the additional components that comprise the entire engine control. The control model consists of a Proportional Integral (PI)-controller regulating the 'Receiving Tank Level Valve' set-point from the 'Process Water Receiving Tank' level sensor feedback. Besides controlling the receiving tank level, the control algorithm also has to ensure that transitions between states in the system is possible, according to signals from engine operators and the WTS control system. This control strategy is formulated in a state-machine running in the EGRCU. The DSE model describes the pressures and flow of all the components illustrated in Fig. 1. The model resembles the preliminary model developed in MATLAB but without details such as pressure build-up in piping, water accumulation in components and pressure loss over valves. The main purpose of the DSE is to test the control strategy, ensuring that all state transitions are possible and that the regulator works correctly. To prepare for HIL simulation, the DSE model is placed in a separate controller called the Engine Simulation Unit (ESU), shown in Fig. 3. The ESU controller is installed in the HIL platform as a representation of the real engine. Data exchange between the EGRCU and ESU imitate the actual communication with the real engine through analog and digital IOs. In SIL, a software implementation of virtual IOs and network simulate the communication.

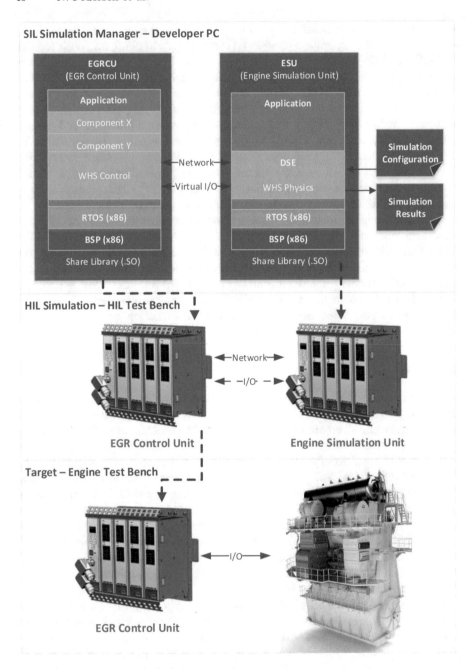

Fig. 3. SIL WHS simulation setup (from [9]).

3.3   Simulation and Verification Process

SIL simulation is achieved by compiling controllers to an x86 platform and into shared libraries. The shared libraries are executed by a simulation manager ensuring temporal execution and correct data exchange between controllers. How the embedded control software has been adapted to enable deterministic simulation has been described in [11] and will not be further explained in this paper. A simulation scenario is provided to the DSE model through a simulation configuration file and the results are delivered in a simulation results file.

In the verification process, when the system has been properly tested in the SIL environment, the models are moved to the HIL platform by cross-compiling to the embedded system. On the HIL test bench additional tests of computation overhead, communication and additional temporal issues are performed.

For final testing, an engine test bench is physically available at the MDT research center in Copenhagen. Only a single test bench is available due to the immense cost and sheer size of the engine. Available time-slots on the test engine are very limited and extremely costly, due to the fuel consumption and amount of operators required, so the proper modelling and testing of the previous steps is desirable.

3.4   Simulation Evaluation

The SIL simulation was used to develop a functioning PI controller that regulates the process water tank level and a state machine for actuating valves and pumps according to a number of states for starting and stopping the WHS system. The system was tested on the HIL test bench, ensuring that the systems worked properly on the controller hardware. Finally, a test session was performed on the engine test bench. This test showed that the PI controller worked as intended, however, an unsuspected situation occurred when stopping the WHS system.

Figure 4 shows the results of running the initial control strategy on the real system. After 100 s the EGR control system ordered the WHS system to prepare for EGR operation. Then after 50 s the state-machine was finished starting different pumps and opening of valves. At this point the Process Water Receiving Tank (PWRT) control is fully engaged. The bottom graph in Fig. 4 shows how the Receiving Tank Level Valve (RTLV) is regulated to redirect water from the process circuit to the WTS for cleaning and stabilizing the PWRT level. The top graph in Fig. 4 shows the water level in the PWRT and how it became stable after a transient period, proving that the PWRT control worked correctly during WHS operation. Vessels are not always required to use EGR, so shut-down of the system should be possible during engine operation. After 600 s a command from the engine operator was ordered, from an operating panel, for the EGR system to shutdown and WHS to stop operation. The state machine started emptying the tank to reach a stable offline level around 20–25% in the tank. At 676 s a behaviour not seen in either the SIL or HIL simulation was observed. When the WHS system started, water in the WMC started to accumulate gradually. At a point in time an equilibrium was achieved due to increased water

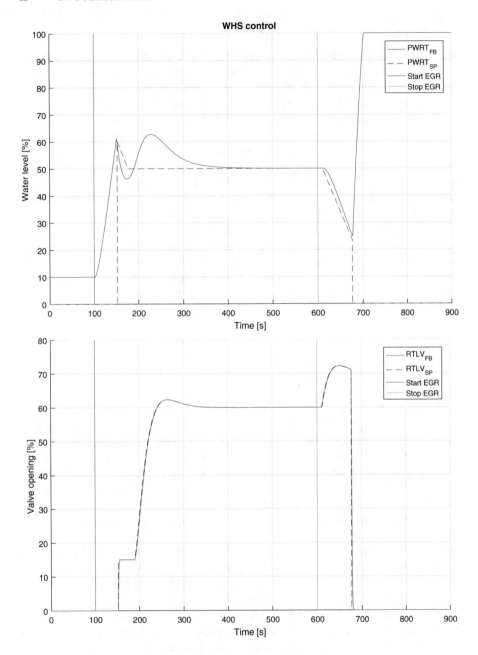

Fig. 4. WHS control results (from [9]).

pressure resulting in a consistent flow through the WMC (without increased water accumulation in the WMC as a consequence). During shutdown, when the desired water level in the PWRT was achieved and RTLV control stopped, the accumulated water in the WMC started to flow to the PWRT tank. As seen in Fig. 4, the amount of residual water in the WMC is so large that it overfills the PWRT.

From the engine test bench it was discovered that the controller actuating the RTLV was working properly, but the state-machine was not properly handling the emptying of the WMC. Engine tests are very costly and MDT would like to investigate if a more efficient development process can be achieved. In the DSE model used for development of the state-machine, the accumulation of water in the WMC had not been modelled. To improve the control strategy of the WHS, a higher-fidelity model should be used. Instead of simply extending the DSE model to include a more detailed WMC model, a co-simulation solution was chosen [1]. The co-simulation should not only include a detailed WMC model but be so generic that changes to the system layout and more advanced models of components can be easily implemented. The argument for the choice of co-simulation is given in the following section.

4   Targeting Co-simulation

The software application framework and DSE is central for development because they are designed for the target platform of the final system, and directly enable validation through both SIL and HIL. Keeping this in mind it is rational to keep the control systems in the framework. However, there are a number of options for enhancing physics modelling that would be beneficial:

- Porting the controller software to a notation that can be used in the MATLAB environment, where it is easier to express the physical model. This would however, just shift the issue to the controller, that then needs to be ported back to the software application framework.
- Enhancing the physical model in DSE, while the standard approach, it is more time consuming than using a dedicated modelling tool like MATLAB, but it enables faster simulation speeds.
- Use a generic solution that enables co-simulation between the control system expressed in the software application framework and a physical modelling tool like MATLAB. This will not require any changes to software development at MDT, but would enable physical models to be created using the desired modelling tool. It would potentially run slower than a complete model expressed in DSE but would be more flexible. This solution would make the representation of the physical dynamics more detailed in SIL simulation. However, the co-simulation model would not be able to run on the HIL platform. The purpose of the HIL test is not to test functionality already verified in the SIL simulation, but to ensure computational overhead and investigate temporal aspects.

The latter approach was chosen because it is generic and it allows well-known modelling tools in the physical domain to be used. To interface between models,

FMI is used to provide a standardised interface. The last constraint on the co-simulation is that it needs to be performed across architectures and platforms. The software application framework is required to run as a Linux 32 bit process. The reason for this is, as previously mentioned, because the framework is developed to build directly to the embedded system which is a 32 bit architecture. It is also a requirement that the physical modelling environment be a Windows 64 bit application. The control developers working in e.g. MATLAB do so in Windows 64 bit and management-wise, introducing co-simulation to the current tool-chain would be preferable. Another reason for the choice of deviation in platform is the lack of 32 bit support for MATLAB on Linux.

It must be possible to run the simulation using Linux 32 bit for the software application framework and Windows 64 bit for MATLAB. Therefore a solution is to use the FMI COE from the INTO-CPS research project since it supports both. However, the co-simulation could not originally span architectures or platforms. Therefore an extension is presented in Subsect. 5.1 that enables co-simulation in a distributed setting, spanning both architectures and platforms.

4.1   Functional Mock-up Interface

FMI is a tool independent standard developed within the MODELISAR project [7]. It supports both model exchange and co-simulation and exists as Version 1, released in 2010 and Version 2, released in 2014. It was developed to improve exchange of simulation models between suppliers and Original Equipment Manufacturers (OEM). The standard describes how simulation units are to be exchanged as ZIP archives called *Functional Mock-up Units (FMUs)* and how the model interface is described in an XML file named *modelDescription.xml*. The functional interface of the model is described as a number of C functions that must be exported by the library that implements the model inside the FMU. Since the FMU only contains a binary implementation of the model it offers some level of intellectual property protection. The focus of this work is on co-simulation, where each FMU is capable of participating in a co-simulation without the need of an external solver, i.e. each FMU includes the required solvers needed for simulation.

4.2   The INTO-CPS Tool Chain

While individual tools and formalisms for the development of controllers, including simulation, testing and code generation, are very mature, the design workflow is only partially integrated. The Horizon 2020 project INTO-CPS [12,13] aimed at closing this gap, by creating an "Integrated Tool-chain for the model-based design of Cyber-Physical Systems" [14]. The chain of tools are connected as illustrated in Fig. 5, moving all the way from requirements to final realisations [15][2]. The underlying idea behind a collection of individual tools which are connected

---

[2] The tool chain is available as open source, owned and further developed by the INTO-CPS Association, see http://into-cps.org/.

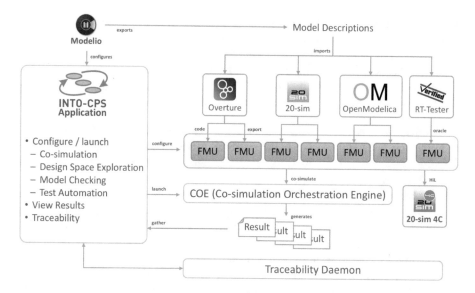

Fig. 5. The INTO-CPS tool chain.

in this fashion is that different stakeholders (potentially from different departments or even different companies) can model their constituent system in the
notation they are familiar with using the tools they already know.

The INTO-CPS Co-Simulation Orchestration Engine: Maestro

One of the core tools of this chain is Maestro, the INTO-CPS Co-Simulation
Orchestration Engine (COE), which is a fully FMI 2.0 co-simulation compliant
Master supporting both fixed and variable step size simulations [16]. It was
decided to use FMI as the interface for the different simulation and testing
tools, since it is a mature standard[3] with an active community.

The COE is developed in a combination of Java and Scala, which makes it
multi-platform and provides the simulation service through HTTP with a REST
architecture[4]. Currently, two methods for time-stepping are implemented; one
for fixed time steps, and one for variable time steps. To determine the step size
when using variable time steps, one must select one or more constraints. The
constraints are the following:

Zero-Crossing: This occurs at a point when a function changes its sign or
crosses a boundary.
Bounded Difference: Keeps the minimal and maximal value of a set of values
within a specified amount
Samping Rate: Ensures that predefined repetitive time instants are chosen
without jitter.

---

[3] http://fmi-standard.org.

[4] The protocol is available at https://into-cps-association.github.io/simulation/coe.
html.

Max Step Size [17]: The concept is to prompt each FMU for the maximum
time step that it can progress and then choose the overall minimum. This
functionality is not part of the FMI Standard, but each tool in INTO-CPS
that generates FMUs support this functionality. The purpose of Max Step
Size is to precisely hit the point in time where events from computational
systems occur and avoid unnecessary synchronisations.

Additionally, the COE supports successive substitution in order to increase
stability and is capable of executing co-simulations using parallelism to improve
the simulation speed [18]. Furthermore, it provides live streaming of variables to
be used for e.g. live plotting or analysis.

The basic iteration principle of FMI co-simulation is to satisfy the dependencies of the FMUs and allow them to progress. This is carried out by 1. getting
inputs, 2. setting outputs, and 3. invoking each FMU to perform a step, after
which the process iterates until a predefined end time is reached. If variable time
step combined with Max Step Size is used, then the FMUs are queried for a step
size between step 2 and 3. The other constraints uses a mix of calculating the
step size between step 2 and 3, and validating the performed step after step 3. An
FMU can reject a given step size because it is too large. In this case the COE
performs a roll back on all FMUs, if they support setting and getting states,
and afterwards uses a smaller step size. Figure 6 presents three FMUs and their
respective ports. The dependencies between these FMUs are therefore: output
a to input b1, output b3 to input c1, output c2 to input b1. The COE uses a
jacobian iteration approach, which means that all outputs are retrieved from the
FMUs (a, b3, and c2 in Fig. 6) followed by setting all inputs (b1, c1, b2), and,
lastly, invoking each FMU (A, B, and C) to perform a step [19]. This approach
allows for parallelising the FMU step calculations, where an improvement in
simulation speed is most likely.

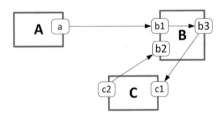

Fig. 6. An example of a simulated CPS with dependencies between FMUs (the rectangles A, B and C) via their respective ports (the rounded squares a, b1, b2, b3, c1
and c2).

The COE has also been extended with a feature called hierarchical FMU that
addresses sharing of FMUs between vendors and thus eventually simulations.
The FMI standard describes how to create an encapsulated FMUs but does
not describe how to share these in a meaningful manner when they comprise
a subsystem of multiple FMUs with pre-configured connections, parameters,

step sizes and other configurations for co-simulation. As a result, it becomes increasingly difficult to configure a co-simulation as more FMUs, which make up a sub-system and thus requires a specific co-simulation configuration, are added. The WHS also presents this challenge because of the externally provided WTS subsystem, had it been provided as a collection of FMUs. An example of creating an hierarchical FMU is shown in https://www.youtube.com/watch? v=MKZb3HkyVtc and described in [16]. The video demonstrates how a fully configured co-simulation using the COE can be exported as an FMU with some inputs and outputs being unconnected as hooks to the full system. This means that the new hierarchical FMU is comprised of an instance of all required FMUs, a co-simulation configuration and finally an instance of the COE that can be controlled through FMI making it a slave of the overall system and a master of the encapsulated co-simulation. As a result, such an hierarchical FMU can be developed and tested by a sub-supplier and used in a subsequent co-simulation by simply connecting the missing inputs and outputs, thereby preserving the co-simulation configuration created by the sub-supplier, who is expected to be the expert of the subsystem.

In addition to the baseline tools incorporated inside the tool chain, a number of other modelling and simulation tools have been tested with the COE. This includes both commercial tools such as Dymola, Modelon, SimulationX and Unity as well as additional open source tools such as 4Diac[5] and AutoFO-CUS3[6]. While the COE is multi-platform it does not directly support mixed-architecture (combinations of 32 bit and 64 bit architectures) or mixed-platform (combinations of e.g. Windows and Linux) simulations as required for the WHS system as discussed next in Sect. 5.

5 WHS Co-simulation

To co-simulate the WHS from Sect. 3 using FMI, it is required that both constituent models must support FMI, and that a suitable orchestration engine that supports FMI and the required platform and architecture combination is available. Since no such simulator was available it was necessary to develop an extension to the COE is described in Subsect. 5.1. To enable FMI for the constituent models, an extension was developed for the MDT software application framework which has been published in [11,20]. The model of the WHS is exported from MATLAB to an FMU using the Modelon FMI Toolbox for MATLAB/Simulink[7]. The complete co-simulation model is shown in Subsect. 5.2, and evaluated in Subsect. 5.3.

5.1 Distributed COE Extension

To enable multi-architecture co-simulation, the challenge of mixing 32 bit and 64 bit code needs to be addressed. Essentially, two processes with inter-process

---

[5] See https://www.eclipse.org/4diac/.

[6] See https://af3.fortiss.org/.

[7] http://www.modelon.com/products/fmi-tools/fmi-toolbox-for-matlabsimulink/.

communication are required by the host system to realize this, where one of them acts as the simulation master. A similar challenge arises when different platforms need to interact in a co-simulation.

An extension to the COE was developed that is capable of both simulating across architectures and platforms. The solution chosen was to utilise an extension point in the COE that allows a custom factory to be used for FMU instantiation. An overview of the extension is realised and shown in Fig. 7. The COE uses the distributed factory to instantiate FMUs that require execution with a different host configuration, either architecture or platform deviation.

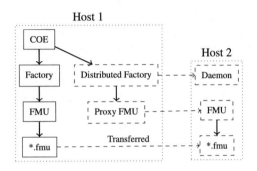

Fig. 7. Distributed extension overview (from [9]).

The extension is realized using Java-Remote Method Invocation (JAVA-RMI) to provide cross-platform communication [21]. It consists of a distribution factory and an FMU proxy that is plugged into the COE. It uses a daemon that must run on the remote host to provide a service that enables the COE to remotely load and control FMUs. The COE configuration is also extended to specify which remote daemon a specific FMU should be executed by. When a co-simulation is started the COE will communicate with the specified remote daemons to configure the co-simulation by first pushing FMUs to the remote daemons that then in turn load and setup a communication channel for the loaded FMUs. These will then be connected to the FMU proxy in the COE, which is responsible for handling remote communication.

5.2   Co-simulation Setup

The co-simulation setup is illustrated in Fig. 8. The master COE is running on the Windows host, and the COE-deamon on the Linux host. A JSON configuration file describes the co-simulation setup to the COE. The configuration file tells the COE where the FMU-archives are located and on which host-IP they should be executed. The configuration file also contains information about connections between the inputs and outputs of the FMUs, parameters and simulation algorithm: variable/fixed time step.

The WHS MATLAB model is code generated into an FMU using the Modelon FMI toolbox for MATLAB/simulink. The toolbox compiles the MATLAB model

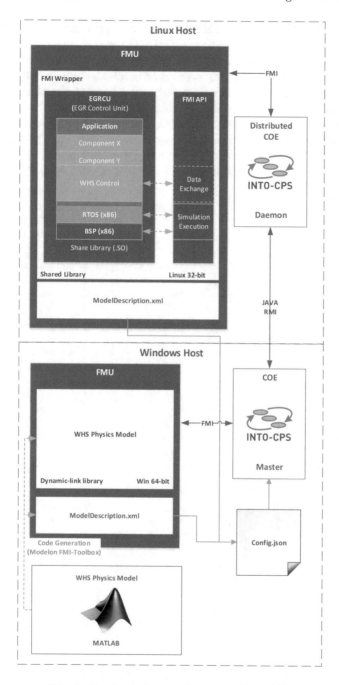

Fig. 8. Co-simulation configuration (from [9]).

to a 64-bit DLL including the FMI-API and auto-generates the model description XML defining the interface to the FMU. The control system FMU has been created by wrapping the FMI Application Programming Interface (API) around the SIL simulation and compiling it to a Linux 32-bit shared library. The simulation can access the RTOS for scheduling and a hook to the clock in the BSP, all described in [11]. Accessing the variables of the WHS control is done through a proxy interface that provides pointers to internal variables to be manipulated. Furthermore, the proxy interface introduces a conversion layer between internal types such as FIXPOINT16 and FMI-types. The SIL simulation only includes the EGRCU controller, which contains the WHS control. The ESU controller,

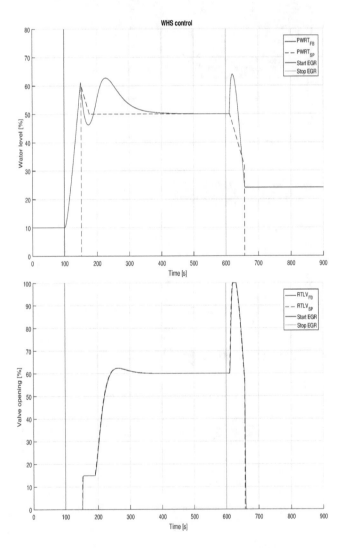

Fig. 9. WHS control results (from [9]).

containing the DSE models, has been replaced with the MATLAB model in the co-simulation.

The simulation is initiated through the COE and the results are delivered in Comma-Separated Values (CSV) format on the Windows host.

5.3 Co-simulation Evaluation

Figure 9 shows the simulation results with the proposed Co-Simulation setup, where the DSE physical model has been replaced with the more detailed MAT-LAB model. Being able to anticipate the behaviour of the accumulated water in the Water Mist Catcher, it is now possible to address it and to modify the state machine accordingly to control the components in a more appropriate way during WHS shutdown. The first 600 s show the same response as in Fig. 1. However, the new state machine now ensures that the WMC is drained before shutting down RTLV control. This prevents the water level in the PWRT from overflowing, but instead stabilise at a desired level of approximately 20–25%.

6   Concluding Remarks

This paper shows how the control development process at MAN Diesel & Turbo could benefit from introducing co-simulation. The conventional approach, where control algorithms and strategy are formulated using simplified models of the physical dynamics, is not always able to properly represent the complexity of the system. Importantly, with the conventional approach, defects are typically not found before moving to the expensive engine test bench. With the co-simulation approach proposed in this paper, higher-fidelity constituent models of physical dynamics, formulated in dedicated tools, can be simulated together with a SIL simulation of the control software, at an earlier stage of development. In the example presented, the accumulation of water in a water mist catcher was neglected in the initial model, essentially resulting in a water tank overflow during shutdown. With the co-simulation, a more detailed model, formulated in MATLAB, could be used for developing a working control strategy. Had the co-simulation been used for initial control development, the issues seen on the test engine would likely have been discovered at an earlier stage, saving money and time.

The main challenge enabling co-simulation at MDT was the deviation in both OS platform and hardware architecture of the simulation tools used. The SIL simulation of the control software is constrained to a 32-bit Linux platform and the MATLAB environment was required to run on a 64-bit Windows platform due to change management concerns. In cooperation with the INTO-CPS-project, the INTO-CPS Co-simulation Orchestration Engine for executing co-simulations complying with the FMI standard was adapted to enable distributed co-simulation. With the distributed extension to the COE it was possible to conduct the co-simulation despite the platform and architecture deviation.

Besides the promising results shown in this paper, additional benefits from the INTO-CPS co-simulation tool chain are anticipated. The extensions developed to the MDT frameworks and development processes enable not only co-simulation of the EGR system, but also of any other system to be developed in the future by MDT, with minimal effort. In the future it will be explored when it makes sense to also make use of more capabilities from the INTO-CPS tool chain. In particular it is imagined that there could be advantages of trying to make models of the supervisory control instead of coding this in C++ initially. For example, using Overture or one of the other baseline tools to develop the models enable the possibility of utilising the Max Step Size functionality described in Subsect. 4.2. Furthermore, using the concept of hierarchical FMU it is possible to split up large co-simulations into subsystems and define co-simulations characteristics of the individual subsystems.

In the EGR Water Handling System presented here, a subsystem called Water Treatment System is delivered by an MDT OEM and neither modelled nor controlled by MDT. One of the main advantages of FMI used by INTO-CPS is that models are exchanged on a binary level offering protection of intellectual property. One of the future ambitions is to be able to share models with OEMs so systems like the WHS and WTS can be simulated together, improving both companies products. Part of the high-fidelity models developed at MDT are very complex and require time to simulate, especially if co-simulated with several other models. One of the additional advantages of the distributed co-simulation is that the simulation process can be parallelised and perhaps distributed to centralised high-performance hardware. This could potentially speed up simulation execution times and enable more advanced system investigations, previously deemed too time consuming. Initial work on using the COE in a cloud setting has already been initiated, in particular in relation to design space exploration in situations where there is large room for different alternative solutions [22].

Acknowledgements. The work presented here is partially supported by the INTO-CPS project funded by the European Commission's Horizon 2020 programme under grant agreement number 664047.

References

1. Gomes, C., Thule, C., Broman, D., Larsen, P.G., Vangheluwe, H.: State of the art. Technical report, Co-simulation (2017)
2. Gomes, C., Thule, C., Broman, D., Larsen, P.G., Vangheluwe, H.: Co-simulation: a survey. ACM Comput. Surv. **51**(3) (2018). https://doi.org/10.1145/3179993. Article No. 49
3. MathWorks: Matlab official website, October 2011. http://www.mathworks.com
4. Dassault Systèmes: 3DS official website, March 2017. https://www.3ds.com/products-services/catia/products/dymola
5. Kleijn, C.: Modelling and simulation of fluid power systems with 20-sim. Int. J. Fluid Power **7**(3) (2006)

6. Fitzgerald, J., Larsen, P.G., Verhoef, M. (eds.): Collaborative Design for Embedded Systems - Co-modelling and Co-simulation. Springer, Heidelberg (2014). https://doi.org/10.1007/978-3-642-54118-6
7. ITEA Office Association: ITEA 3 project 07006 MODELISAR, December 2015. https://itea3.org/project/modelisar.html. Accessed 12 June 2015
8. MAN Diesel & Turbo: Emission project guide, MAN BW two-stroke marine engines. Technical report, MAN Diesel & Turbo (2016)
9. Pedersen, N., Lausdahl, K., Sanchez, E.V., Larsen, P.G., Madsen, J.: Distributed co-simulation of embedded control software with Exhaust Gas Recirculation water handling system using INTO-CPS. In: Proceedings of the 7th International Conference on Simulation and Modeling Methodologies, Technologies and Applications (SIMULTECH 2017), Madrid, Spain, pp. 73–82, July 2017. ISBN: 978-989-758-265-3
10. International Maritime Organization IMO: MARPOL ANNEX VI and NTC 2008 with guidelines for implementation - supplement. Technical report, September 2015
11. Pedersen, N., Bojsen, T., Madsen, J., Vejlgaard-Laursen, M.: FMI for co-simulation of embedded control software. In: Linköping Electronic Conference Proceedings, no. 124, pp. 70–77 (2016)
12. Fitzgerald, J., Gamble, C., Larsen, P.G., Pierce, K., Woodcock, J.: Cyber-Physical Systems design: formal foundations, methods and integrated tool chains. In: FormaliSE: FME Workshop on Formal Methods in Software Engineering, Florence, Italy, May 2015. ICSE 2015
13. Fitzgerald, J., Gamble, C., Payne, R., Larsen, P.G., Basagiannis, S., Mady, A.E.-D.: Collaborative model-based systems engineering for Cyber-Physical Systems – a case study in building automation. In: INCOSE 2016, Edinburgh, Scotland, July 2016
14. Larsen, P.G. Fitzgerald, J., Woodcock, J., Fritzson, P. Brauer, J., Kleijn, C., Lecomte, T., Pfeil, M., Green, O., Basagiannis, S., Sadovykh, A.: The INTO-CPS project. In: Integrated Tool Chain for Model-Integrated Tool Chain for Model-Based Design of Cyber-Physical Systems, CPS Data Workshop, Vienna, Austria (2016)
15. Bandur, V., Larsen, P.G., Lausdahl, K., Thule, C., Terkelsen, A.F., Gamble, C., Pop, A., Brosse, E., Brauer, J., Lapschies, F., Groothuis, M., Kleijn, C., Couto, L.D.: INTO-CPS tool chain user manual. Technical report, INTO-CPS Deliverable, D4.3a, December 2017
16. Thule, C., Lausdahl, K., Larsen, P.G., Meisl, G.: Maestro: the INTO-CPS co-simulation orchestration engine (2018). To be submitted to Simulation Modelling Practice and Theory
17. Broman, D., Brooks, C., Greenberg, L., Lee, E.A., Masin, M., Tripakis, S., Wetter, M.: Determinate composition of FMUs for co-simulation. In: 2013 Proceedings of the International Conference on Embedded Software (EMSOFT), pp. 1–12 (2013)
18. Thule, C., Larsen, P.G.: Investigating concurrency in the co-simulation orchestration engine for INTO-CPS. In: Petrenko, A.K., Kamkin, A.S., Terekhov, A.N. (eds.) Preliminary Proceedings of the 10th Anniversary Spring/Summer Young Researchers' Colloquium on Software Engineering (SYRCoSE 2016), Krasnovidovo, Russia, 30 May–1 June 2016, pp. 223–228. ISP RAS, May 2016
19. Bastian, J.. Clauss, C., Wolf, S., Schneider, P.: Master for co-simulation using FMI. In: 8th International Modelica Conference (2011)

20. Pedersen, N., Madsen, J., Vejlgaard-Laursen, M.: Co-simulation of distributed engine control system and network model using FMI and SCNSL. In: 10th IFAC Conference on Manoeuvring and Control of Marine Craft MCMC 2015, vol. 48, no. 16, pp. 261–266 (2015)
21. Java remotemethodinvocation specification 1.5.0 (2004). http://java.sun.com/j2se/1.5/pdf/rmi-spec-1.5.0.pdf
22. Gamble, C., Payne, R., Fitzgerald, J., Soudjani, S., Foldager, F.F., Larsen, P.G.: Automated exploration of parameter spaces as a method for tuning a predictive digital twin (2018). Submitted for publication

Model Oriented System Design Applied to Commercial Aircraft Secondary Flight Control Systems

Isabella Panella[1] and Graham Hardwick[2(✉)]

[1] Systems Engineering Manager Actuation Systems, UTC Actuation Systems,
Stafford Road, Wolverhampton WV10 7EH, UK
isabella.panella@utas.utc.com
[2] Chief of Systems Performance Wolverhampton, UTC Actuation Systems,
Stafford road, Wolverhampton WV10 7EH, UK
graham.hardwick@Utas.utc.com

Abstract. Modelling and simulation is at the core of future technology developments such as Big Data analytics, Artificial Intelligence, and More Electric Aircraft. The work herein presented explores the adoption of modelling and simulation methodologies based on a Model Based Systems Design approach to support the definition of performance envelopes and verification procedures for the Power Drive Unit (PDU) of a Fowler flap system commonly used on commercial aircraft. Specifically, the work presented will introduce the modelling and simulation of the flap system and the modelling and simulation process used to assess the influence of PDU gearbox ratio using the Simulink flap system model. Three types of analysis are carried out: time histories are provided to understand the key parameters in the system performance; sensitivity study to identify optimal gear ratio for the PDU; and finally the performance envelope of the PDU is evaluated.

Keywords: More Electric Aircraft · Flap system · Model based design

1 Introduction

The aerospace landscape is evolving into a more connected, more intelligent, and more electric one. The growing application of Artificial Intelligence (AI) technologies and Big Data analytics, the advances in sensors technologies and communication infrastructure are more than ever included in the design process of commercial aircraft and with them the associated challenges of guaranteeing the systems integrity, safety, reliability, and maintainability.

Increasing commercial investments within the global market have seen significant expansion of robotics and other intelligent systems to address consumer and industrial applications. At the same time, autonomy is being embedded in a growing array of software systems to enhance speed and consistency of decision-making, among other benefits, as reported in [1].

© Springer Nature Switzerland AG 2019
M. S. Obaidat et al. (Eds.): SIMULTECH 2017, AISC 873, pp. 55–76, 2019.
https://doi.org/10.1007/978-3-030-01470-4_4

As reported in [1], p. 18, an autonomous system must be capable to independently compose and select a course of action among multiple options in order to accomplish its goals based on its knowledge and understanding of the world, itself, and the situation.

Moreover, in the past 20 years, the commercial aircraft landscape has seen an increase focus on environmental issues. As reported in [2], higher energy efficiency, lower nitrogen oxide (NOx) emissions, and lower audible noise for aircraft have become critical in the design of future commercial aircraft.

The demand to optimize aircraft performance, decrease operating and maintenance costs, increase dispatch reliability, and reduce gas emissions have driven the industry to support the development of more electric aircraft (MEA), and ultimately an all-electric aircraft.

More Electric Aircraft are seen as a means to improve aircraft weight, fuel consumption, total life cycle costs, maintainability and aircraft availability.

The MEA implementation supports the utilization of electric power for all non-propulsive systems.

There are two energy sources on conventional aircraft: the primary sources are the engines and the secondary sources are electric, pneumatic, and hydraulic systems powered by the primary source.

As reported in [3], traditionally these secondary sources are driven by a combination of different secondary power such as hydraulic, pneumatic, mechanical and electrical.

Advances in the field of power electronics, specifically the development of high density electric motors, power generation and conversion systems, as well as in the field of electro-hydraulic actuators have made MEA aircraft a reality, with the most outstanding examples represented by the Boeing 787, Airbus 380, and Lockheed F35.

Modelling and simulation is fundamental in the evolution and inclusion of new technologies in the aircraft design process. It enables to understand and evaluate the behavior of the platforms, their safety constraints and performance envelopes, to support the verification process of the system, as well as to assess the risks associated with their development, integration challenges. Moreover, it enables the introduction of new technologies and their evaluation and effectiveness when applied to aircraft systems.

In this chapter, the authors explore the adoption of modelling and simulation methodologies to support the definition of performance envelopes and verification procedures for a commercial aircraft secondary flight control system, also known as a high lift system. Specifically, the work presented will introduce the modelling and simulation of a flap system

The chapter is organized as follows.

In Sect. 2, an overview of the high lift system is provided, whilst in 3 the aircraft design approach is linked to the systems engineering model based design methodology.

In Sect. 4, the high lift functional architecture is drawn and discussed, followed by the detailing of the mathematical model (Sect. 5). The results of the work are presented in Sect. 6 and conclusions and recommendations are reported in Sect. 7.

2 High Lift Systems

UTC Aerospace Systems design, manufacture, and integrate secondary flight control systems for a variety of commercial aircraft, from wide body to single aisle configuration, from business jets to the A380.

System modelling tools within the business are utilized from the proposal/concept stage all the way through system design, development, manufacturing, system integration, and in service-conditions.

Commercial aircraft utilize secondary flight control surfaces, such as flaps and slats to modify the wing profile in order to increase aerodynamic lift for a given air speed. This allows aircraft landing speeds/distances to be reduced.

The lift provided by a wing follows Bernoulli's law:

$$L = 1/2\rho V^2 S C_L$$

Whereby L is the lift provided, ρ is the air density, V is the speed of the airflow around the wing, S is the wing surface area, and C_L is the lift coefficient. At any given speed, the increase of lift of an aircraft can be achieved by increasing:

(a) The wing surface area (S), and/or
(b) The lift coefficient (C_L)

High-lift systems enable to increase the area of the wing and to change the shape of the wing aerofoil to support a change in lift coefficient. There are two types of high-lift systems: the leading-edge slats and the trailing edge flap (Fig. 1).

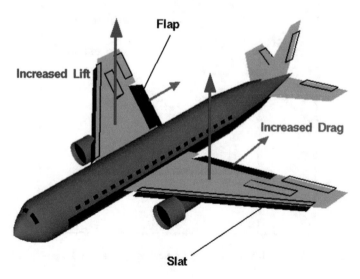

Fig. 1. Slat and Flap [4].

Slats are usually deployed to allow the wing to operate at a higher angle of attack and decrease stall speed. When deployed, the slat systems extend the wing area, which in turn enables to produce a higher coefficient of lift (C_L) for a given angle of attack (α) and speed (V). Therefore, the slat deployment enables an aircraft to fly at slower speeds, or take off and land in shorter distances. They are usually deployed while landing or performing maneuvers which take the aircraft close to stall, whilst they are retracted in cruise conditions.

Flaps systems are usually mounted on the wing trailing edges of a fixed-wing aircraft, which are used to lower the minimum speed at which the aircraft can be safely flown, and to increase the angle of descent for landing.

When deployed the control systems enable the camber or curvature of the wing to be increased, raising the maximum lift coefficient (C_L) or the upper limit to the lift (L) a wing can generate. This enables the aircraft to fly at reduced air speeds and reduce the stall speed of the aircraft. An increase in wing camber also generates an increase in wing drag (D). Therefore, the use of flap systems is beneficial during approach and landing as it also reduces the air speed of the aircraft.

A representative example of the effects of deploying the flap and the slat to the lift coefficient is provided in Fig. 2, taken from Rudolph.

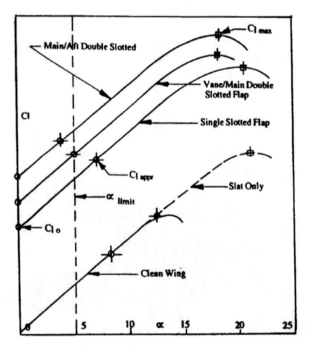

Fig. 2. High-lift performance for landing ([5], Fig. 1.34, p. 43).

In this work, the authors are focusing on the modelling and simulation of a flap system, specifically a Fowler flap system. Fowler flap systems are a special form of slotted flap system. Fowler flaps increase the curvature of the wing and therefore the lift coefficient (C_L), but they also increase the wing area (S). They are slotted to allow air to flow from the lower to the upper side of the wing and therefore to reduce the danger of a stall of the airflow on the upper side.

3 The Aircraft Design Process and System Engineering Approach

In aircraft design, there are three well known phases; conceptual, preliminary, and detailed design.

In the conceptual design phase, the customers' requirements are explored and basic configuration arrangement, sizing, weight, and performance are defined. This requires significant modelling and simulation to be performed as in conceptual design the requirements are extremely fluid and can be questions and modified if needed. The aircraft sizing is carried out and it is driven by design requirements such as payload and range as well as operational requirements, such as range, maneuver required and so on (Fig. 3).

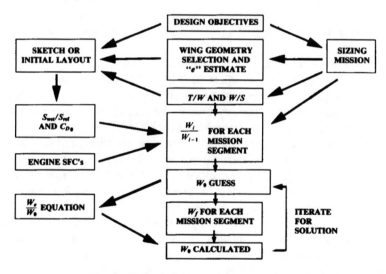

Fig. 3. Refined sizing method ([6], p. 107).

The iterative nature of this phase is described in the Design Wheel reported in Fig. 4:

In the preliminary design phase, the configuration is frozen and the design moves into the design off the major subsystems, the development of test and evaluation procedures, and the development of actual cost estimates.

Finally, in the detailed design phase, the full development of the aircraft takes place and the production process is defined.

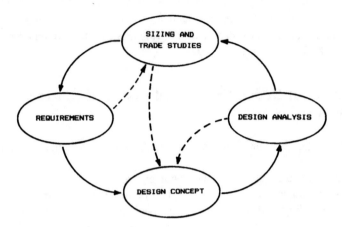

Fig. 4. Design wheel ([6], p. 3).

As it can be seen, the aircraft design approach described matches the model based design approach formalized by the International Council on Systems Engineering (INCOSE). In our work, we focus the modelling and simulation efforts to the validation and verification of the conceptual and preliminary design stages of our Fowler flap system.

The validation and verification system design process is a legacy approach in system design and within the aerospace industry has been endorsed within the standards defined in [7, 8]. Systems engineering utilises Model Based Design within the systems Validation and Verification cycle as a core instrument to guarantee robustness and integrity within the systems, as reported in INCOSE.

The validation and verification processes based on modelling and simulation have been further represented in the following diagram (Fig. 5).

The picture captures the iterative cycles of validation and verification for each stage of the systems engineering life-cycle. This process is followed for the implementation of the functional high lift system. As described in [11], the work herein presented details the iterative nature of the preliminary design stage of a product life cycle, which corresponds to the first level of iteration depicted in Fig. 5. The first iteration level is also referred to as functional level of modelling and simulation and it enables an initial customers requirements decomposition through a process of requirements allocation and elicitation implemented through modelling and simulation, specifically through:

- Sensitivity studies
- Trade off analysis
- Test case development
- Decomposition of requirements

The engineering product development schedule is superimposed with the validation and verification process to highlight the stages where MBSE can add value.

In general, model based design emphasises the use of models through the entire life cycle such as developing test cases and aiding verification activities, generating

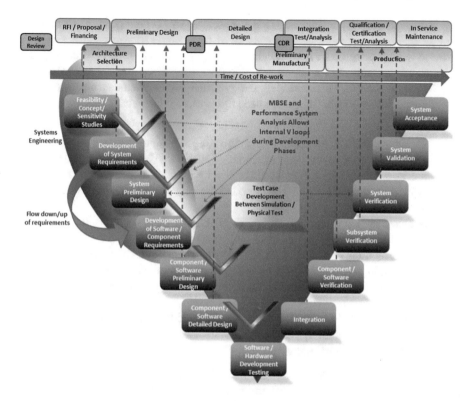

Fig. 5. Validation and verification system design process [11].

prototype control code, supporting solving problems, as well as for hardware in the loop activities and also supporting certification activities.

4 High Lift Functional Architecture

The following describes the High Lift Functional architecture. The architecture shown in Fig. 6 is the result of a previous studies presented in [9, 11].

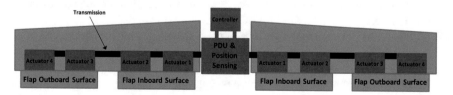

Fig. 6. Generic high-lift architecture ([9, 11]).

The functional architecture provided as described by [11] is a generic medium sized commercial aircraft flap system, characterized by a single transmission line and distributed actuators spaced symmetrically with respect to the aircraft centerline. This architecture presents only a flap system and does not include the slat actuators.

The study in [11] demonstrated the physical layout includes elements of the functionalities that a high lift system needs to present which are:

- Central source Power Drive Unit (PDU) which provides the power drive actuation to the system and has position sensing capability; This interfaces with the secondary flight controller for control and monitoring functions. The PDU is dual channel for redundancy. Hydraulic valves are controlled by the flight controller and regulate the flow to the hydraulic motors. The motors drive a mechanical gearbox that drives the transmission.
- Transmission shafts connect the PDU to the actuators on the wing and hence enabling synchronous movement of both wings.
- Mechanical wing actuation with four rotary geared actuators (RGA) per wing; These contain gearboxes which provide mechanical advantage to the transmission drive torques. This enables the transmission to drive aerodynamic loads with large torques.
- The secondary flight control system interfaces with the PDU, sensors and safety devices which arrest the system during failure case scenarios. It also interfaces with the main aircraft flight controller.

The engineering product development schedule is superimposed with the validation and verification process to highlight the stages where MBSE can add value.

In general, model based design emphasises the use of models through the entire life cycle such as developing test cases and aiding verification activities, generating prototype control code, supporting solving problems, as well as for hardware in the loop activities and also supporting certification activities.

5   Mathematical Model

Due to the long timescales of commercial aircraft project lifecycles the models need to be:

1. Appropriate to application. For example pseudo dynamic modelling is used for static size case development or fully dynamic so that transient dynamics can be interrogated.
2. Modular and documented. Due to programs spanning over decades, technologies and modelling capability will evolve and the model will need to be modular and well documented in order to support future updates.
3. Refer to requirements. Model requires appropriate referencing to requirements where appropriate.
4. Version controlled so that configuration of the model can be accessed and co-ordinated with the correct design build standard of the hardware.
5. Verified using unit, subsystem and system physical tests.

Herein, the mathematical model of a High Lift System as described in [11] is provided and mapped into a Simulink model.

The study in [11] utilised the model to perform a sensitivity study to support optimal design point for the selection of the PDU, considering as the key parameter its gear ratio.

The first step is to define the model requirements. Requirements enable the definition of the overall performance requirement for the system, operational and environmental conditions, as well as regulatory considerations are needed to ensure safety. They set the "boundary conditions" for the systems, which will need to be validated and verified.

It is important to consider another constraint when implementing a simulation model. The model complexity is proportional to run time. Therefore, the model fidelity needs to be traded with the speed of simulation that we want to achieve.

Once the requirements are defined, specific performance descriptions need to be allocated to the functional blocks, and model outputs are linked to the systems requirements, such as:

- Flap system deployment times;
- Flap system hydraulic flow rate;
- Dynamic and steady state transmission loads;
- PDU normal operating velocity.

Dynamic modelling has been achieved through the application of first order differential equations.

These equations are represented in the form of a state-space model within the Simulink modelling environment, which utilises a non-stiff variable step ordinary differential solver (ODE). The model contains continuous states but the governing equations are non-linear and hence the system is a non-linear time invariant system. State space modelling is a control systems technique to represent the dynamic behaviour of a system as reported in [10].

5.1   Model Architecture

As explained in [11] the functional architecture is mapped into the simulation environment through the application of mathematical equations capturing the individual functions' behavior, according to mechanical, hydraulic and electrical physics laws.

Examples of the equations used to create the blocks are provided in the following sections.

The principle of operation is the following. Mechanical transmission blocks connect the PDU to the actuators. The drive to the PDU is generated by transferring hydraulic power into mechanical using hydraulic motors.

Figure 7 highlights the system architecture mapped into a Simulink model as provided by [11]. This contains the following subsystems:

- "Controller" is the Secondary flight controller (which contains I/O to the aircraft controller);
- "Power Drive Unit" is the Dual channel power drive unit;

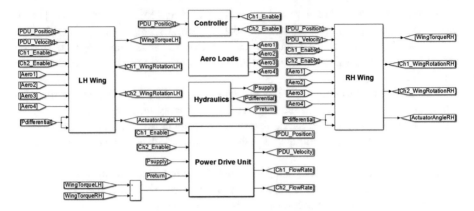

Fig. 7. Model architecture of a generic secondary flight control system in Simulink [11].

- "LH & RH Wing" are the two mechanical wings containing the high lift system;
- "Aero Loads and Hydraulics" are the interface definition of the aerodynamic loads and hydraulic system.

As described in ref [11] the PDU and the controller generically represented in the functional architecture (Fig. 6) are translated into the Simulink diagram (Fig. 7) and are connected using "GoTo" blocks for example:

- Ch1(2)_Enable – controller to PDU enable electrical signal.
- PDU_Position – PDU to controller position
- Aero1(n) – Aerodynamic loads between the interface and wings
- Psupply – Hydraulic supply pressure between interface and PDU

5.2   High Lift Transmission

The high lift transmission wing subsystems are modelled as mechanical blocks that include component stiffness, inertia, damping, transmission efficiency and drag. Reference [11] provides the detailed description which is cited in this section. These blocks are defined as "LH Wing" and RH Wing" (Fig. 7). A portion of these blocks have been expanded in Fig. 8.

Each inertia element contains one-dimensional rotational states (accelerations/ velocities). For example the equation to convert relative transmission shaft deflection (dx) and velocity (dv) into a transmission torques is modelled as follows (where k_{trans} and c_{trans} are the respective torsional stiffness and damping of the transmission):

$$Shaft\,Torque = k_{trans} * dx + c_{trans} * dv \qquad [12] \qquad (1)$$

Figure 8 which is cited in [11] provides the Simulink model of the transmission system including shafts and flap panels showing the connections between the Simulink

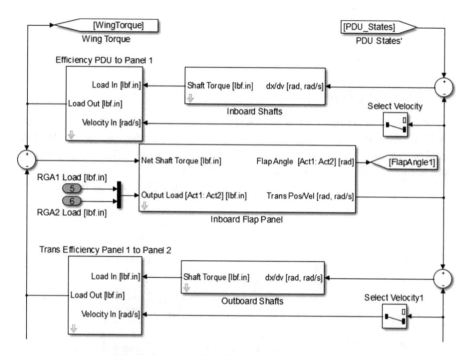

Fig. 8. Model of high lift system in the wing [11].

blocks. The shaft torque is multiplied by a transmission efficiency η_{trans} that calculates the downstream torque for example:

$$Output\ Torque = Input\ Torque * \eta_{trans} \qquad [12] \qquad (2)$$

All shaft torques attached to the panel are summated and then integrated with the aerodynamic loads into the flap panel dynamic model. This process is repeated down the transmission line. The transmission torque and aerodynamic loads are inputs to the flap panel block.

The net torque acting on the flap panel (T_{flap}) is calculated as follows. The aerodynamic loads (T_{aero}) are converted into the transmission torque reference frame using the actuator gear ratio (G_{act}) and mechanical efficiency (η_{act}). This is summed with the torque from the transmission shafts (T_{trans}) and drag (T_{drag}) is deducted which is described in Eq. 3:

$$T_{flap} = (\eta_{act} * T_{aero}/G_{act}) + T_{trans} - T_{drag} \qquad [12] \qquad (3)$$

The model contains additional complexities for example determining the direction of the drag torque dependant on the direction of rotation.

The net torque on the flap panel (T_{flap}) is converted into panel acceleration (A_{flap}) by dividing by the inertia of the panel and transmission components (I_{act}).

$$A_{flap} = (\eta_{act} * T_{flap}/I) \qquad [12] \qquad (4)$$

The velocities and displacements are then determined by integrator blocks.

5.3 Power Drive Unit

As cited by [11] the power drive unit model incorporates two hydraulic channels that contain a constant displacement motor and control valve to activate the motor. More complex hydraulic control arrangements can be used to control the power drive unit more precisely such as electro hydraulic servo valves.

This block is defined as "Power Drive Unit" in Fig. 7 [11] and has been expanded in Fig. 9 [11].

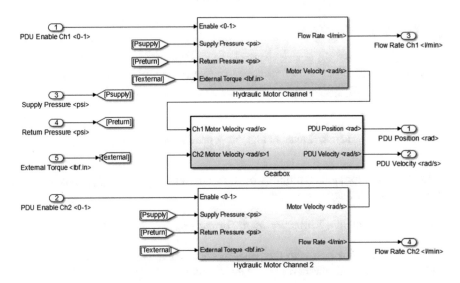

Fig. 9. Power drive unit model [11].

The PDU enable signal controls both the brake in the PDU and the control valve. The control valve dynamics are represented by a first order transfer function using the time constant T_c:

$$Valve\ Transfer\ Function = 1/(1 + Tc.(s)) \qquad [12] \qquad (5)$$

Movement of the control valve determines the pressure drop across (ΔP) the motor. The pressure drop is converted to a motor torque (T_m) by multiplying by the motor displacement (K_{mot}) and incorporating drag (T_{drag_m}) and motor efficiency (η_{mot}) as shown by Eq. 6:

$$Tm = \Delta P * K_{mot} * \eta_{mot} - T_{drag\_m} \qquad [12] \qquad (6)$$

Hydraulic motor acceleration is calculated by dividing the motor torque by the motor inertia. Integrating the acceleration provides the angular velocity of the motor. Both motor speeds are transferred through a gearbox where the PDU output shaft position and velocity states are passed to the wing.

5.4 Secondary Flight System Controller

The secondary flight control opens the control valve and releases the system brakes when the position demand is not equal to the present PDU position. When the system reaches target position the control valve is closed and all brakes are engaged.

Multiple brakes are often used in high-lift systems due to the transmission disconnect failure scenarios. When activated the brakes prevent excessive asymmetry between the left and right hand wing. An actual high-lift controller will typically have a number of sensors to monitor certain failure conditions and arrest the system before they become catastrophic. However, for the purposes of this paper these are omitted.

5.5 Model Parameters

A list of the model parameters is provided in the table below. Those are set via the use of Matlab functions. This allows multiple model parameters to be run through batches. All parameters used in the model are arbitrary and do not relate to a particular aircraft. The model outputs are used to assess the operating speed of the system and resulting deployment times. The transmission loads generated are also analyzed (Table 1).

Table 1. Table of model parameters [11].

K_{hyd}	Hydraulic fluid bulk modulus
ρ_{hyd}	Hydraulic fluid density
I_{mot}	Hydraulic motor inertia
K_{mot}	Hydraulic motor displacement
T_s	Control valve time constant
A_{valve}	Control valve area
C_{valve}	Coefficient of discharge for control valve
P_{PDU}	PDU brake disengagement pressure
G_{PDU}	PDU gearbox ratio
η_{mot}	Motor efficiency
T_{drag_m}	Motor drag
k_{trans}	Transmission torsional stiffness
c_{trans}	Transmission torsional damping

(*continued*)

Table 1. (continued)

K_{hyd}	Hydraulic fluid bulk modulus
η_{trans}	Transmission dynamic efficiency
T_{drag}	Transmission system drag
I_{act}	Actuator inertia
T_{OBB}	Outboard brake response time
P_{OBB}	Outboard brake disengagement pressure
G_{sensor}	Gear ratio position sensor
T_{aero}	Aerodynamic loads at all actuators
G_{act}	Gear ratios of rotary geared actuators
η_{act}	Actuator efficiency

5.6   Model Verification

The mathematical models provided herein have been verified against physical test data. This has provided the confidence to use the model outputs to support the design and development of numerous high lift aircraft systems.

Model verification has been performed at numerous stages of the system engineering process for example at component level (actuator, PDU etc.) and against full system rig and simulated results.

Excellent model correlation has been achieved at both individual component and full system level over a range of environmental temperatures, aerodynamic loads across multiple programs.

Models have been further utilized in software and hardware in the loop environments with suppliers and customers.

Models of secondary flight control systems developed in the Simulink environment have been rapid prototyped to support hardware in the loop simulation.

This has provided the following benefits:

- Development time of software algorithms significantly improved;
- Control models tested in simulation prior to integration on the test rigs improve systems' robustness and easiness of integration.
- Use of the mathematical model as a precise implementation of the controller and thereby reducing misinterpretation of formal requirements.

In recent years there has been a significant shift in the aerospace industry to move from rig tests to modelling and simulation environments. Simulations allow testing of the system in extreme cases that test rigs may not be able to perform due to:

- Too dangerous/expensive to perform;
- Not possible to perform such as all system tolerances being maximised/minimised;
- Limitations of the test rigs.

6 Results

This section summarises the results of simulations that assess the influence of PDU gearbox ratio (GPDU) using the flap system model as cited in [11]. The PDU gearbox ratio is a critical parameter within the high lift system design that is defined during the preliminary design stage. GPDU is defined as the ratio between the hydraulic motor velocity and PDU gearbox output velocity and is implemented by the use of multiple spur gear trains and or planetary gearboxes.

A ratio greater than one is typically required to amplify the hydraulic motor torque to keep the motor size down and hence reduce weight. This allows the PDU to drive the transmission systems against the external aerodynamic loads and internally generated friction.

Firstly time histories are provided to define the fundamental system metrics that are used to measure the system performance i.e.:

- PDU Output Velocity
- Deployment Times
- Hydraulic Flow Rate
- System Position
- Torques in the transmission system

Then a sensitivity study is performed where the system metrics are evaluated with respect to changes in the PDU gearbox ratio. This provides the relative influence that each metric has for a change in ratio and also provides indications of which gear ratio should be used to optimise these metrics.

Finally the performance envelope of the PDU is evaluated by using the:

- PDU output torque with respect to output speed
- PDU flow rate with respect to output speed

This analysis indicates how the performance envelope changes with respect to gear ratio and can be used to assess what gear ratio should be chosen depending on the demanded performance requirements.

6.1 Time Histories

The basic output from the model is logged in time histories. Figure 10 highlights the PDU output shaft velocity during deployment.

This shows that when the system proceeds at a normal operating speed of 184 rpm. It takes approximately 36.9 s to fully deploy the system from the stowed condition.

Both hydraulic motors consume approximately 12.6 l/min flow rate during operation as provided by Fig. 11:

The following time history shows the system position at both the PDU and the wing tips. It can be seen that the wing tips marginally lag the PDU. The lag is due to stiffness of the transmission system and it is the expected systems behavior (Fig. 12).

Finally the transmission torque from each wing can be plotted. This shows an initial transient peak torque at 97 Nm reducing to a steady state torque of 93 Nm per wing (Fig. 13).

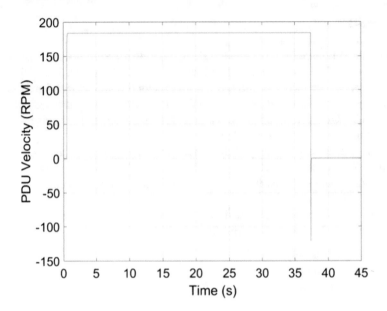

Fig. 10. Time history of PDU output shaft velocity [11].

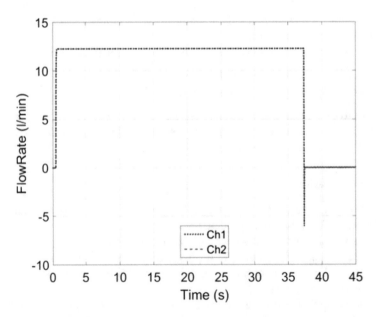

Fig. 11. Hydraulic motor flow rate consumption [11].

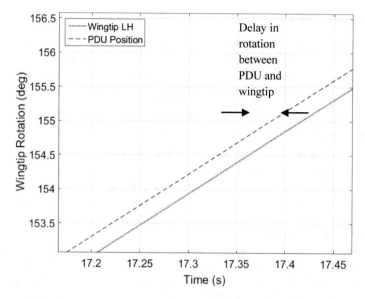

Fig. 12. System position at the PDU and wing tip sensors [11].

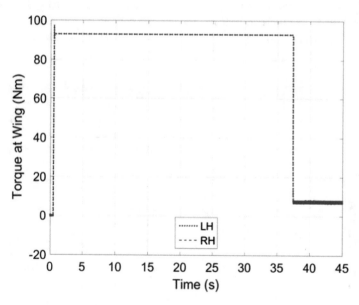

Fig. 13. Wing torques during deployment [11].

6.2   Sensitivity Study

As studied in [11] the model parameters have been fed into the model as ranges in order to assess the systems sensitivity. The example provided shows how the gear ratios of the PDU gearbox can be modified to achieve various system requirements such as:

- Minimise running and peak transmission torques;
- Minimising the hydraulic flow rate consumption;
- Reducing deployment time.

Figure 14 provides the results from a batch of simulations where the PDU gear ratio was varied between 20:1 to 30:1.

The results from the simulations indicate the optimal gear ratio to be 26:1 which maximizes operating speeds and hence reducing deployment times. An optimum occurs because of the trade-off between the mechanical advantage of the PDU and the drag which increases as function of speed.

At low gear ratios the PDU has a lower mechanical advantage and hence the PDU's torque capability is lower, which implies lower operating speeds. At high gear ratios the motor speed is higher for a given PDU output speed and the speed induced drag in the motor increases and limits the PDU speed.

As studied in [11] Fig. 15 presents the sensitivity analysis of the variation of the PDU ratio vs. the normal operating velocity (blue line) and the PDU gear ratio vs. the deployment time. The optimal gear ratio is the one defined by the value for which we have operating velocity and minimum deployment time. In this example, the optimum

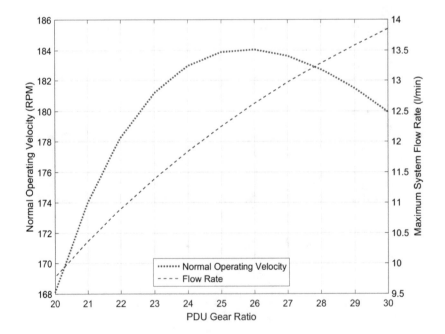

Fig. 14. Operating velocity and flow rate vs gear ratio [11].

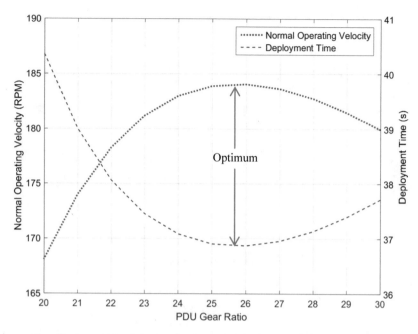

Fig. 15. Normal operating velocity and deployment time vs gear ratio [11].

PDU gear ratio for this system is 26:1. The transmission torques generated by the model can be used to as an input into stress sizing simulations. Figure 16 shows that the running torques in the transmission are highest at 26:1 due to the highest operating speeds. The peak transmission torque has a marginally different trend due to the dynamic transmission characteristics.

A hydraulic high lift system is typically sized based upon cold temperature operating and breakout performance. The reason behind it is that the drag in the transmission is at the highest while the PDU capability is the lowest.

6.3   Performance Envelope

The performance envelope of the PDU can be evaluated using the mathematical model. For example Fig. 17 provides the PDU Output Torque vs Velocity curves for various PDU gear ratios. The external load on the PDU was increased and the corresponding output speed recorded. This provides pseudo steady state power curves that the PDU can generate based upon the system boundary conditions. The boundary conditions are fixed due to being specified by the aircraft manufacturer such as:

- Hydraulic supply and return pressures
- Hydraulic and environmental temperatures

Figure 17 highlights the characteristic reduction in PDU output velocity as the external torque is increased. This is due to the fixed input power available (fixed pressure supply and regulated maximum flow rate available) and corresponding

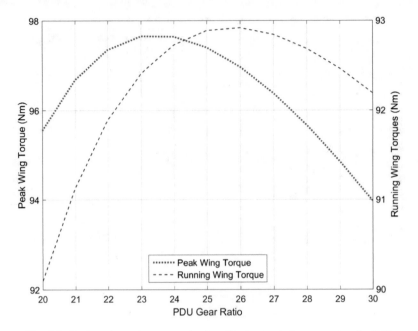

Fig. 16. Peak and running torque in the wing transmission vs gear ratio [11].

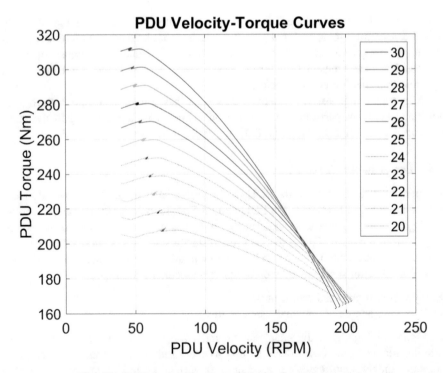

Fig. 17. PDU Output Torque vs Velocity for each gear ratio.

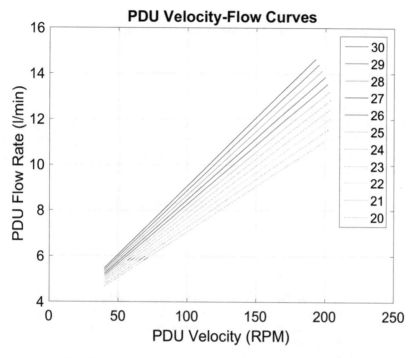

Fig. 18. PDU Output Torque Vs Velocity for each gear ratio.

hydraulic and mechanical efficiencies of the PDU. Below <170 rpm the torque/velocity envelope is increased with a larger gear ratio due to the increased mechanical advantage available i.e. increasing the gear ratio allows a higher output torque for a set speed.

At velocities >170 rpm the envelope degrades with increased gear ratio due to the output speed reducing more for a given motor speed. This can be summarised using the following expressions:

$$\text{Power (fixed)} \propto T_{PDU} \times V_{PDU} \propto G \times T_{Motor} \times V_{Motor}/G$$

where:

TPDU = PDU output torque
VPDU = PDU output velocity
TMotor = Hydraulic motor torque
VMotor = Hydraulic motor velocity
G = PDU gear ratio (between output shaft and motor)

As the input power to the PDU is fixed then any increase in output torque will be accommodated with a reduction in velocity. Hence an increase in PDU gear ratio will increase torque but will reduce the output velocity.

Another trade off with increased gear ratio is that a higher flow rates from the aircraft are required which may not be available as Fig. 18 shows. Finally the PDU stall torque also proportionally increases with gear ratio which needs consideration.

Therefore higher gear ratios will require the transmission to be designed to accommodate the increased torques which has implications on system weight.

7   Conclusions

This work has described the modelling process of a commercial aircraft high-lift transmission and power drive system. An example model was provided that was developed in the Simulink modelling environment and results of a sensitivity analysis have been provided. Model validation has been discussed together with the integration of models into other engineering environments.

This paper provides a generic functional model and demonstrates the analysis techniques used within the design and development of high lift systems that are in-service.

References

1. Defence Science Board, "Defence Science Board Summer Study on Autonomy," June 2016
2. Huang, H.: Challenges in More Electric Aircraft (MEA). IEEE Newslett. August 2015. https://tec.ieee.org/newsletter/july-august-2015/challenges-in-more-electric-aircraft-mea, http://www.moreelectricaircraft.com/
3. Slat and Flap. https://www.grc.nasa.gov/www/k-12/airplane/flap.html
4. Rudolph, P.K.C.: High-lift systems on commercial subsonic airliners, NASA CR4746, September 1996
5. Raymer, D.: Aircraft Design: A conceptual Approach, 2nd edn. AIAA Education Series (1992)
6. ARP4754A, Guidelines for Development of Civil Aircraft and Systems, SAE International (2010)
7. Walden, D.: INCOSE Systems Engineering Handbook. INCOSE, 4th edn. Wiley, New York (2015)
8. SEBoK. Guide to the Systems Engineering Body of Knowledge – Validation and verification diagram (2017). sebokwiki.org
9. Hardwick, G., Hanna, S., Panella, I.: Functional modelling of high lift systems. In: NAFEMS World Congress 2017. NAFEMS (2017). awaiting publication
10. Zadeh, Desoer: Linear System Theory – The State Space Approach. McGraw Hill, New York (1963)
11. Hardwick, S., Panella, I.: Dynamic modelling of commercial aircraft secondary flight control systems. In: 7th International Conference on Simulation and Modeling Methodologies, Technologies and Applications, SIMULTECH 2017. INSTICC (2017)

Simulation, Modeling and Technologies for Drones Coordination Techniques in Precision Agriculture

F. De Rango[✉], G. Potrino, M. Tropea, A. F. Santamaria, and N. Palmieri

University of Calabria, Via P. Bucci Rende, Cosenza, Italy
{derango,m.tropea,afsantamaria,n.palmieri}@dimes.unical.it,
peppepotrino@live.com

Abstract. The aim of this chapter is to provide an application of UAV to support the agricultural domain in monitoring the land for checking and countering the presence of parasites that can damage the crop. However, to properly manage a UAVs team, equipped with multiple sensors and actuators, it is necessary to test these technologies and design proper strategies and coordination techniques able to efficiently manage the team. At this purpose, the chapter proposes a simulator suitable for the agriculture domain in order to design novel coordination and control techniques of a UAVs team.

1 Drones Overview

In the modern age, the use of technology is playing an important role in supporting human activities. Man is concentrating more and more on intellectual work, trying to make practical activities more and more automated in order to increase their efficiency [1]. In this regard, the use of drones is increasingly becoming a key aspect of this automation process. This is because a drone has many advantages including agility, efficiency and reduced risk, especially in dangerous missions. The use of drones is therefore becoming increasingly important and for this reason we want to propose a new use in a new context, measuring the costs and efficiency, in order to establish the actual convenience and risks to be faced. Among the various existing contexts, the agricultural sector is chosen. This choice derives from the fact that agricultural activities are becoming more and more automated as they are strenuous activities for humans and therefore it is better if carried out by machines. Moreover, they are activities that need more and more efficiency, which can not be given by simple human strength. The increase in the world population has also caused the reduction of cultivable areas due to urbanization, but this has also increased the demand for products. There is therefore a need to increase production using little land and this can not be done except by applying intensive agriculture instead of old extensive agriculture. Obviously, it aims to make this agriculture sustainable also trying to preserve environmental resources. Among the various aspects of agriculture

© Springer Nature Switzerland AG 2019
M. S. Obaidat et al. (Eds.): SIMULTECH 2017, AISC 873, pp. 77–101, 2019.
https://doi.org/10.1007/978-3-030-01470-4_5

in this work we treat the fight against parasites. In fact, they are able to destroy entire plantations, eliminating production. This is becoming a real problem in the last period. For example, we remember the whole plantations of olive trees destroyed in Puglia following the attack of the Xylella or the destruction of many palm trees as a result of attacks by "Red Punteruolo".

Drones, flying devices lacking a human pilot on-board, have attracted major public attention (Fig. 1).

Fig. 1. Drone.

The different types of drones can be differentiated in terms of the type (fixed-wing, multi-rotor, etc.), the degree of autonomy, the size and weight, and the power source. These specifications are important, for example for the drone's cruising range, the maximum flight duration, and the loading capacity [2].

Today, drone technology is no longer confined to use in military, industry and meteorology. Small UAV toys, capable of capturing live videos and images, can be purchased today for few hundred dollars from various toy retailers.

A drone is a remotely controlled aircraft that is capable of capturing images and video sequences of a targeted region and transferring them to a remote server for storage and further processing. A drone is usually controlled by a handheld device such as a radio controller, a mobile phone or a tablet.

Today, both technological and legal factors restrict what can be achieved and what can be allowed safely. For example, the US Federal Aviation Administration (FAA) requires drones to operate within line-of-sight (LOS) of a pilot who's in control, and also requires drones to be registered. A probable forecast on use of drones in both public and commercial fields is shown in Fig. 2.

It is possible to do some classifications for drones on the basis of their characteristics like weight, type of control, type of wings, and so on. These main characteristics are summarized in Figs. 3 and 4 on the basis of data gathered from in [4,5].

In recent times, technological development led to a progressive reduction of costs, overall dimensions, consumption and weight of electronic components [3]. With performance improvement of electric motors and the rapid increase specific energy (capacity/ratio weight) available from ion batteries of lithium, the fixed-wing configuration is almost completely replaced by multi rotor configuration.

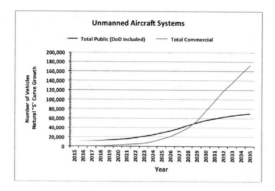

Fig. 2. Total Unmanned Aircraft Systems Forecast 2015–2035 data forecast [4,5].

UAS Description	Weight (Pounds)	Overall Size (Feet)	Mission Altitude (Feet Above the Surface)	Mission Speed (Miles per Hour)	Mission Radius (Miles)	Mission Endurance (Hours)
Nano	<1	<1	<400	<25	<1	<1
Micro	1 to 4.5	<3	<3,000	10 to 25	1 to 5	1
Small UAS	4.5 to 55	<10	<10,000	50 to 75	5 to 25	1 to 4
Ultralight Aircraft*	55 to 255	<30	<15,000	75 to 150	25 to 75	4 to 6
Light Sport Aircraft*	255 to 1320	<45	<18,000	75 to 150	50 to 100	6 to 12
Small Aircraft*	1,320 to 12,500	<60	<25,000	100 to 200	100 to 200	24 to 36
Medium Aircraft*	12,500 to 41,000	TBD	<100,000	TBD	TBD	TBD

Fig. 3. Classification for weight. Data Source: [4,5].

The success of these platforms for remote sensing applications such as aerofo-
togrammetry, analysis of the thermal dispersion of the ground and/or buildings
and the analysis of air quality is linked to the availability of reliable avionics and
low-cost that allowed to implement stabilization systems and piloting the aircraft
that allows an operator without particular piloting experience of use the drone
in security too in case of meteorological conditions adverse. Moreover, compared
to the traditional fixed-wing aircraft, the multirotor aircraft presents a configuration more compact, greater ease of use and the ability to operate in tight
spaces, thanks to the vertical flight capacity. Finally the steady flight capacity
(hovering) allows greater precision on the data acquired by the sensor, while the
possibility of piloting remote from ground station allows the accomplishment
of missions in environments hostile and/or contaminated in total safety for the
human operator.

In this chapter, it is supposed to use drones in the agriculture domain in a situation where it is necessary to monitor plants or products in a land as suggested in several works [6–9]. Often, farmers need to face many issues such as parasites or sudden climate changes that can severely affect the agricultural products quality. At this purpose, new technologies such as drones equipped with specific sensors, cameras and fertilizers can support farmers to face the threats and, in the specific case, to kill parasites that can destroy plants in the cultivated field. Let us suppose to have a cultivated area where a certain number of plants are distributed. In this area, it is possible that some parasites decide to go in order to attack the perceived plants and to reproduce themselves. In this case, it can be suitable to react to this attack in a timely and efficient manner. It is not known where exactly parasites are because they can be distributed in the land too and they can reproduce and move in a distributed manner. In this specific case, a drones team can be very useful in overseeing the overall area in order to localize through cameras the attacked plants or to see parasites moving among plants. In real situation, unfortunately, drones are powered by a battery with limited lifetime with a consequent risk to drain all energy an stop to move in the area. Moreover, the pesticide is also limited considering the heavy constraints of the drones that can support not to high weights on-board. Due to these restrictions, the objective to coordinate drones among the cultivated area avoiding to pass more times on the same pieces of land. At the same time, the coordination task in order to distribute the surveillance task preserving the energy and accounting for the pesticide consumption can be a challenging task. In order to propose coordination techniques of drones for precision agriculture domain, it is necessary to have a simulator where parasites and drones mobility modules are implemented. Moreover, basic communication modules to allow drones to communicate maps and topologies in order to reduce the drones movements, to avoid to waste pesticide and drain energy are necessary. There is no simulator till now, at the best of our knowledge, that provide these modules and this testifies the novelty of the research area although an increasing interest has been shown on the industry and research field. In [10], one of the first simulator has been designed to consider the effect of parasites, plants (trees) and drones in precision agriculture. This book chapter extends that work considering technologies that can be applied in agriculture with reference to sensors and cameras on board to drones. Moreover, the problem statement with all variables included in the simulator to evaluate coordination techniques have been added and explained.

2   Precision Agriculture

In this paragraph, a brief introduction on some interesting application in the field of precision agriculture is faced. Smart farming and Precision agriculture acquire even more interests in the last few years. Industrial and research investments are increasing to achieve a better integration of *Internet of Things* (IoT) technologies offering new products and services. Thus, it will be possible to cut down the overall cost of products, increasing quality and quantity of the harvesting.

Remote human pilot	Real-time control by remote pilot
Remote human operator	Human provides the flight parameters to invoke the built-in functions for vehicle control
Semi-autonomous	Human controlled initiation and termination, autonomous mission execution
Autonomous	Automated operation after human initiation
Swarm control	Cooperative mission accomplishment via control among the vehicles

Fig. 4. Classification for control. Data Source: [5].

Novel technologies in the field of communications permit to offer new solutions for continuous monitoring of the croplands, livestock, water system with the final goal to reduce the human activities. Thus, new systems based on embedded technologies and *Machine to Machine* (M2M) communications are being used in the precision agriculture domain. In fact, in this field it is important to continuous monitoring all climatic conditions. The communication protocols may help to reduce the overall amount of data that are generated to better use the network resources. These solutions help to improve the knowledge of the environment and to improve the environmental and agricultural sustainability to improve crop traceability and to increase overall yield [11–16].

With the increase of population around the world, the demand of food is increased. However, the amount of areas available for food production get smaller day after day. Solutions that may help to increase production and to exploit those areas in a better way are considered. These goals drive farmers to improve the efficiency of their crops [17]. Thus the research and industrial interests covering different area to help farms to optimize processes. The management of the resources is one of the most important topic in the precision agriculture because the better the resources are utilized the better are the achieved results. One of the most important resource is represented by the availability of water. It is clear that to optimize the use of this resource efficient solutions must be adopted. Commonly irrigation control technology waste lot of water. In a smart irrigation control system, distribution of water is done avoiding water wastages adopting cooperation strategies between devices that have in charge monitoring and actuating tasks. Typically the irrigation control system does not consider weather forecasting, status of plants and environment but the activation is made on the basis of technician perceptions. An IoT based Crop-Field Monitoring and Irrigation system full automatized require a better knowledge of the environment. More information can be collected about the environment to adopt a better solution for the crops and their yields. These solutions are composed of several smart devices that communicate between them to increase their knowledge of the environment and to optimize their decision making algorithms. Moreover, a software framework can complete the entire system for globally manage it and to adopt global strategies that may regard the choice of the area to utilize and how these must be managed. In this region, covering large area in rural zone, the communication availability represents an open issues. Fortunately, in these last years cellular networks are spreading their coverage and offer new technologies that permit to have more network resources to exploit such as low delay and high bandwidth. M2M communications present different protocols that can be adopted. *Message Queue Telemetry Transport* (MQTT) and *Constrained Application Protocol* (CoAP) are deeply utilized in these kinds of applications. In particular, MQTT is utilized because it is easy to use and offer good performances. It is based on a publishing/subscribe mechanism where a broker manages connections and topics. Smart devices based on embedded technologies may help to improve results by acting smart way applying customized solutions. Good results have been obtained in terms of reducing the extra man power efforts, water requirement and fertilizer requirement.

An example of smart devices designed for accomplishing tasks for water management is shown in Fig. 5. Here sensors device monitor crops and send data exploiting M2M protocols. These data are collected by remote services that provide to analyze and store these data. Moreover, at the same time, data about environment such as weather conditions, water availability are collected as well. A decision making algorithm is performed to regulate water flow in case of water needs.

2.1 Video Analysis Systems for Supporting Precision Farming Activities

The problem of monitoring the condition and the growth is one of the most important topics for precision agriculture and livestock farming. In fact, to acquire more information and increase the knowledge of the environments image based measurements and smart data mining are required. Aerial images of crops

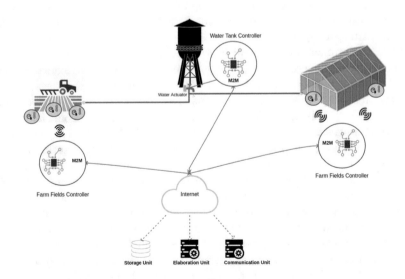

Fig. 5. Example of network architecture based on M2M communication.

fields and livestocks farming can be achieved using *Unmanned Aerial Vehicle* (UAV). To check the overall health, vitality of the vegetation can be evaluated using a well-known methods called *Normalized Difference Vegetation Index* (NDVI). This index is useful for evaluation but it cannot be enough to apply an efficient solution to increase production level and quality of their products. Efficient policies can be applied differencing zones. Taking under consideration the NDVI and the classification algorithms it is possible to divide crop-fields in several zones and grouping them by finding zones with similar aspects. This can help farmers to decide the action to perform in each zone [17]. NDVI information is a good measure to predict the planting yield variation [18]. Following the results achieved by analyzing captured frames it is possible to determine the vitality of vegetation. From these two last useful works, it is possible to evaluate NDVI index by using the Eq. 1, if the red channel is available, or with Eq. 2, if only the blue spectrum is present.

$$NDVI = \frac{NIR - Red}{NIR + Red} \tag{1}$$

$$NDVI = \frac{NIR - Blue}{NIR + Blue} \tag{2}$$

Another important field of interest and application scenario in which video analysis may help the developing of plantations is the identification of weeds in the crops fields. The correct identification of this kind of vegetation is an open issue. The more utilized approach to solve weeds infestation is to use herbicides. This method brings up several side effects. Reducing cost and improving crop yield is highly needed. Precision farming technologies are deeply developed and improved during last two decades. In particular, weed is a plant that needs of water, nutrients and space to develop that reduces crops yields. The use of herbicides can be reduced developing alternative systems that work in a real-time way. These systems use optical sensors and video-camera devices for analyzing the area and finding infestation. In [19] for example a machine vision system for precision sprayer is developed. It differentiates between plant and weed using gabor filter and blob detection. In this field also the UAV may help to find weeds and plantations problems. In [20], a terrestrial UAV is equipped with a smart camera and IR sensors. Here data are sent towards the computational module for the classification step. In case of weed the UAV is authorized to spray herbicide.

In Fig. 6 an example of terrestrial drone for weeds recognition is shown. This drone is equipped with several sensors to acquire environmental measurements such as temperature and humidity. Moreover, it is equipped with a smart camera for acquiring real-time images of the crop-lands. Images are sent to the elaboration unit that provides to analyze video, extract features and classify recognized objects. Here the main task is to identify weeds and no weeds plant and their density. This analysis is then sent to the decision making algorithm that chooses if the drone sprays herbicides or not.

Many applications for precision agriculture for livestocks was birth in the last decade. The rise of interest in this field is given by the good results achieved.

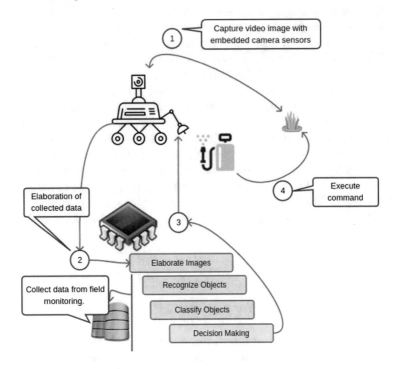

Fig. 6. Example of terrestrial drone architecture for precision agriculture.

The continuous monitoring related to the growth of animals make possible to optimized farm processes in terms of feed, water and health care. An example is supplied by [21] in which a dynamic monitoring and analyzing system is proposed for object detection and growth assessment of pigs. Captured images are acquired on the flight from a special 3d-camera to analyze objects exploiting the depth of the acquired images. After the depth images acquired a watershed algorithm is adapted to segment each individual animal. Furthermore, due to the resembling color and uncontrollable living behaviors pigs are difficult to separate with each other. Therefore, new sensors like time-of-flight camera are preferred as it can provide the 3-D position information in depth image rather than the RGB image [21].

2.2 Use of UAV in Precision Agriculture

In the past years, UAVs improve their contribution in the extensive agriculture to better monitoring crop-fields [22]. Solution based on this technology can be improved exploiting M2M and IoT technologies to increase the knowledge of the environment and to create more sophisticated algorithms in decision making steps [23–25]. For example, the problem of pesticide pollution is still present [26]. Several improvements can be done to reach new goals. For example it is important to evaluate the energy consumption of UAV for long time missions and provide smart solutions that promote energy harvesting technologies. The UAV systems will work in autonomous way but several issues are still opened such as normative regulations of flighting objects and some technical issues [27]. The UAVs are widely used for area monitoring, spraying and agricultural insurance survey. Area scanning is an area of interest commonly covered by the use of UAV equipped with high resolution cameras [28–30]. Several solutions propose a service based on flight planning. Here drones start their activities using a GPS sensor to move in the environments. Several check points have been chosen in the map and a prefixed route is defined by flight control service.

3   Coordination Techniques Based on Routing

A coordination technique is a process to organize a group. In drones field, coordination refers to lead a drones group to reach a common goal. In this case coordination is permitted by packets exchange. Communication is permitted by a wireless interface because drones are mobile nodes. This network type is called Mobile Ad Hoc Network (MANET) [31,32]. This network type is built in dynamic environments. The main MANET problem is the routing management caused by drone mobility and limited drone energy. A dynamic network requires periodically packets exchange which gives updated routing tables. Then, routing algorithms should:

- Assure reasonably small routing tables;
- Chose the best path to reach other nodes;
- Help to have updated routing tables;
- Reach the optimality in a short time sending a small number of packets;
- Offer multiple path to reach a destination considering the various costs.

3.1   Routing Protocols

We can classify routing protocol in the following way:

- Proactive: In proactive algorithms, MANET mobile devices exchange theirself packets periodically. For this reason, routing is immediately available. The idea is to maintain updated every node routing table. Proactive protocols have the following characteristics:
  - Packets are exchanged periodically;
  - Tables usage;
  - Tables updating.
  Proactive protocols have the following advantages:
  - They permit to have immediately available routing of informative packets;
  - Routing tables contain the real network structure because they are updated periodically.
  Proactive protocols have following disadvantages:
  - Excessive packets exchange;
  - High signaling traffic.

- Reactive: Reactive protocols make paths among nodes on-demand. This is an advantage because MANET networks are very dynamic. Reactive protocols send a packet from a source to a destination in the following way:
  - Route discovery: this procedure determinates various paths between the source node and the destination node;
  - Route maintenance;
  - Route deletion.
  Reactive protocols, unlike Proactive protocols, have following advantages:
  - They make paths only on-demand;
  - They delete the high network traffic;
  - There are no tables to store routing data.
- Hybrid: Hybrid protocols merge proactive protocols advantages with reactive protocols advantages. Some applications of Hybrid algorithms are Location-Aided Routing (LAR) and Zone Routing Protocol (ZRP) [33].
- Hierarchical: Hierarchical protocols reduce packets overhead of the proactive protocols. They divide nodes in classes and maintain the relative entry class of each node in the table. The network is divided in cluster and a leader cluster is elected (likewise of the Base Station in cellular networks). This approach makes a centralized structure which is more scalable. In each cluster, for having always available paths, proactive techniques are used. In inter-cluster communication reactive techniques are used [34].
- Location-based: These routing protocols are designed to maintain routing tables using geographical node position (GPS). Routing is optimized to work in a specified area (Routing zone) which contains the Expected Zone which is the zone in which it expects to find the destination node [35].
- Power-aware: In these techniques every node is conscious of its bounded energy. This consciousness permits the node to decide when to stay switched on or switched off or to choose a less expensive path. An example of power aware technique is Power-Aware Routing Optimized (PARO) [36].

3.2   Routing Type

Routing algorithms has the aim to fill and to maintain updated routing tables. This can take place in two ways:

- Static routing: In static routing each node has a routing table manually configured. In many cases there is an administrator which configure the routing table. If the network changes topology, the administrator must manually update the routing table. For this reason, this technique is applied to static small networks. In this case, in fact, it is very efficient because packets exchange to updated routing table is deleted. Among static routing advantages there is bandwidth, CPU and RAM saving.
- Dynamic routing: In dynamic routing, tables are created and updated by a routing algorithm in execution on the node. This is made by a packets exchange among nodes which increase bandwidth, CPU and RAM usage. These types of algorithms can adapt at network change finding more paths to the same destination. It can be classified also in centralized or o distributed dynamic routing.

3.3 Problem Statement for UAVs Coordination

In our work the routing algorithm is applied to coordinate drones movements in order to perform adequate count measure against parasites attacks.

3.4 Drones

Each drone is equipped with a battery that provides limited energy, a tank of limited capacity that contains the pesticide, and a wireless module used to communicate with the nearby drones. The energy of the battery is dissipated in three ways: by moving the drone, with the use of the pump that sprays the pesticide and with the transmission of data (Fig. 7).

The drones can be mainly in two conditions: moving at constant speed or stopping spraying the pesticide on infected plants. The speed of drone varies between a minimum and a maximum based on what it is doing. If the drone was recruited by a another drone, its speed increases in reaching the destination, based on its current charge. In fact, the higher is the traveling speed the higher is the energy consumption of the UAV. Each drone, when it has an empty tank or has a critical battery level, returns to the base to recharge both. Drones can use the recruitment techniques in case of necessity, or if they have found parasites but the pesticide tank is empty or the battery has a low charge. The drones can be at a height between the height of the plants and the maximum height reached. Each drone has its own height. Each drone is at a height different than every other drone in order to avoid collisions. Each drone, in detecting parasites, has a certain visual angle also to see more surface as it moves away from the ground. Obviously, greater is its height and greater is the error of detection of parasites, which causes waste. In order to achieve their goal, drones cooperate with each other by exchanging information via wireless. The exchange of information is coordinated through the application of specific routing protocols. In this chapter we want to use various approaches including one based on the local flooding and one based on the use of the Link-State. Therefore, we want to compare the two approaches and test the changes in performance as they vary some parameters. The exchange of information mainly aims at two aspects:

- coordination of drone movements, in order to avoid waste. In this case there is a flooding exchange that guides the neighboring drones into the choice of the next move;
- recruit other drones in case of need for help.

The dissipation of the energy of the drones is as follows:

- a drone dissipates energy to keep in flight and to move. This dissipation is linked in a linear manner to the height of the drone. The higher is the altitude the higher is the energy demand of the UAV because of a more rarefied air;
- drone dissipates energy to use the pesticide pump. This consumption was linked to the ml of pesticide to be sprayed;
- a drone dissipates energy to use the wireless module to transmit packets.

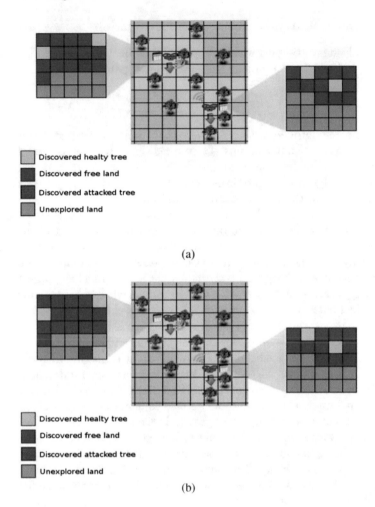

Discovered healty tree
Discovered free land
Discovered attacked tree
Unexplored land

(a)

Discovered healty tree
Discovered free land
Discovered attacked tree
Unexplored land

(b)

Fig. 7. Typical situation with plants, visited cells, unexplored cells and infected plants: (a) before map exchange; (b) after map exchange. Source data: [37].

The drones can search for parasites by choosing random directions or using a distributed search algorithm, subsequently described.

We consider a team of UAVs, that performs a mission of monitoring a land for destroying parasites in the plants. Once the parasites are detected by a UAV, it tries to destroy them through a certain quantity of pesticide that the UAVs carry. However, the pesticide resources of each drone is limited in quantity. Once a drone notices parasites in a certain position of the area, it becomes a leader and it has to form a coalition in this region based on its current resource level such as pesticide and energy resources. Moreover it informs other drones about the found parasites, broadcasting some information. The other drones, that have the required resources, will evaluate their availability to reach the leader or not.

It should be noticed that the coalitions formed are temporary by nature; once the detected parasites are destroyed the coalition members can perform other tasks.

Plants disseminated on the field attract the parasites. It is assumed that each parasite does not communicate with other parasites to perform a strategy but it moves on the basis of its local scope. It is oriented towards a single direction and it is limited by its local scope that is assumed in our case to be its three cells in front. During the movement the parasite can go in one of the adjacent and visible cells.

Parasites Mobility Model. Plants disseminated on the field attract the parasites. It is assumed that each parasite does not communicate with other parasites to perform a strategy but it moves on the basis of its local scope. It is assumed that the parasites is not influenced by the presence of other parasites and it is oriented in a direction through a restricted view, it is able to see up to three cells in front him and in this way it is able to move in one of the visible and adjacent cells as it depicted in Fig. 8. In Fig. 8(a) an example of parasite movement is depicted. Parasite is oriented toward North-West (NW) but it decides to move on North (N), then, it moved to North and new cells can be seen now in it local scope. Blue cells are cells that can see.

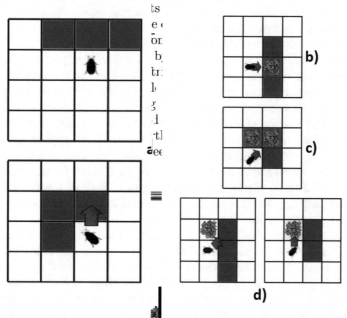

Fig. 8. (a) Parasite movement; (b) only one plant is visible; (c) more cells are visible; no cells are visible. Source data: [37].

During the movement the parasite can go in one direction and three main cases can happen:

- Case 1: Only in one of the visible cells is present a plant: the parasite moves directly on the cell which contain a plant such as shown in Fig. 8(b);
- Case 2: There are more cells which contain a plant: the parasite moves randomly in one of the cells which contain a plant such as shown in Fig. 8(c);
- Case 3: No one of the visible cells contains a plant such as shown in Fig. 8(d):
  - parasite choses randomly a direction;
  - if the new one is not equal to the current one, the parasite turns in the new direction and if turning in the new position it sees a plant, it will go on it;
  - otherwise it will continue to go in the direction chosen before.

4   UAVs Simulator to Coordinate Drones in Precision Agriculture

4.1   UAVS Simulator

We consider a team of UAVs/drones, that performs a mission of monitoring a land for destroying parasites in the plants. Once the parasites are detected by a UAV, it tries to destroy them through a certain quantity of pesticide that the UAVs carry. However, the pesticide resources of each drones are limited in quantity. Once a drone (called leader) notices parasites in a region, it has to form a coalition based on its current resource level such as pesticide and energy resources and it broadcasts some information (i.e, its location, type and number of its capabilities) to the other UAVs. The other drones, that have the required resources, will evaluate their availability to reach the leader or not. It should be noticed that the coalitions formed are temporary by nature; once the detected parasites are destroyed, the coalition members can perform other tasks. Drones movement is modeled through the use of a local map that is stored on-board. This map collects info about already visited fields, if there are plants in the near field (cell) and if some plants are infected or not. Each of these info last for a time on the basis of the time set by the simulator. This temporal dependence of plants and insects reflect the real situation where an insect can attach again the plant or a plant can be planted again on the field. The map is implemented by a cells square matrix and each cell represents a portion of field. The drone is assumed to be in the middle of the cell. The info related to the cell can be the presence or not of the plant, the presence of parasites and consequently if the plant can be infected or not. Moreover, the info in the cell can also express if a plant is under care or if the cell has not been explored yet. Drone updates its map after any movement removing the old info on the map and replacing them with more updated info. This local map helps the drone about the next field (cell) to visit. It selects one of the unexplored cells performs the field selection. The cell selection probability is uniformly distributed among all unexplored cells. However, after visiting some cells, the selection probability can be distributed again among the remaining cells. When all cells in the local map cannot be visited because they have been already visited from a few time or because these cells are occupied by a health plant, the cell selection probability can be changed. Each drone can exchange its map with all neighbor drones (drones within the drones transmission range). This possibility can assure that a drone can know the situation of neighbor cells also without going to explore them and this speed up the convergence of the overall field exploration. Moreover, we consider many plants disseminated on the field that can attract the parasites. It is assumed that each parasite does not communicate with other parasites to perform a strategy but it moves on the basis of its local scope. It is oriented towards a single direction and it is limited by its local scope that is assumed in our case to be its three cells in front. During the movement the parasite can go in one of the adjacent and visible cells.

4.2   Simulator Indicators in the GUI

Each drone is equipped with a battery providing the energy to fly and move in the interested area. It is also equipped with a pesticide tank to spray on the parasite to kill them and a wireless module to communicate with neighbor drones. The communication range of each UAV is assumed to be limited. However, a UAV can reach another UAV by a sequence of communication links. The fuel tank is considered in milliliters (ml) whereas the battery power in Watts (W). All these parameters are selected before assessing the simulation through a GUI in the front-end of the simulator. The battery power can be dissipated in three ways: drone movements, pesticide spraying and communication, which is due to radiation, signal processing as well as other circuitry. It is assumed that to eliminate a parasite is necessary a fixed amount of pesticide is defined in the GUI simulator interface. It is also assumed that drones can move at a constant speed from a cell to another one or they can stop to spray the pesticide on the tree attached by parasites. This assumption could be removed extending the mobility model and without affecting the validity of the overall proposed simulator model. Simulator allows also defining the minimum pesticide quantity (it is defined by GUI). Moreover, when the fuel tank is below a minimum level (defined in GUI) the drone needs to come back at its base station or when the drone terminated its pesticide. It is possible in the simulator to set also the minimum time necessary to recharge the drone and refill the fuel tank. In order to provide a friendly graphical interface, all drones are represented with some indicators that can change the color on the basis of the quantity of remaining pesticide or the battery level.

For better understanding how UAV levels work, we reported the Fig. 9, which is already presented in [10,37]. Moreover, considering definitions done regarding used color, in [10,37] in the Fig. 10(a), it is shown the circle colored in the most internal part to indicate the residual pesticide quantity of each drone i (S_i). The residual battery levels of each drone i (P_i), instead, is represented with colors indicated in Fig. 10(b). Indicators are also related to the tree health in order to represent the resistance level to the parasites attack (Fig. 10(c)). More

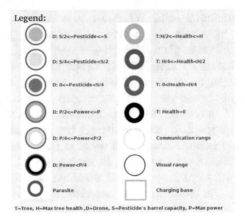

Fig. 9. Example of UAVs level as presented in [10,37].

◉	$S_i > \frac{1}{2} S$ (tank capacity)
○	$\frac{1}{4} S < S_i \leq \frac{1}{2} S$ (tank capacity)
◉	$0 < S_i \leq \frac{1}{4} S$ (tank capacity)

a)

○	$P_i > \frac{1}{2} P$ (Initial battery capacity)
○	$\frac{1}{4} S < P_i \leq \frac{1}{2} P$ (Initial battery capacity)
○	$0 < P_i \leq \frac{1}{4} P$ (Initial battery capacity)

b)

○	$H_i > \frac{1}{2} H$ (Initial health state of the tree)
○	$\frac{1}{4} H < H_i \leq \frac{1}{2} H$ (Initial health state of the tree)
○	$0 < H_i \leq \frac{1}{4} H$ (Initial health state of the tree)
○	$H_i = 0$ (Three completely damaged)

c)

Fig. 10. Here the reference configuration is reported has already presented in [10,37]. In particular, (a) Indicator of pesticide remaining quantity; (b) Indicator of battery remaining quantity; (c) Tree health indicators.

specifically, when a tree has been attached to parasites, its health decreases on the basis of number of parasites attaching the tree and the time of the parasites remain into the tree. We refers to H_i as the current health state of each $i - th$ tree and H terms is its initial health state.

4.3   Simulator Front-End

From the main window of the simulator, it is possible to set the number of drones to deploy in the interested area, the number of parasites uniformly distributed in the area, the number of base stations from which drones leave or where they come back to recharge battery and pesticide. Moreover, as indicates in [37], it is possible to define also the tree lifetime and the damage produced by parasites to tree in order to evaluate the efficiency and efficacy of the coordination strategy among to kill all parasites and save the trees; the capacity of pesticide measured in ml; the maximum and minimum level of battery of a drone; the distribution of trees in the considered field; the health state of the tree in order to take into account the parasites attach; the number of drones charge stations; the minimum amount of pesticide (in ml) to kill a single parasite; the time necessary to the parasite to move by 1 m; TTL for help request, that represents the maximum number of hops that a help request can propagate in the network. The simulator previews a GUI in which it is possible to select and define some parameters (some are those cited previous in the text) use during the simulations.

4.4   Proposed Recruitment Protocols

In this section, we show the protocols used in order to perform a comparison. We show a comparison between a reactive protocol such reactive flooding and a proactive ones such the link state routing.

Drones Coordination Based on Reactive Flooding. In this recruitment method, the drone that needs help sends a help packet in broadcast and starts a timeout, this step is shown in Fig. 11. Reachable drones receive the message and, if they are available, they send a response message that will be stored from the destination drone, else they forward in broadcast to their neighbor drones the request until it reaches the maximum hop number. When timeout expires, the drone that needs help reads received responses and chooses the drone with owns the maximum pesticide quantity, this mechanism is shown in Fig. 12. If two drones have equal pesticide quantity, drone with more energy is chosen. If two drones have also equal energy, nearest drone is chosen. Among these, it chooses the drone that sends the response first. The drone that needs help sends a recruitment request to the best drone chosen and comes back to the base station to recharge. Drone that receives recruitment request, if available, goes to the part of field in which was situated the drone that needed help. If the drone that needs help does not receive response at timeout expiration (there are no reachable drones), it stores its actual coordinate for coming back after recharge.

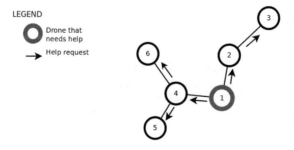

Fig. 11. Drone 1 sends a help request.

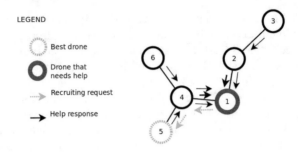

Fig. 12. Drone 1 sends a recruitment request to drone 5.

Packets used in this recruitment protocol are:

– Help packet: this packet is sent to request help or to recruit a drone chosen between those that have sent a response. This packet weights 28 bytes and it is composed in the following way:
  - Destination IP;
  - Source IP;
  - TTL;
  - X coordinate;
  - Y coordinate.

Destination IP field refers to destination drone. When a help request is sent, this field must contain the broadcast address network. When a recruitment request is sent, this field must contain the IP address of the recruited drone. Source IP field refers to drone that needs help. TTL field refers to the maximum hop number. X and Y coordinate fields refer to coordinate of the area that contains parasites.

– Help response: this packet is sent by a free drone that receive a help packet containing a broadcast address as destination. It weighs 52 bytes and it is composed in the following way:
  - Destination IP;
  - Source IP;
  - TTL;
  - Pesticide energy;
  - X coordinate;
  - Y coordinate;
  - Z coordinate.

Destination IP and Source IP fields refer respectively to receiver and sender. TTL field refers to maximum hop number. Pesticide field refers to *ml* of pesticide owns by sender. Energy field refers to joules owns by sender. *X*, *Y* and *Z* coordinate fields refer to the position of sender (recruiter).

Link State. This recruitment method uses Link-State protocol to update and to maintain routing tables of nodes. Each node sends a LSA packet when some information change in its memory. The Fig. 13 shows the LSA packet forwarding mechanism. A LSA packet is sent in the following case: a drone changes position, a drone destroys some parasites, known network graph is updated. Then, each drone knows the network graph reachable and for each graph node it knows its quantity of remained pesticide. Each drone has a database in which it stores, for each reachable node, its address IP, the sequence number of the last received LSA packet and the remained pesticide quantity. This database is used to choose the best drone to recruit in case of needs. The drone with the greatest pesticide quantity is chosen. The path in the network graph to reach the best drone is computed with Dijkstra algorithm. Each node that receives a recruitment packet forwards it to the destination node computing the next hop with Dijkstra algorithm. Receiver node stores this packet and, if available, goes to the part of field in which was situated the drone that needed help, this behavior is shown in the Fig. 14.

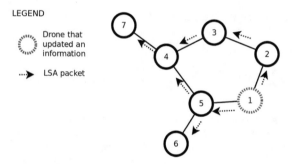

Fig. 13. Drone 1 sends a LSA packet.

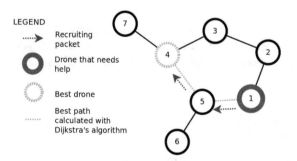

Fig. 14. Drone 1 sends a recruitment request to drone 4.

Drone's characteristics:		Parasite's characteristics:	
Weight of drone (g):	1500	Moving delay: (ms per m)	2000
		Parasites in a level of infestation:	1
Height of drone (m):	0.6	Damage inflicted by parasite to tree(every s):	1
Max reached height (m):	50.0	Needed pesticide to kill a parasite(ml per par.):	5
Visual angle (°):	140		
Min speed (m/s):	4	**Tree's characteristics:**	
Max speed (m/s):	8	% trees in land:	50
Recharging delay (ms):	15000	Trees's health:	2500
Spray delay (ms per mL):	600	Max trees's height (m):	3.0
Max loop cycle time(ms):	2400000		

Consumption:		Drone's barrels:	
		Pesticide(ml):	100
Spray consumption(j per ml):	10	Max power (j):	400000
Density of pesticide (g/cm^3)	1.5	Min power (j):	1000

Communication:		Other:	
TTL (hop's number):	5	Number of charging bases :	8
Bandwidth (bps):	11534336	Number of simulations:	1
Max range (m):	150.0		
Propagation delay (m/s):	300000	**Used protocol in help request:**	
Processing delay (ms):	4	○ Reactive Flooding ○ Link-State	

Reactive Flooding:	
Timeout to receive help response (ms):	2000

Distributed Search Drone's parameters:	
Memory size to local map (Byte):	3025
DSM trasmission period (ms):	1000
Deadline DSM (ms):	30000

☑ Show communication/visual range
☑ Variable drone's speed

Fig. 15. GUI to set the simulation parameters relative to Drone's parameters, Parasites parameters, plant's characteristic, wireless communication parameters, energy values, search task parameters, pesticide fuel parameters, protocol setting, flooding value. This configuration steps are also available in previous works [10,37]. The simulator configuration phase is not changed under the point of view of User Interface.

Packets used in this recruitment protocol are:

- LSA packet: this packet is sent to update routing tables in each node. If a node receives a duplicated packet, it will be deleted. This packet, without the network graph, weights 24 bytes and it is composed in the following way:
  - Source IP;
  - Prop. IP;
  - TTL;
  - Sequence number;
  - Pesticide;
  - Graph.
  Source IP field refers to source drone. Propagator IP field refers to drone that forwards LSA packet. TTL field refers to the maximum hop number. Sequence number field refers to the packet sequence number. Pesticide field refers to ml of pesticide owns by sender. Graph field refers to the network graph in which each node represents a drone IP and each edge represents that two nodes are directly connected.

– Final help packet: this packet is sent by the drone that needs help. It is sent only to a neighbor computed by Dijkstra algorithm. The next hop is computed in every node. This packet weights 24 bytes and it is composed in the following way:
  • Destination IP;
  • Source IP;
  • X coordinate;
  • Y coordinate.
Destination IP field refers to recruited drone. Source IP field refers to drone that needed help. X and Y coordinate fields refer to coordinate of the area that contains parasites.

4.5   Performance Evaluation Through Simulator

Simulations are made by using parameters defined in the *Fig.* 15, Here the base simulator already presented in [10, 37] has been improved to carry out new results in the considered environment. In the following some simulation results related to consumed energy, killed parasites and pesticide consumption are shown. We tested the parameters and strategy implemented in the simulator in order to analyze the performance of the team in exploring the area and saving the plants.

Figure 16(a) shows the level of consumed power for both the compared protocols. It is possible to note that reactive flooding protocol has better performance respect to the link state one. This is due to the maps exchange that is exploited with the reactive flooding allowing the reduction of the drones movements in the explored areas. The Fig. 16(b) shows the number of recruiting requests in the time. It is possible to view that this number is greater using a protocol based

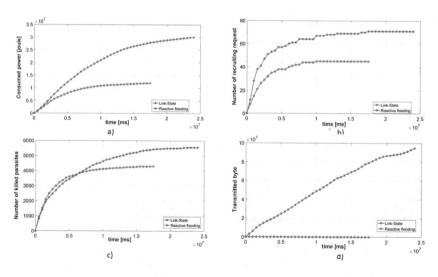

Fig. 16. (a) Consumed power; (b) Number of recruiting requests; (c) Number of killed parasites; Transmitted bytes.

on link state paradigm respect to the Reactive Flooding protocol through the difference between the two protocols is little. In the Fig. 16(c) is depicted the number of killed parasites comparing the two considered protocols. In this graph it is possible to observe that the reactive flooding protocol permits a better performance of the system allowing a greater number of killed parasites respect to the behavior of the protocol based on link state algorithm. At last, the Fig. 16(d) shows how the reactive flooding protocol allows a very low number of transmitted bytes compared with the link state based protocol that has the need of sending much more packets in order to exchange information between drones.

5   Conclusions

In this paper, we propose a novel simulator able to test the efficiency and evaluate the costs of a drones fleet, equipped with multiple sensors and actuators, in precision agriculture. The paper proposes novel coordination and control techniques of UAVs team. Moreover, in this work the attention is focused on the fight against parasites that represent a treat because they are able to destroy entire plantations and, then, they can destroy an entire production. The main idea raises from real world observations and needs. Few years ago, plantations of olive trees were destroyed in Puglia following the attack of the Xylella. Moreover, many palm trees fall because of "Red Punteruolo" attacks. In order to evaluate the goodness of our approach we design a simulator that allows us to carry out some considerations in terms of costs and risks. This step is important before of using real technologies. Consumed energy, killed parasites and pesticide consumption results are discussed through several simulation campaigns. We evaluate parameters and strategies focusing our attention on the capability of the team to explore interested area and saving plants. Results demonstrate the goodness of the proposal and performances in terms of networking saving resources in message exchange processes. Moreover, we save energy and prolong the work cycle of the UAVs. This allows device to accomplish more tasks, contributing to kill more parasites in the considered applicative domain.

References

1. Hall, R.J.: An Internet of Drones. IEEE Internet Comput. **20**(3), 68–73 (2016)
2. Vergouw, B., Nagel, H., Bondt, G., Custers, B.: Drone technology: types, payloads, applications, frequency spectrum issues and future developments. In: Custers, B. (ed.) The Future of Drone Use, pp. 21–45. TMC Asser Press, The Hague (2016)
3. Stöcker, C., Bennett, R., Nex, F., Gerke, M., Zevenbergen, J.: Review of the current state of UAV regulations. Remote. Sens. **9**(5), 459 (2017)
4. Bedford, M.A.: Unmanned Aircraft System (UAS) Service Demand 2015–2035
5. Zeng, Y., Zhang, R., Lim, T.J.: Wireless communications with unmanned aerial vehicles: opportunities and challenges. arXiv preprint arXiv:1602.03602 (2016)
6. Zhang, C., Kovacs, J.M.: The application of small unmanned aerial systems for precision agriculture: a review. Precis. Agric. **13**(6), 693–712 (2012)

7. Primicerio, J., Di Gennaro, S.F., Fiorillo, E., Genesio, L., Lugato, E., Matese, A., Vaccari, F.P.: A flexible unmanned aerial vehicle for precision agriculture. Precis. Agric. **13**(4), 517–523 (2012)
8. Costa, F.G., Ueyama, J., Braun, T., Pessin, G., Osorio, F.S., Vargas, P.A.: The use of unmanned aerial vehicles and wireless sensor network in agriculture applications. In: IEEE International Geoscience and Remote Sensing Symposium (2012)
9. Tripicchio, P., Dabisias, G.: Towards smart farming and sustainable agriculture with drones. In: International Conference on Intelligent Environments (2015)
10. De Rango, F., Palmieri, N., Tropea, M., Potrino, G.: UAVs team and its application in agriculture: a simulation environment. In: SIMULTECH 2017, pp. 374–379 (2017)
11. Pooja, S., Uday, D.V., Nagesh, U.B., Talekar, S.G.: Application of MQTT protocol for real time weather monitoring and precision farming. In: 2017 International Conference on Electrical, Electronics, Communication, Computer, and Optimization Techniques (ICEECCOT), Mysuru, pp. 1–6 (2017). https://doi.org/10.1109/ICEECCOT.2017.8284616
12. MIT Technology Review: Agricultural Drones. Relatively cheap drones with advanced sensors and imaging capabilities are giving farmers new ways to increase yields and reduce crop damage (2015). http://www.technologyreview.com/featuredstory/526491/agricultural-drones/
13. Vincente-Guijalba, F., Martinez-Marin, T., Lopez-Sanchez, M.: Dynamical approach for real-time monitoring of agricultural crops. IEEE Trans. Geosci. Remote. Sens. **53**(6), 3278–3293 (2015)
14. Stehr, N.J.: Drones: the newest technology for precision agriculture. Nat. Sci. Educ. **44**(1), 89–91 (2015)
15. Abdullahi, H.S., Mahieddine, F., Sheriff, R.E.: Technology impact on agricultural productivity: a review of precision agriculture using unmanned aerial vehicles. In: International Conference on Wireless and Satellite Systems, pp. 388–400. Springer, Cham, July 2015
16. Karpowicz, J.: Above the field with UAVs in precision agriculture. Commercial UAV Expo: Las Vegas, NV, USA (2016)
17. Rezende Silva, G., Cunha Escarpinati, M., Duarte Abdala, D., Rezende Souza, I.: Definition of management zones through image processing for precision agriculture. In: 2017 Workshop of Computer Vision (WVC), Natal, pp. 150–154 (2017). https://doi.org/10.1109/WVC.2017.00033
18. Teal, R.K., Tubana, B., Girma, K., Freeman, K.W., Arnall, D.B., Walsh, O., Raun, W.R.: In-season prediction of corn grain yield potential using normalized difference vegetation index contribution from the Oklahoma agricultural experiment station. Agron. J. **98**, 1488–1494 (2006). https://doi.org/10.2134/agronj2006.0103
19. Bossu, J., Gee, C., Truchetet, F.: Development of machine vision system for real time precision sprayer. Electron. Lett. Comput. Vis. Image Anal. **7**(3), 54–66 (2008)
20. Arakeri, M.P., Vijaya Kumar, B.P., Barsaiya, S., Sairam, H.V.: Computer vision based robotic weed control system for precision agriculture. In: 2017 International Conference on Advances in Computing, Communications and Informatics (ICACCI), Udupi, pp. 1201–1205 (2017). https://doi.org/10.1109/ICACCI.2017.8126005

21. Sun, H., Zhu, Q., Ren, J., Barclay, D., Thomson, W.: Combining image analysis and smart data mining for precision agriculture in livestock farming. In: 2017 IEEE International Conference on Internet of Things (iThings) and IEEE Green Computing and Communications (GreenCom) and IEEE Cyber, Physical and Social Computing (CPSCom) and IEEE Smart Data (Smart-Data), Exeter, pp. 1065–1069 (2017). https://doi.org/10.1109/iThings-GreenCom-CPSCom-SmartData.2017.162

22. Pederi, Y.A., Cheporniuk, H.S.: Unmanned aerial vehicles and new technological methods of monitoring and crop protection in precision agriculture. In: Proceedings of 3rd IEEE International Conference Actual Problems of Unmanned Aerial Vehicles Developments (APUAVD), pp. 289–301 (2015)

23. Dlodlo, N., Kalhezi, J.: The Internet of Things in agriculture for sustainable rural development. In: International Conference on Emerging Trends in Networks and Computer Communications (ETNCC), pp. 13–18 (2015)

24. Vasisht, D., Kapetanovic, Z., Won, J., Jin, X., Chandra, R., Sinha, S.N., Stratman, S., et al.: FarmBeats: an IoT platform for data-driven agriculture. In: NSDI, pp. 515–529, March 2017

25. Ye, J., Liu, Q., Fang, Y.: A precision agriculture management system based on Internet of Things and WebGIS. In: 21st International Conference on Geoinformatics, pp. 1–5 (2013)

26. Devraj, R.J., Deep, V.: Expert systems for management of insect-pests in pulse crop. In: 2nd International Conference on Computing for Sustainable Global Development (INDIACom), pp. 1144–1150 (2015)

27. Dai, B., He, Y., Gu, F., Yang, L., Han, J., Xu, W.: A vision-based autonomous aerial spray system for precision agriculture. In: 2017 IEEE International Conference on Robotics and Biomimetics (ROBIO), Macau, Macao, pp. 507–513 (2017). https://doi.org/10.1109/ROBIO.2017.8324467

28. Colomina, J., Molina, P.: Unmanned aerial systems for photogrammetry and remote sensing: a review. J. Photogramm. Remote. Sens. (ISPRS) **92**, 79–97 (2014)

29. Hassan-Esfahani, L., Tores-Rua, A., Ticlavilca, A.M., Jensen, A., McKee, M.: Topsoil moisture estimation for precision agriculture using unmanned aerial vehicle multispectral imagery. In: IEEE International Geoscience and Remote Sensing Symposium (2014)

30. Anthony, D., Elbaum, S., Lorenz, A., Detweiler, C.: On crop height estimation with UAVs. In: IEEE/RSJ International Conference on Intelligent Robots and System (2014)

31. Macker, J.: Mobile ad hoc networking (MANET): routing protocol performance issues and evaluation considerations. RFC 2501 (1999)

32. Varghese, V.T., Sashidar, K., Rekha, P.: A status quo of WSN systems for agriculture. In: International Conference on Advances in Computing, Communications and Informatics (ICACCI), pp. 1775–1781 (2015)

33. Alam, M., Khan, A.H., Khan, I.R.: Swarm intelligence in MANETs: a survey. Int. J. Emerg. Res. Manag. Technol. **5**(5), 141–150 (2016)

34. Manap, Z., Ali, B.M., Ng, C.K., Noordin, N.K., Sali, A.: A review on hierarchical routing protocols for wireless sensor networks. Wirel. Pers. Commun. **72**(2), 1077–1104 (2013)

35. Blazevic, L., Le Boudec, J.Y., Giordano, S.: A location-based routing method for mobile ad hoc networks. IEEE Trans. Mob. Comput. **4**(2), 97–110 (2005)
36. Jung, E.S., Vaidya, N.H.: Power aware routing using power control in ad hoc networks. ACM SIGMOBILE Mob. Comput. Commun. Rev. **9**(3), 7–18 (2005)
37. Rango, F.D., Palmieri, N., Santamaria, A.F., Potrino, G.: A simulator for UAVs management in agriculture domain. In: 2017 International Symposium on Performance Evaluation of Computer and Telecommunication Systems (SPECTS), pp. 1–8 (2017)

Particle Based Blood Pressure Simulation in the Aorta with the Model Generated from CT Images

Nobuhiko Mukai[✉], Kazuhiro Aoyama, Takuya Natsume, and Youngha Chang

Graduate School of Engineering, Tokyo City University, Tokyo 158-8557, Japan
{mukai,aoyama,natsume,chang}@vgl.cs.tcu.ac.jp

Abstract. We have performed the blood pressure simulation in the aorta and the left ventricle with the model generated from CT (Computerized Tomography) data. There were some works related to the aorta; however, some simulations performed the aortic valve behavior with artificial models, and others investigated the blood flow only in the aorta with models generated from MRI (Magnetic Resonance Imaging) data. Then, we demonstrated the simulation of the blood flow and the pressure change in the aorta and the left ventricle with a model generated from CT data. In the simulation, as the blood flowed into the left ventricle through the mitral valve, the pressure in the left ventricle increased, and the aortic valve finally opened by the high pressure in the left ventricle. As the result of the simulation, we confirmed that the pressure change in the left ventricle was similar to a real data; however, the blood pressure in the aorta was lower than the real data. Therefore, in this paper, we demonstrate that the blood pressure change in the aorta becomes similar to the real data by setting some initial pressure in the aorta and forming a closed region of the aorta.

Keywords: Computer graphics · Physics based simulation
Particle method · Medical application · Aorta · Blood pressure

1 Introduction

In our hearts, there is a valve called "aortic valve", which is located between the aorta and the left ventricle. If the valve becomes malfunction, the blood does not flow correctly and some kind of surgery is necessary. The surgery is mainly divided into two types: AVR (Aortic Valvular Replacement) and AVP (Aortic ValvuloPlasty). AVR replaces the dysfunctional live valve with a prosthetic one, and the surgery is not so difficult; however, it is necessary to take a medicine called "warfarin" to prevent blood from coagulation. On the other hand, AVP recovers the valvular function by repairing the dysfunctional live valve, and it is not necessary to take warfarin since there is no extraneous materials in the aorta; however, the surgery is very complexed and difficult so that the preoperative

© Springer Nature Switzerland AG 2019
M. S. Obaidat et al. (Eds.): SIMULTECH 2017, AISC 873, pp. 102–113, 2019.
https://doi.org/10.1007/978-3-030-01470-4_6

computer simulation is mandatory. For the simulations related to the aorta, there are some previous researches.

Hart et al. suggested 2D fluid-structure interaction model [1]. Then, they expanded the model into 3D and presented the maximum principle Cauchy stresses in the leaflets of the aortic valve [2]. Cheng et al. also researched the fluid velocity distribution and the wall shear stress on a bi-leaflet mechanical heart valve [3]. In the simulation, there are two kinds of materials: blood and aorta, which are fluid and solid body, respectively so that fluid-solid interaction should be considered. Loon et al. employed Navier-Stokes equation for the blood flow and hyper-elastic Neo-Hookean model for the solid deformation, respectively [4]. On the other hand, Carmody et al. utilized FEM (Finite Element Method) for the simulation of fluid-structure interaction [5]. We also performed the aortic valve simulation with a particle method by considering heart's pulsation [6]. In addition, Hsu et al. employed Lagrangian-Eulerian methods for fluid-structure interaction [7,8], and they generated artificial models for the simulations; however, the models were not generated from medical data such as CT or MRI.

On the other hand, Seo et al. simulated the flow characteristics in the aortic arch with a model generated from CT images [9]. In addition, Wendel et al. generated an aortic model from MRI to research the behavior of the aortic valve [10], and we also generated a simulation model from CT data [11]. The simulation models generated from real data are realistic and useful. The models, however, had only the aorta including the aortic valve, and did not include the left ventricle located next to the aorta. Then, Le and Sotiropoulos employed a model including both of the aorta and the left ventricle for the simulation of fluid-structure interaction between the blood flow and a mechanical heart valve behavior [12].

In the previous studies, some researches employed artificial models to simulate the aortic valve behavior, while others utilized models generated from real data such as CT or MRI; however, the target valves for the simulation were prosthetic instead of live valve. Therefore, we investigated the pressure change in the aorta and the left ventricle with a particle model generated from real CT data [13]. In the simulation, the aortic valve opened by the high pressure in the left ventricle, and the pressure change in the left ventricle was similar to a real data. The pressure in the aorta, however, was lower than the real data. Then, in this paper, we report the blood pressure change in the aorta becomes similar to the real data by setting some initial pressure in the aorta and forming a closed region of the aorta.

2   Simulation Model

In the simulation, we have to treat two kinds of materials: blood and the aorta including the aortic valve and the left ventricle, which are fluid and solid bodies, respectively. Then, we have to consider the fluid-solid interaction. In general, FEM and particle methods are used for the analysis of fluid and solid body, respectively. In the simulation, however, the aortic valve opens and closes according to the blood pressure so that the blood flow is broken by the aortic valve behavior. In other words, the topology of blood changes. Therefore, we employ a particle method for the simulation of both fluid and solid body since particle methods are suitable for topological change rather than FEM. Then, the simulation model should be generated from a medical data and constructed with particles. Figure 1 shows the CT images of a heart that has been used for the model generation. The CT data is constructed with 114 images, and the images are numbered from the top to the bottom. The image format is "bitmap" and the resolution is 512×512.

On the one hand, Fig. 2 shows one image of the vertical cross section in the heart, which depicts the aorta, the aortic wall, the aortic valve, the Valsalva's sinus, and the left ventricle. Pseudo colors are mapped for easy recognition.

For the generation of a particle model, the target regions should be extracted from the volume data in Fig. 1; however, the image resolution is 512×512 and the number of the images is 114 so that the volume data would be about 30M voxels. Then, if each voxel is set as one particle, the model becomes composed of 30M particles, which are too many for a normal PC (Personal Computer) to

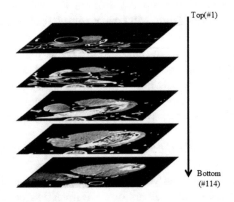

Fig. 1. CT images used for the model generation [13].

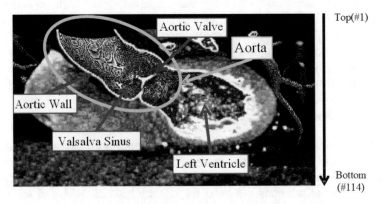

Fig. 2. Vertical cross section image in the heart [13].

handle with. Then, some data reduction is necessary. In the reduction, if the volume data is reduced by simple thinning process, some breaks or holes are generated in the model. In addition, in particle methods, it is said that some extra particles are necessary outside of the model to prevent high speed particles from moving out of the model. Then, the particle model generation algorithm is as follows.

<Particle model generation algorithm>

1. The resolution of the CT image is reduced from 512×512 to 64×64.
2. Every four image plane is selected for the model generation.
3. The reduced and selected image is binarized with a threshold and the target pixels are manually extracted as a closed region.
4. The closed region is filled with pixels, and the two outermost pixels are extracted as the aortic wall elements.
5. The two extra pixels are added outside of the model as dummy pixels to prevent particles from moving out of the model.
6. A 3D particle model is generated by integrating all 2D pixel data generated above.

The generated model has three dimensional shape, and Fig. 3(a) depicts a cross section of the model. The lower right and upper left parts are the left ventricle and the aorta, respectively. The central hole in the left ventricle is the mitral valve, and the blue part between the left ventricle and the aorta is the aortic valve. Figure 3(b) is a vertical cross section of the model. The lower part is the left ventricle and the upper part is the aorta. The right part of the left ventricle is the mitral valve. Actually, the aortic valve is not so clear on the CT image so that the model has manually been generated by referencing to the shape of the valsalva sinus shown in Fig. 2. Figure 3(c) shows the aortic valve model, which has three leaflets that are differentiated in color.

3   Simulation Method

We employ a particle method for the simulation since the topology of blood often changes according to the opening and the closing of the aortic valve. In general, there are two types of particle methods: SPH (Smoothed Particle Hydro-dynamics) and MPS (Moving Particle Semi-implicit). SPH is usually used for compressible fluid, while MPS is used for incompressible fluid. Blood is generally treated as incompressible fluid so that we adopt MPS [14] for the simulation.

Two kinds of governing equations are used for continuous body simulations: Cauchy's equation of motion and equation of continuity, which are written as follows (Eqs. (1) and (2)).

$$\rho \frac{D\mathbf{v}}{Dt} = \nabla \cdot \sigma + \mathbf{b} \tag{1}$$

$$\frac{D\rho}{Dt} + \rho \nabla \cdot \mathbf{v} = 0 \tag{2}$$

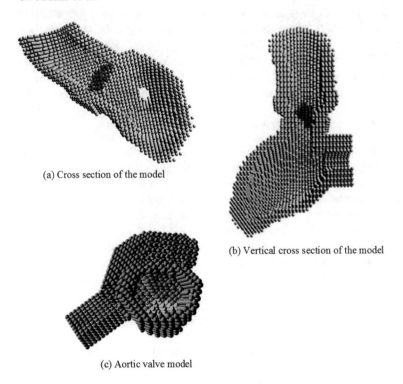

(a) Cross section of the model

(b) Vertical cross section of the model

(c) Aortic valve model

Fig. 3. Particle model generated from CT images [13].

where, ρ is density, $\mathbf{v}$ is velocity, t is time, σ is stress tensor, and $\mathbf{b}$ is body force acceleration such as gravity. In addition, the constitutive equation of elastic body is described as the following (Eqs. (3) and (4)).

$$\sigma^e = \lambda \mathrm{tr}\,(\varepsilon)\mathbf{I} + 2\mu\varepsilon \tag{3}$$

$$\varepsilon = \frac{1}{2}\left\{\nabla\mathbf{u} + (\nabla\mathbf{u})^T\right\} \tag{4}$$

where, σ^e is stress of elastic body, ε is strain tensor, $\mathbf{I}$ is unit tensor, $\mathbf{u}$ is displacement, λ and μ are lame constants, which are expressed as follows (Eqs. (5) and (6)).

$$\lambda = \frac{\nu E}{(1+\nu)(1-2\nu)} \tag{5}$$

$$\mu = \frac{E}{2(1+\nu)} \tag{6}$$

where, ν is Poisson's ratio and E is Young's module.

By substituting Eqs. (3) and (4) for Cauchy's equation (Eq. (1)), the next Cauchy-Navier equation (Eq. (7)) is obtained, which equation is used to analyze the behavior of the aortic valve.

$$\rho\frac{D^2\mathbf{u}}{Dt^2} = (\lambda + \mu)\,\nabla\,(\nabla \cdot \mathbf{u}) + \mu\nabla^2\mathbf{u} + \mathbf{b} \tag{7}$$

On the other hand, the constitutive equation of fluid is written as the following (Eqs. (8) and (9)).

$$\sigma^f = -p\mathbf{I} + 2\eta\mathbf{D} \tag{8}$$

$$\mathbf{D} = \frac{1}{2}\left\{\nabla\mathbf{v} + (\nabla\mathbf{v})^T\right\} \tag{9}$$

where, σ^f is stress of fluid, p is pressure, $\mathbf{I}$ is unit tensor, η is viscosity, $\mathbf{D}$ is tensor of strain velocity, and $\mathbf{v}$ is velocity. By substituting Eqs. (8) and (9) for Eq. (1), Navier-Stokes equation (Eq. (10)) is obtained as follows, which is used to analyze the behavior of blood.

$$\rho\frac{D\mathbf{v}}{Dt} = -\nabla p + \eta\nabla^2\mathbf{v} + \mathbf{b} \tag{10}$$

4   Heart Pulsation

In this simulation, we investigate how blood particles flow from the left ventricle to the aorta and how the pressure changes in the left ventricle and the aorta by analyzing the particle behavior. Figure 4 shows the pressure change in the left ventricle and the aorta of a human heart [15–18].

In Fig. 4, the aortic pressure is higher than that in the left ventricle at the atrial systole stage, when the aortic valve closes. At the isovolumetric contraction stage, blood flows into the left ventricle through the mitral valve and the left ventricle is filled with blood, while the left ventricle shrinks isovolumetrically so that the pressure in the left ventricle rapidly increases up to the same level as the aortic pressure. At the rapid ejection stage, the pressure in the left

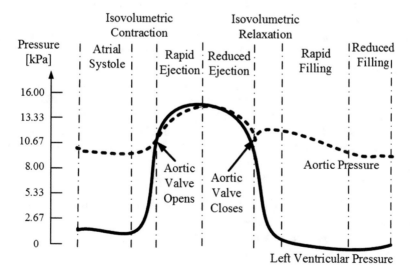

Fig. 4. Pressure change in one heart pulsation [13].

ventricle becomes slightly higher than that in the aorta so that the aortic valve opens. During blood flows from the left ventricle to the aorta, the pressure in the left ventricle is almost the same level as the pressure in the aorta although the pressure in the left ventricle is slightly higher than that in the aorta. After some blood has flown from the left ventricle to the aorta, the pressure in the left ventricle gradually decreases at the reduced ejection stage. As the result, the aortic valve closes with the help of reflection from peripheral blood vessels. After the aortic valve closes, the left ventricle expands isovolumetrically at the isovolumetric relaxation stage, and the pressure in the left ventricle rapidly decreases. If the aortic valve closes correctly, no blood flows back from the aorta to the left ventricle and the pressure difference between the aorta and the left ventricle increases. Finally, at the rapid and reduced filling stage, blood flows into the left atrium and the stage goes back to the atrial systole stage.

5   Simulation Results

The simulation was performed with a normal PC, which has i7-3770K CPU (Central Processing Unit) and GeForce GTX570 GPU (Graphics Processing Unit). The simulation time for 1[step] corresponds to 0.1[ms] in real time, and the particle radius was 2 [mm].

Figure 5 shows the visualization of the blood flow and the pressure change in the left ventricle and the aorta. At the beginning of the simulation, there is no blood particle in the aorta, while some blood particles flow into the left ventricle through the mitral valve (Fig. 5(a)). Then, the pressure in the left ventricle gradually increases (Fig. 5(b) and (c)). When the pressure in the left ventricle becomes over about 10 [kPa], the aortic valve opens and some blood particles flow into the aorta (Fig. 5(d)). Thereafter, the pressure in the left ventricle continues to increase (Fig. 5(e)) and reaches about 15 [kPa] (Fig. 5(f)).

On the other hand, Fig. 6 shows the diagram of the pressure change in the left ventricle and the aorta, where the simulation results are overlaid on Fig. 4. The figure says that the simulated pressure in the left ventricle corresponds well to the literature value; however, the simulated pressure in the aorta does not correspond to the literature value. One reason is that particles flown from the left ventricle to the aorta spread out since the aorta does not form a closed region. The other is that the initial pressure in the aorta is zero although the aorta in the actual heart has some pressure even at the beginning (atrial systole stage).

Then, we have amended the simulation model as follows.

– A lid, which is elastic body, is added at the end of the aorta model to prevent particles in the aorta from spreading out.
– Some particles are set in the aorta at the beginning in accordance with the literature value.

Figure 7 shows the modified aortic model with a lid, with which some particles stay in the aorta and the pressure in the aorta can be kept. However, unless any

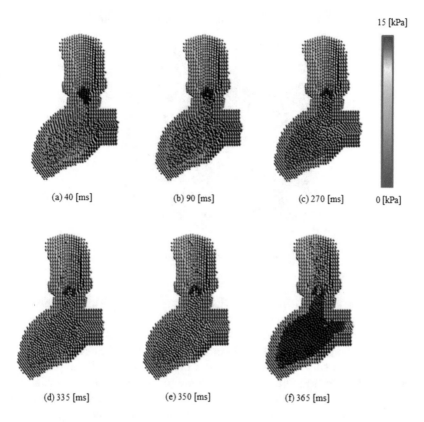

Fig. 5. Visualization of blood flow and pressure change [13].

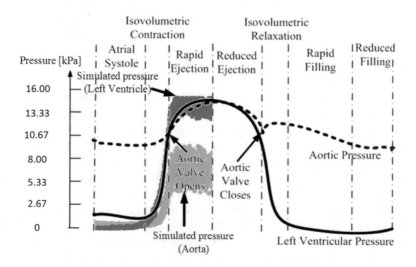

Fig. 6. Pressure change in the simulation results [13].

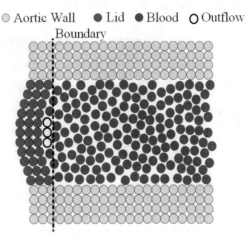

Fig. 7. Aortic model with lid.

particles can flow out of the aorta, the pressure in the aorta increases forever so that some particles should flow out of the aorta. The dotted line in Fig. 7 shows an imaginary boundary. If particles move beyond the boundary, they are removed since they are considered to outflow forward peripheral blood vessels.

Figure 8 shows the visualization of the blood flow and the pressure change in the left ventricle and the aorta with some initial value in the aorta. At the beginning, there are some blood particles in the aorta, and the initial value is set as 9 [kPa] (Fig. 8(a)) in accordance with the literature value (Fig. 4). As some blood particles flow into the left ventricle through the mitral valve, the pressure in the left ventricle increases (Fig. 8(b) and (c)), and both pressures in the aorta and the left ventricle become almost the same (Fig. 8(d)). When the pressure in the left ventricle becomes higher than that in the aorta, the aortic valve opens and some blood particles flow into the aorta (Fig. 8(e)). Thereafter, the pressure in the left ventricle continues to increase (Fig. 8(f)).

On the other hand, Fig. 9 shows the diagram of the pressure change in the left ventricle and the aorta with the initial value. This time, the figure says that the simulated pressure in the aorta corresponds well to the literature value, while the simulated pressure in the left ventricle does not and it is higher than that in the aorta. There are two reasons for the fact that the pressure in the aorta corresponds well to the literature value. One is that some initial value in the aorta is set at the beginning and particles do not spread out of the model because the aorta forms a closed region. The other is that some blood particles are removed when they move beyond the boundary because they are considered to outflow forward peripheral blood vessels. On the other hand, the reason that the pressure in the left ventricle does not correspond well to the literature value is that blood particles continue to flow into the left ventricle through the mitral valve. Then, the mitral valve should close after enough blood particles flow into the left ventricle.

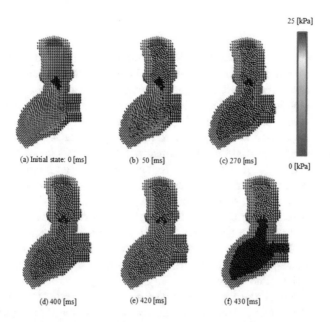

Fig. 8. Pressure change with the initial pressure in the aorta.

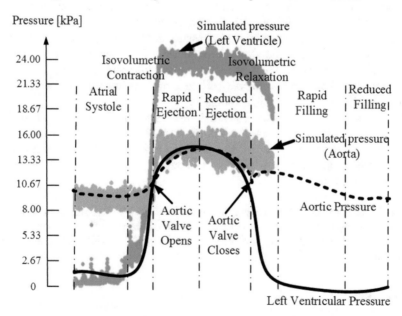

Fig. 9. Pressure change with the initial value in the aorta.

6   Conclusions

Computer simulation of the aortic valvular behavior is mandatory for the surgery called AVP so that some researches have been developed. However, some of the previous researches used artificial models, and others were performed for AVR even if they used realistic models generated from real data. Then, we have demonstrated the blood flow and the pressure change in the aorta and the left ventricle with a particle model generated from real CT data. For the simulation, we have employed a particle method since particle methods are very useful for the topological change of blood flow. As the result of the simulation, we have found that the pressure change in the left ventricle corresponded well to the literature value; however, the pressure change in the aorta did not. There are two reasons for it. One is that the aorta does not form a closed region so that blood particles flown into the aorta spread out and the pressure in the aorta does not go up. The other is that there is no blood particle at the beginning of the simulation although the literature shows there is some pressure even at the beginning (atrial systole stage).

Therefore, we have tried to perform the blood flow simulation with a particle method by setting initial pressure in the aorta at the beginning and forming a closed region of the aorta although some particles are removed if they go beyond a boundary since they are considered to outflow forward peripheral blood vessels. As the result of the simulation, the pressure change in the aorta has corresponded well to the literature value, while the pressure in the left ventricle has been over the literature value since blood particles continued to flow into the left ventricle through the mitral vale. Then, we have to consider the interlock between the aortic value and the mitral valve, and to close the mitral vale when enough blood particles flow into the left ventricle.

In addition, we have treated the aortic valve and the aortic wall as solid body; however, they should be treated as elastic body. Moreover, we did not consider the time scale in the diagram of the pressure change in the aorta and the left ventricle. Then, in the future, we plan to consider these issues and perform the better simulations.

Acknowledgements. We greatly appreciate Dr. Shuichiro Takanashi, who is a chief director of Sakakibara Heart Institute, for providing us the CT data and some useful advices. This work has also been supported by JSPS KAKENHI Grant Number 15K00176.

References

1. Hart, J.D., Peters, G.W.M., Schreurs, P.J.G., Baaijens, F.P.T.: A two-dimensional fluid-structure interaction model of the aortic value. J. Biomech. **33**, 1079–1088 (2000)
2. Hart, J.D., Peters, G.W.M., Schreurs, P.J.G., Baaijens, F.P.T.: A three-dimensional computational analysis of fluid-structure interaction in the aortic valve. J. Biomech. **36**, 103–112 (2003)

3. Cheng, R., Lai, Y.G., Chandran, K.B.: Three-dimensional fluid-structure interaction simulation of bileaflet mechanical heart valve flow dynamics. Ann. Biomed. Eng. **32**, 1471–1483 (2004)
4. Loon, R.V., Anderson, P.D., Baaijens, F.P.T., van de Vosse, F.N.: A three-dimensional fluid-structure interaction method for heart valve modelling. C. R. Mecanique **333**, 856–866 (2005)
5. Carmody, C.J., Burriesci, G., Howard, I.C., Patterson, E.A.: An approach to the simulation of fluid-structure interaction in the aortic valve. J. Biomech. **39**, 158–169 (2006)
6. Mukai, N., Abe, Y., Chang, Y., Niki, K., Takanashi, S.: Particle based simulation of the aortic valve by considering heart's pulsation. In: Medicine Meets Virtual Reality, pp. 285–289. IOS Press (2014)
7. Hsu, M.C., Kamensky, D., Bazilevs, Y., Sackes, M.S., Hughes, T.J.R.: Fluid-structure interaction analysis of bioprosthetic heart valves: significance of arterial wall deformation. Comput. Mech. **54**, 1055–1071 (2014)
8. Hsu, M.C., et al.: Dynamic and fluid-structure interaction simulations of bioprosthetic heart valves using parametric design with T-splines and Fung-type material models. Comput. Mech. **55**, 1211–1225 (2015)
9. Seo, T., Jeong, S.H., Kim, D.H., Seo, D.: The blood flow simulation of human aortic arch model with major branches. In: International Conference on Biomedical Engineering and Informatics, pp. 923–926 (2011)
10. Wendell, D.C., et al.: Including aortic valve morphology in computational fluid dynamics simulations: initial findings and application to aortic coarctation. Med. Eng. Phys. **35**, 723–735 (2013)
11. Mukai, N., Takahashi, T., Chang, Y.: Particle-based simulation on aortic valve behavior with CG model generated from CT. VISIGRAPP **2016**, 248–253 (2016)
12. Le, T.B., Sotiropoulos, F.: Fluid-structure interaction of an aortic heart valve prosthesis driven by an animated anatomic left ventricle. J. Comput. Phys. **244**, 41–62 (2013)
13. Mukai, N., Okamoto, Y., Aoyama, K., Chang, Y.: Blood flow and pressure change simulation in the aorta with the model generated from CT data. In: SIMULTECH 2017, pp. 392–397 (2019)
14. Koshizuka, S.: Particle Method. Maruzen, Tokyo (2005)
15. Izawa, Y.: Medical Note: Cardiovascular Disease. Nishimura, Tokyo (2009)
16. Levick, J.R.: An Introduction to Cardiovascular Physiology. Medical Science International, Tokyo (2011)
17. Klabunde, R.E.: Color Atlas of Physiology, 2nd edn. Lippincott Williams & Wilkins, Baltimore (2012)
18. Silbernagl, S., Despopoulos, A.: Color Atlas of Physiology, 6th edn. Georg Thieme Verlag, Stuttgart (2009)

Formal Methods

Stochastic Simulation of Meteorological Non-Gaussian Joint Time-Series

Nina Kargapolova[1,2(✉)]

[1] Institute of Computational Mathematics and Mathematical Geophysics SB RAS,
Novosibirsk 630090, Russia
nkargapolova@gmail.com
[2] Novosibirsk State University, Novosibirsk 630090, Russia

Abstract. A numerical stochastic model of joint non-stationary non-Gaussian time-series of daily precipitation, daily minimum and maximum air temperature is proposed in this paper. The model is constructed on the assumption that these weather elements are non-stationary non-Gaussian random processes with time-dependent one-dimensional distributions. This assumption takes into account the diurnal and seasonal variation of real meteorological processes. The input parameters of the model (one-dimensional distributions and correlation structure of the joint time-series) are determined from the data of long-term real observations at weather stations. On the basis of simulated trajectories, some statistical properties of rare and extreme weather events (e.g. sharp temperature drops, extended periods of high temperature and precipitation absence) were studied.

Keywords: Stochastic simulation · Non-stationary random process
Non-Gaussian process · Air temperature · Daily precipitation
Extreme weather event

1 Introduction

Stochastic simulation is one of the main tools for solving the problems of statistical Meteorology and Climatology. Using the Monte Carlo method, both the properties of specific meteorological processes and their complexes are studied (see, for example, [1,2,5,9]). Depending on the problem being solved, time-series of meteorological elements of different time scales are simulated (with hours, days, decades, etc. as a time-step). The type of simulated random processes (stationary or non-stationary, Gaussian or non-Gaussian, etc.) is determined by the properties of real meteorological processes and by the selected time step. In recent decades a lot of scientific groups all over the world work at development of so-called "stochastic weather generator". At its core, "generators" are software packages that allow numerically simulate long sequences of random numbers having statistical properties, repeating the basic properties of real meteorological series. Most often series of surface air temperature, daily minimum and maximum

© Springer Nature Switzerland AG 2019
M. S. Obaidat et al. (Eds.): SIMULTECH 2017, AISC 873, pp. 117–127, 2019.
https://doi.org/10.1007/978-3-030-01470-4_7

temperatures, precipitation and solar radiation are simulated ([3,6,14–16]). Not only single-site time series, but also spatial and spatio-temporal meteorological random fields are simulated with the use of "weather generators" ([5,7,10]).

In this paper a stochastic parametric simulation model that provides daily values for precipitation indicators, maximum and minimum temperature at a single site on a yearlong time-interval is presented. The model is based on the assumption that these weather elements are non-stationary random processes and their one-dimensional distributions vary from day to day. A latent Gaussian process and its non-linear transformation (so called threshold transformation) are used for simulation of precipitation indicators. Parameters of the model are chosen for each location on the basis of real data from a weather station situated in this location. An extension of the model, that let to simulate amount of precipitation is also proposed.

2   Real Data Analysis

As input parameters of a numerical stochastic model of joint time-series of precipitation indicator, daily minimum and maximum air temperature in this paper one-dimensional distributions of weather elements and a set of correlation coefficients that fully describes auto- and cross-correlation structure of joint time-series are used. To define the models' parameters data of real meteorological observations from 1976 to 2009 were used. Real data were collected at weather stations located in different climatic zones (for example, weather stations in the cities of Sochi (subtropical zone), Ekaterinburg (temperate continental zone), Tomsk (continental-cyclonic zone). Pogranichniy (moderately monsoon zone), etc.). For the sake of convenience, year-long time-series that start on the January, 1st are considered and data for February 29 is not taken into consideration. Since the most noticeable feature of the time-series at such time interval is the seasonal variation, estimation of the model parameters was done on an assumption that real weather processes are non-stationary. In order to reduce statistical errors on the model parameters' estimations, resulting from a small sample size, a five-day moving averaging window was used.

To construct a stochastic model, the use of sample one-dimensional distributions is not advisable, since the sample distributions don't have any tails, and therefore do not allow to estimate the probability of occurrence of extreme values of meteorological elements. In this connection, it is necessary to approximate the sample distributions by some analytic densities, which, on the one hand, do not greatly alter the form of the distribution and its moments, and on the other, possess tails. Numerical experiments show that Gaussian distributions $F_j^I(x)$, $F_j^A(x)$ approximate closely (in sense of the Pearson χ^2-criterion) sample histograms of minimum and maximum air temperature time-series $I_1, I_2, \ldots, I_N$ and $A_1, A_2, \ldots, A_N$ respectively for all days $j = \overline{1, N}$, $N = 365$ at all considered weather stations.

Daily precipitation indicator E_j in a day number j $\left(j = \overline{1, N}\right)$ is defined as 1 if amount of precipitation during this day in more of equal than 0.1 mm and as 0 otherwise. This means that daily precipitation indicator, as a random variable in each day number j, has the Bernoulli distribution $(P(E_j = 1) = p_j)$.

For simulation of a joint time-series in this paper sample correlation matrices were used. Analysis of real data shows that for all meteorological stations correlation functions $corr(i, i + h)$ as functions of shift h decrease rapidly. As an illustration, Fig. 1 shows examples of sample autocorrelation functions $corr(1, 1 + h)$ as functions of the shift h. Examples of a behavior of correlation functions $corr(i, i + h)$ as functions of the first argument i, are shown in Fig. 2.

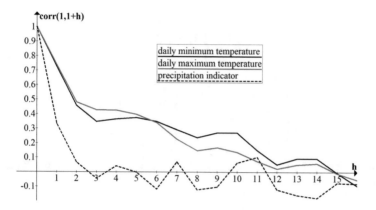

Fig. 1. Autocorrelation coefficients $corr(I_1, I_{1+h})$, $corr(A_1, A_{1+h})$, $corr(E_1, E_{1+h})$. Novosibirsk, Russia.

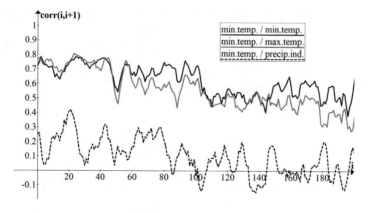

Fig. 2. Coefficients $corr(I_i, I_{i+1})$, $corr(I_i, A_{i+1})$, $corr(I_i, E_{i+1})$. Novosibirsk, Russia.

3   Model Description

In this section a model of precipitation indicator, daily minimum and maximum air temperature joint time-series is presented. This model was first published in [4]. Each simulated model trajectory is a matrix $M = \left(\boldsymbol{I}^T, \; \boldsymbol{A}^T, \; \boldsymbol{E}^T \right)$, where a column-vector $\boldsymbol{I}^T = (I_1, I_2, \ldots, I_N)^T$ is a vector whose component I_j is daily minimum air temperature in a day number j, $\boldsymbol{A}^T = (A_1, A_2, \ldots, A_N)^T$ is a vector of daily maximum temperatures and a column-vector $\boldsymbol{E}^T = (E_1, E_2, \ldots, E_N)^T$ is a vector of daily precipitation indicators.

Elements of a joint time-series M are calculated with the help of a transformation

$$
\begin{aligned}
I_j &= \sigma_j^I \xi_j^I + \mu_j^I, \\
A_j &= \sigma_j^A \xi_j^A + \mu_j^A, \\
E_j &= \begin{cases} 1, & \xi_j^E \le c_j, \\ 0, & \xi_j^E > c_j, \end{cases}
\end{aligned}
\tag{1}
$$

where vectors

$$
\begin{pmatrix} \mu_1^I \\ \mu_2^I \\ \vdots \\ \mu_N^I \end{pmatrix}, \quad
\begin{pmatrix} \mu_1^A \\ \mu_2^A \\ \vdots \\ \mu_N^A \end{pmatrix}, \quad
\begin{pmatrix} \sigma_1^I \\ \sigma_2^I \\ \vdots \\ \sigma_N^I \end{pmatrix}, \quad
\begin{pmatrix} \sigma_1^A \\ \sigma_2^A \\ \vdots \\ \sigma_N^A \end{pmatrix}
$$

are vectors of mean and standard deviation. Threshold values c_j are defined from the equations

$$
P\left(E_j = 1\right) = \frac{1}{\sqrt{2\pi}} \int_{-\infty}^{c_j} \exp\left(-\frac{t^2}{2}\right) dt = p_j, \; j = \overline{1, N}.
$$

Values of $\mu_j^I, \mu_j^A, \sigma_j^I, \sigma_j^A, p_j$ are estimated on a basis of real data from a weather station. It should be noted that for all $j = \overline{1, N}$ equality $c_j = 0$ is true if and only if $p_j = 0.5$, inequalities $c_j > 0$ and $p_j > 0.5$ are equivalent. Hereafter it is supposed that $p_i \ne 0$, $p_i \ne 1$, $i = \overline{1, N}$. Variables $\left(\xi_j^I, \xi_j^A, \xi_j^E\right)$ are components of a joint Gaussian process $\Omega = \left(\left(\boldsymbol{\xi}^I\right)^T, \left(\boldsymbol{\xi}^A\right)^T, \left(\boldsymbol{\xi}^E\right)^T \right)$ with zero mean and specific correlation matrix

$$
G = \begin{pmatrix} G_{II} & G_{IA} & G_{IE} \\ G_{AI} & G_{AA} & G_{AE} \\ G_{EI} & G_{EA} & G_{EE} \end{pmatrix}.
$$

Matrix G must be such that a process M has a sample correlation matrix

$$
R = \begin{pmatrix} R_{II} & R_{IA} & R_{IE} \\ R_{AI} & R_{AA} & R_{AE} \\ R_{EI} & R_{EA} & R_{EE} \end{pmatrix}.
$$

Method of matrix G calculation is described below. Dimension of matrixes G and R are $1095 \times 1095 (3N \times 3N)$. Element $r_{XY}(i, j)$ of a matrix block R_{XY} is a correlation coefficient between X_i and Y_j $(X, Y \in \{I, A, E\}, \; i, j \in \{1, 2, \ldots, N\})$.

Element $g_{XY}(i,j)$ is corresponding to $r_{XY}(i,j)$ correlation coefficient of a Gaussian process.

Let's take a closer look at the matrix G and find equations that define this matrix when the matrix R is given. In [11] a special case of such equations was considered. General case was first considered in [4]. Normalisation of two correlated Gaussian random variables doesn't change a correlation coefficient between them, which implies

$$
\begin{aligned}
G_{II} &= R_{II}, \; G_{AA} = R_{AA}, \\
G_{IA} &= R_{IA}, \; G_{AI} = R_{AI}.
\end{aligned}
\tag{2}
$$

Definition of a correlation coefficient leads to equations

$$
r_{IE}(i,j) = \frac{EI_i E_j - EI_i EE_j}{\sqrt{DI_i}\sqrt{DE_j}} = -\frac{g_{IE}(i,j)}{\sqrt{2\pi p_j(1-p_j)}} \exp(-c_j^2/2)
$$
$$
i,j = \overline{1,N}.
\tag{3}
$$

These equations fully define the matrix G_{IE}. Matrices G_{EI}, G_{AE}, G_{EA} are defined in a similar way. Since

$$
r_{EE}(i,j) = \frac{P(E_i = 1, E_j = 1) - p_i p_j}{\sqrt{p_i(1-p_i)}\sqrt{p_j(1-p_j)}},
$$
$$
P(E_i = 1, E_j = 1) = P\left(\xi_i^E \le c_i, \xi_j^E \le c_j\right), \quad i,j = \overline{1,N}
$$

following equalities hold for $i,j = \overline{1,N}$ & $i \neq j$,

$$
r_{EE}(i,j) = \frac{F(c_i, c_j, g_{EE}(i,j)) - p_i p_j}{\sqrt{p_i(1-p_i)}\sqrt{p_j(1-p_j)}},
\tag{4}
$$

where

$$
F(h,k,\rho) = \frac{1}{2\pi\sqrt{1-\rho^2}} \times \int_{-\infty}^{h}\int_{-\infty}^{k} \exp\left(-\frac{1}{2(1-\rho^2)}\left(x^2 - 2\rho x y + y^2\right)\right) dx dy.
$$

Obviously,

$$
r_{EE}(i,i) = g_{EE}(i,i) = 1, \quad i = \overline{1,N}.
$$

It should be noted that equations for $r_{EE}(i,j)$, $i,j = \overline{1,N}$ & $i \neq j$, don't have any analytical solutions, but it is possible to solve them numerically. Given above Eqs. (2)–(4) fully define matrix G and let to formulate a simulation algorithm:

Step 1. Estimate $\mu_j^I, \mu_j^A, \sigma_j^I, \sigma_j^A, p_j$, $j = \overline{1,N}$ on a basis of real data.

Step 2. Using given above equations and equalities, define matrix G.

Step 3. Simulate required number of trajectories of a joint Gaussian process Ω with zero mean and correlation matrix G.

Step 4. Using (1) transform simulated trajectories of a Gaussian processes into trajectories of a non-Gaussian process M.

Remark 1. Due to a physical sense of daily minimum and maximum temperatures, an inequality $I_j \leq A_j$ must be true for all $j = \overline{1, N}$, but transformation (1) doesn't guarantee it. This means that one must eliminate from consideration all trajectories in which this inequality violates. In practice, it is typical that $\mu_j^I \ll \mu_j^A$ and σ_j^I, σ_j^A are relatively small, so usually there are few trajectories with $I_j > A_j$.

Remark 2. It may happen that matrix G is not positively defined. In this case it is necessary to regularize the matrix G (several regularisation algorithms are described in [12]). In this paper, regularization was performed by replacing the negative eigenvalues of the specified matrix by small positive ones. This has led to the fact that the correlation structure of the simulated process differs from the correlation structure of real one, but this difference is relatively small.

Remark 3. There are a lot of algorithms for simulation of a Gaussian process with the given correlation matrix. The most common are algorithms based on Cholesky and spectral decomposition of the correlation matrix. In this paper the last one was used.

Remark 4. Numerical solution of equations for $r_{EE}(i, j)$ is a time-consuming problem. There is a way to reduce computational time. So-called Owen's formulas ([13]) give a representation of function $F(h, k, \rho)$ via one-dimensional integrals:

$$F\left(c_i, c_j, g_{EE}\left(i, j\right)\right) = \frac{1}{2}\Phi\left(c_i\right) + \frac{1}{2}\Phi\left(c_j\right) - T\left(c_i, a_1\right) - T\left(c_j, a_2\right) + \frac{1}{2},$$

if $c_i \geq 0$, $c_j \geq 0$ or $c_i < 0$, $c_j < 0$, and

$$F\left(c_i, c_j, g_{EE}\left(i, j\right)\right) = \frac{1}{2}\Phi\left(c_i\right) + \frac{1}{2}\Phi\left(c_j\right) - T\left(c_i, a_1\right) - T\left(c_j, a_2\right),$$

if $c_i < 0$, $c_j \geq 0$ or $c_i \geq 0$, $c_j < 0$, where

$$\Phi\left(c_i\right) = \frac{1}{\sqrt{2\pi}} \int\limits_0^{c_i} \exp\left(-t^2/2\right) dt,$$

$$T\left(c, a\right) = \frac{1}{2\pi} \int\limits_0^a \exp\left(-\frac{c^2\left(1+t^2\right)}{2}\right) \frac{dt}{1+t^2},$$

$$a_1 = \frac{c_j - c_i g_{EE}(i,j)}{c_i \sqrt{1 - g_{EE}^2(i,j)}}, \quad a_2 = \frac{c_i - c_j g_{EE}(i,j)}{c_j \sqrt{1 - g_{EE}^2(i,j)}}.$$

This representation together with the fact that

$$p_i = \frac{1}{2} + \Phi\left(c_i\right), \quad i = \overline{1, N}$$

let to replace equations for $r_{EE}(i,j)$ with equations ([4])

$$
r_{EE}(i,j) = \begin{cases}
\dfrac{\frac{1}{2}p_i+\frac{1}{2}p_j-p_ip_j}{\sqrt{p_i(1-p_i)}\sqrt{p_j(1-p_j)}} - \dfrac{T(c_i,a_1)+T(c_j,a_2)}{\sqrt{p_i(1-p_i)}\sqrt{p_j(1-p_j)}}, & if\ c_ic_j>0, \\[2ex]
\dfrac{\frac{1}{2}p_i+\frac{1}{2}p_j-\frac{1}{2}-p_ip_j}{\sqrt{p_i(1-p_i)}\sqrt{p_j(1-p_j)}} - \dfrac{T(c_i,a_1)+T(c_j,a_2)}{\sqrt{p_i(1-p_i)}\sqrt{p_j(1-p_j)}}, & if\ c_ic_j<0, \\[2ex]
-\dfrac{2T\left(c_j,-\dfrac{g_{EE}(i,j)}{\sqrt{1-g^2_{EE}(i,j)}}\right)}{\sqrt{p_j(1-p_j)}}, & if\ c_i=0,\ c_j\neq0, \\[3ex]
-\dfrac{2T\left(c_i,-\dfrac{g_{EE}(i,j)}{\sqrt{1-g^2_{EE}(i,j)}}\right)}{\sqrt{p_i(1-p_i)}}, & if\ c_i\neq0,\ c_j=0, \\[3ex]
\dfrac{2\arcsin g_{EE}(i,j)}{\pi}, & if\ c_i=c_j=0.
\end{cases}
$$

Numerical experiments show that computational time required for solution of last equations for $r_{EE}(i,j)$ is approximately 4 times less than computational time required for solution of equations for $r_{EE}(i,j)$ with double integrals.

4 Extended Model

For solution of some applied problems it's not always enough to simulate precipitation indicators, that show either day is a dry day or a wet one, – precipitation amount is also very important characteristic. The model presented in this paper may be easily extended, and simulation of precipitation indicators $\boldsymbol{E}^T$ may be replaced by simulation of daily precipitation amount $\boldsymbol{D}^T = (D_1, D_2, \ldots, D_N)^T$ in a form of a multiplicative process

$$
D_j = E_jC_j, \quad j = \overline{1,N},
$$

where $\boldsymbol{C}^T = (C_1, C_2, \ldots, C_N)^T$ is a non-Gaussian random process describing amount of daily precipitation on the assumption of their presence. This means that instead of a process $(\boldsymbol{I}^T,\ \boldsymbol{A}^T,\ \boldsymbol{E}^T)$ with correlation matrix R a process $(\boldsymbol{I}^T,\ \boldsymbol{A}^T,\ \boldsymbol{E}^T,\ \boldsymbol{C}^T)$ have to be simulated with a correlation matrix

$$
\widetilde{R} = \begin{pmatrix}
R_{II} & R_{IA} & R_{IE} & R_{IC} \\
R_{AI} & R_{AA} & R_{AE} & R_{AC} \\
R_{EI} & R_{EA} & R_{EE} & O \\
R_{CI} & R_{CA} & O & R_{CC}
\end{pmatrix},
$$

where O is $N \times N$ zero matrix. Elements $g_{IC}(i,j)$, $g_{AC}(i,j)$, $g_{CI}(i,j)$, $g_{CA}(i,j)$, $g_{CC}(i,j)$ of a corresponding matrix $\widetilde{G}$ of a standard Gaussian process are found within the framework of inverse distribution function method from the equations

$$
r_{XY}(i,j) = \int\limits_{-\infty}^{\infty}\int\limits_{-\infty}^{\infty} F_i^{-1}(\Phi(x))\, F_j^{-1}(\Phi(y))\, \phi(x,y,g_{XY}(i,j))\, dxdy,
$$

where

$$\phi\left(x, y, g_{XY}\left(i, j\right)\right) = \left[2\pi\sqrt{1 - \left(g_{XY}\left(i, j\right)\right)^2} \exp\left(\frac{2g_{XY}\left(i, j\right)xy - x^2 - y^2}{2\left(1 - \left(g_{XY}\left(i, j\right)\right)^2\right)}\right)\right]$$

is a distribution density of a bivariate Gaussian vector with zero mean, variance equal to 1 and correlation coefficient $g_{XY}\left(i, j\right)$ between components number i and j, $\Phi\left(\cdot\right)$ is a CDF of a standard normal distribution, $F_i\left(\cdot\right)$, $F_j\left(\cdot\right)$ are CDFs of X_i, Y_j. As approximations F_j^C, $j = \overline{1, N}$ of sample one-dimensional distributions of precipitation mixtures of two Gamma-distributions were used. Parameters of these mixtures were chosen using an algorithm, proposed in [8]. This algorithm let to choose such parameters of a mixture F_j^C that mathematical expectation, variance and skewness of a random variable with a CDF F_j^C are equal to corresponding sample characteristics and function F_j^C is the closest (among all mixtures of two Gamma-distributions) to a sample CDF in a sense of the Pearson's functional.

5   Numerical Experiments

Described above stochastic models were used for simulation of joint meteorological non-stationary time-series on more than 50 weather stations situated in different climatic zones in Russia. For models verification, it is necessary to compare simulated and real data based estimations of such characteristics, which, on the one hand, are reliably estimated by real data, and on the other are not input parameters of the model. Verification of both models shows that the models give satisfactory results for most of the considered stations. Here are several examples of a process characteristics that were used for the model verification.

First characteristic that was used for the extended model verification was "average numbers of days in a month, when minimum temperature is below $0\,^\circ C$ and maximum temperature is above $0\,^\circ C$ $\left(I_j < 0, A_j > 0\right)$" This characteristic is not the model input parameter, so it can be used for verification. Table 1 presents values of this characteristic, estimated over real and simulated data. It can be seen from Table 1, that the model reproduces this characteristic accurately (up to a statistical mistake). Hereafter for all estimations on basis of simulated data 10^6 trajectories were used.

Another characteristic that was considered was "average number of days in a month with a minimum/maximum daily temperature below/above given level $l\,^\circ C$". Table 2 shows corresponding estimations.

Since the model is adequate to real weather processes, it may be used for study of rare/extreme events. Here are two examples. One of the weather event that may be dangerous both to individuals and to agricultural industry is long-term combination of high air temperature and absence of precipitation. Such combination may negatively influence on individuals' health and may cause soil drying up. Table 3 presents average number of time-intervals lasting k days,

Table 1. Average number of days with $I_j < 0, A_j > 0$. St. Petersburg, Russia([4]).

Month	Average number of days	
	Real data	Simulated data
October	4.79	4.89
November	8.94	9.08
December	5.70	5.47
January	9.34	9.43
February	7.71	7.44
March	16.34	16.80

Table 2. Average number of days in a month with a minimum daily temperature above given level $l\,°C$. Sochi, Russia. December.

l	Average number of days	
	Real data	Simulated data
4	21.30	20.77
6	15.41	14.95
8	9.60	9.56
10	4.13	4.16
12	1.78	1.75
14	0.62	0.61

Table 3. Average number of summer time-intervals lasting k days, with absence of precipitation and daily minimum temperature above $20\,°C$. Astrakhan, Russia ([4]).

k	Average number of time-intervals	
	Real data	Simulated data
1	5.38	5.72
2	2.09	2.20
4	0.75	0.76
5	0.61	0.59
6	0.09	0.53
8	0.09	0.37
9	0.33	0.32
10	0.11	0.09

when daily minimum temperature was above $20\,°C$ and there were no precipitations. Averaging was done over summer months. Described in the paper model (the model without an extension) reproduces this characteristic for short time-intervals (up to $k = 5$) satisfactory, so model results for longer time-intervals may be considered as reasonable.

The last rare weather event, considered in this paper, is "appearance of k consequent wet days with total precipitation amount above level l mm". Persistent heavy rainfall can lead to floods and to rotting of plants roots. Table 4 shows probabilities of $k = 5$ consequent wet days with total precipitation amount above level l mm, estimated over real data and simulated trajectories of the extended model. Here 10^8 simulated trajectories were used. These number of trajectories let to estimate probabilities of heavy rainfalls for high levels l for which an estimate from a small sample of real data yields a zero result, although the event is possible.

Table 4. Probabilities of $k = 5$ consequent wet days with total precipitation amount above level l mm. Novosibirsk, Russia. July.

l	Probabilities	
	Real data	Simulated data
10	0.18	0.192
15	0.10	0.098
20	0.06	0.058
25	0.04	0.042
30	0.03	0.029
35	0.02	0.018
40	0.01	0.012
45	0.01	0.009
70	0.00	0.006
90	0.00	0.002

6   Conclusions

In this paper 2 models for simulation of meteorological time-series were considered. It was also shown that simulated trajectories may be used for study of rare/extreme events. It is planned to combine an approach presented in this paper with ideas from [10] and to simulate spatio-temporal joint random fields of precipitation amount and daily minimum and maximum air temperature.

Acknowledgements. This work was supported by the Russian Foundation for Basis Research (grant No 18-01-00149-a), the President of the Russian Federation (grant No MK-659.2017.1).

References

1. Ailliot, P., Allard, D., Monbet, V., Naveau, P.: Stochastic weather generators: an overview of weather type models. J. de la Société Française de Stat. **156**(1), 101–113 (2015)
2. Derenok, K.V., Ogorodnikov, V.A.: Numerical simulation of significant long-term decreases in air temperature. Russ. J. Num. Anal. Math. Modelling. **23**(3), 223–277 (2008)
3. Furrer, E.M., Katz, R.W.: Generalized linear modeling approach to stochastic weather generators. Clim. Res. **34**(2), 129–144 (2007)
4. Kargapolova, N.: Stochastic simulation of non-stationary meteorological time-series: daily precipitation indicators, maximum and minimum air temperature simulation using latent and transformed Gaussian processes. In: Proceedings of the 7th International Conference on Simulation and Modeling Methodologies, Technologies and Applications, pp. 173–179 (2017)
5. Kargapolova, N.A.: Stochastic "weather generators". Introduction to stochastic simulation of meteorological processes: lecture course. PPC NSU, Novosibirsk (2017)
6. Kargapolova, N.A., Ogorodnikov, V.A.: Inhomogeneous Markov chains with periodic matrices of transition probabilities and their application to simulation of meteorological processes. Russ. J. Num. Anal. Math. Modelling. **27**(3), 213–228 (2012)
7. Kleiber, W., Katz, R.W., Rajagopalan, B.: Daily spatiotemporal precipitation simulation using latent and transformed Gaussian processes. Water Resour. Res. **48**(1), 1–17 (2012)
8. Marchenko, A.S., Siomochkin, A.G.: Models of one-dimensional and joint distributions of non-negative random variables. Meteorol. Hydrol. **3**, 50–56 (1982). [in Russian]
9. Ogorodnikov, V.A., Kargapolova N.A., Sereseva, O.V.: Numerical stochastic models of meteorological processes and fields and some their applications. In: Topics in Statistical Simulation. Springer Proceedings in Mathematics & Statistics, pp. 409-417 (2014)
10. Ogorodnikov, V.A., Kargopolova, N.A., Seresseva, O.V.: Numerical stochastic model of spatial fields of daily sums of liquid precipitation. Russ. J. Num. Anal. Math. Modellin. **28**(2), 187–200 (2013)
11. Ogorodnikov, V.A., Khlebnikova, E.I., Kosyak, S.S.: Numerical stochastic simulation of joint non-Gaussian meteorological series. Russ. J. Num. Anal. Math. Modelling. **24**(5), 467–480 (2009)
12. Ogorodnikov, V.A., Prigarin, S.M.: Numerical Modelling of Random Processes and Fields: Algorithms and Applications. VSP, Utrecht (1996)
13. Owen, D.B.: Table for computing bivariate normal probabilities. Ann. Math. Statist. **27**(4), 1075–2000 (1956)
14. Richardson, C.W.: Stochastic simulation of daily precipitation, temperature and solar radiation. Water Resour. Res. **17**(1), 182–190 (1981)
15. Richardson, C.W., Wright, D.A.: WGEN: A Model for Generating Daily Weather Variables. U.S. Department of Agriculture, Agricultural Research Service, ARS-8 (1984)
16. Semenov, M.A., Barrow, E.M.: LARS-WG: A Stochastic Weather Generator for Use in Climate Impact Studies. Version 3.0. User Manual (2002)

Approaches to Fault Detection for Heating Systems Using CP Tensor Decompositions

Erik Sewe[1,2]([✉]), Georg Pangalos[3], and Gerwald Lichtenberg[4]

[1] PLENUM Ingenieurgesellschaft für Planung Energie Umwelt mbH,
Hamburg, Germany
`erik.sewe@plenuming.de`
[2] Hamburg Ministry of Environment and Energy, Hamburg, Germany
[3] Application Center Power Electronics for Renewable Energy Systems,
Fraunhofer Institute for Silicon Technology ISIT, Hamburg, Germany
`georg.pangalos@isit.fraunhofer.de`
[4] Faculty Life Sciences, Hamburg University of Applied Sciences, Hamburg, Germany
`gerwald.lichtenberg@haw-hamburg.de`

Abstract. Two new signal-based and one model-based fault detection methods using canonical polyadic (CP) tensor decomposition algorithms are presented, and application examples of heating systems are given for all methods. The first signal-based fault detection method uses the factor matrices of a data tensor directly, the second calculates expected values from the decomposed tensor and compares these with measured values to generate the residuals. The third fault detection method is based on multi-linear models represented by parameter tensors with elements computed by subspace parameter identification algorithms and data for different but structured operating regimes. In case of missing data or model parameters in tensor representation, an approximation method based on a special CP tensor decomposition algorithm for incomplete tensors is proposed, called the decompose-and-unfold method. As long as all relevant dynamics has been recorded, this method approximates – also from incomplete data – models for all operating regimes, which can be used for residual generation and fault detection, e.g. by parity equations.

Keywords: Fault detection · Multi-linear systems
Operting regimes · Tensor decomposition
Nonlinear parameter identification · Parity equations · Heating systems

1 Introduction

Data often can be arranged in a systematic way by multi-index structures which are mathematically described by tensors. Tensor decomposition methods are applicable for data compression, signal processing or data mining [1]. In this

© Springer Nature Switzerland AG 2019
M. S. Obaidat et al. (Eds.): SIMULTECH 2017, AISC 873, pp. 128–152, 2019.
https://doi.org/10.1007/978-3-030-01470-4_8

contribution, data compression is not the main focus – although a side aspect. In the first part, factor matrices resulting from tensor decomposition of measurement data – e.g. from heating systems – are used for signal-based fault detection. In the second part, tensors are used to represent a set of linear time-invariant models which are the basis of the presented model-based fault detection method.

The analysis of the operation of heating systems, with multiple, possibly different heat generation units and/or different types of heat consumption, usually shows potential for optimization, see [2]. An investigation of a multitude of malfunctions of heating systems done by PLENUM shows that in many cases component-failure, poorly tuned controller parameters and user interaction result in problems of heat supply, excess consumption, wear or data loss. Frequently occurring problems are operation errors of combined heat and power plants and gas boilers. Faults vary from wrong parameter values up to complete loss of the heat production. A further common fault is the operation of a system, which is not adapted to its needs. On the one hand, components of the heating system might be in operation without being used, which results in an unnecessary heat consumption. On the other hand components might be in failure or switched off, although needed, which results in a bottleneck in supply.

The methods of fault detection used for heating systems are not best practice according to today's technical knowledge and options, [3]. A fault is defined as a deviation of at least one characteristic property or parameter of the system from its nominal value [4]. A fault is observable, if its presence – independent of its size or type – is resulting in a change of the nominal output of the system [5]. Fault detection can be done in an automated procedure, if for a component or a (sub-)system

- the expert knowledge is applicable in general and in particular e.g. for the definition of rules and parameters,
- there is sufficient knowledge of the physical process or sufficient measurement data to construct a nominal model,
- or if the measurements are taken in different similar systems and compared with each other.

Expert-based fault detection uses the experience and knowledge of a human with the drawback of high personnel costs. Also because of lack of attention or high time pressure faults can be overlooked. Expert-based fault detection should not be used for regular checks or large amounts of data. For this reason a continuous automated fault detection is desirable.

Signal-based fault detection uses measurement data to specify and evaluate indicators. It is perfectly suitable for simple checks and to be aware of deviations quickly. Currently this is mostly done on a simple level checks, meaning thresholds are checked and ordinary rules are applied. Unfortunately, this reaches its limits of transferability and expandability, see [6]. In many cases the measurements of a component or (sub-)system of a heating system are prone to a structure, i.e. a regular behavior in dependence of a disturbance or another component. A novel approach is to examine this structure using tensor decomposition methods [7]. In a further step the factor matrices resulting from tensor decomposition can be used for fault detection, as proposed in this contribution, see Sect. 3.1. Usually signal-based fault detection is only applicable for static processes, if a dynamical process is under investigation this might not be sufficient, [8], leading to model-based fault detection.

For model-based fault detection a model of the heating system, of a subsystem or of a component is constructed and used for fault detection. It is possible to check the dynamical processes and the interaction of different components as long as an adequate model exists. The modeling, including the parametrization, is based either on measurement data of the nominal behavior, i.e. black-box modeling, on physical equations and constants, i.e. white-box modeling or a combination of both, where the parameters of the physical equations are estimated by the measurement data, i.e. grey-box modeling.

A model-based fault detection can provide an enhanced detection performance, but is associated with many obstacles, see [9]. The behavior of heating systems and buildings is difficult to predict. Systems are unique and the creation of exact models is difficult, time consuming and therefore expensive. Detailed building plans and schemes are rarely available, the intended function often poorly documented. Measurement data is intermittent or of poor quality. Some variables are not directly measurable. A high number of disturbances occurs. Additionally, (sub-)system are nonlinear.

The basis of model-based fault detection is a model of the system, that captures the main system dynamics in a precision, that is sufficient for the task. Models of heating systems can be derived by defining heat power balances, see [10]. The heat power balances have a multilinear property, arising from the multiplication of temperature spread and flow rate. Since both – temperature spread and flow rate – are states of the system or linearly depending on a state of the system (depending on the realization of the model) a direct description as a linear state-space model is not possible. Furthermore, discrete signals like boiler switching (on/off) turn the problem into a hybrid one. One possibility to model these hybrid heating systems is to use tensor systems, see e.g. [11,12]. This approach is quite new, such that so far no methods for system identification exist, only standard nonlinear parameter identification algorithms – which do not take care of the multilinear structure – could be applied. For that reason a different approach is used here to model the system. The model of the chosen approach could be represented as a switched system, see [13]. By quantizing one of the signals – temperature spread or flow rate – no multiplication of these signals is needed, because for each quantized system the signal which was not quantized is multiplied with a constant [14]. Depending on the size of the quantization interval, inaccuracies between the original and the quantized system arise. For each interval of the quantized signal and for each combination of the discrete signals one linear model is estimated using standard techniques, e.g. the N4SID algorithm, [15], resulting in several linear time-invariant systems, i.e. a multi-linear time-invariant model. Note, that multi-linear is used to denote multiple linear models, whereas multilinear – without dash – is used to denote models where the multiplication of all states and inputs – and all combinations with all numbers of multiplicants – are admissible. In multilinear systems no square terms are allowed, this could be modeled by polynomial models. Note that for polynomial systems black box identification methods exist, see [16]. The heating systems under investigation have the aforementioned multilinear property that could be modeled in the polynomial framework but the hybrid property (the discrete signals) are beyond the scope. In addition, the identification is, opposed to the one step identification of N4SID an iterative identification which also needs to be initialized, [17].

This contribution shows signal-based and model-based methods used for fault detection in heating systems. Preliminaries on tensors and tensor decomposition methods are given in Sect. 2. Section 3 shows signal based fault detection, Sect. 3.1 explains the method, two application examples are given in Sect. 3.2. In Sect. 4 a model based method and structure for estimating the parameter sets and fault detection are explained. The complete system can be stored in a tensor structure, see Sect. 4.1. Identification of parameter sets is explained in Sect. 4.2. Missing parameter sets for operating points not included in training data can be estimated by a decompose-and-unfold method, see Sect. 4.3. Fault detection is done using parity equations, Sect. 4.4. In Sect. 4.5 an application example is given. In Sect. 5 conclusions are drawn.

2 Preliminaries on Tensors

The preliminaries on tensors are based on Cichocki et al. [18] and Kolda et al. [1]. This section is outlines the summary published by Müller et al. in [19].

A tensor of order n

$$\mathsf{X} \in \mathbb{R}^{I_1 \times I_2 \times \cdots \times I_n} \tag{1}$$

is an n-dimensional array. The elements $x(i_1, i_2, \ldots, i_n)$ are indexed with $i_j \in \{1, 2, \ldots, I_j\}$ in each dimension $j = 1, 2, \ldots, n$. For simplicity the index vector

$$\mathbf{I} = (i_1, i_2, \ldots, i_n) \tag{2}$$

is defined. A first order tensor is therefore a vector and a second order tensor is a matrix.

A tensor of order three or higher will be denoted as higher order tensor. A third order tensor $\mathsf{X} \in \mathbb{R}^{I_1 \times I_2 \times I_3}$ (notation introduced by [18]) is shown in Fig. 1 [1]. The elements in X are indexed by $x(i_1, i_2, i_3)$. To index rows the MATLAB notation is used, e.g. the first dimension with specific i_2 and i_3 is indexed by $\mathbf{x}(:, i_2, i_3)$, where the colon stands for all values of $i_1 = 1, 2, \ldots, I_1$. To index slices this is analogously done, e.g. the first two dimensions with constant i_3 are indexed by $\mathbf{X}(:, :, i_3)$.

The *outer product*, [18],

$$\mathsf{Z} = \mathsf{X} \circ \mathsf{Y} \in \mathbb{R}^{I_1 \times I_2 \times \cdots \times I_n \times J_1 \times J_2 \times \cdots \times J_m} \tag{3}$$

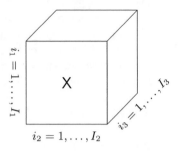

Fig. 1. Third order tensor [1].

of two tensors $\mathsf{X} \in \mathbb{R}^{I_1 \times I_2 \times \cdots \times I_n}$ and $\mathsf{Y} \in \mathbb{R}^{J_1 \times J_2 \times \cdots \times J_m}$ is a tensor of order $n + m$ with the elements

$$z(i_1, i_2, \ldots, i_n, j_1, j_2, \ldots, j_m) = x(i_1, i_2, \ldots, i_n)y(j_1, j_2, \ldots, j_m). \tag{4}$$

The outer product is associative and can be applied in sequence by

$$\bigcirc_{i=1}^{N} \mathsf{X}_i = \mathsf{X}_1 \circ \mathsf{X}_2 \circ \cdots \circ \mathsf{X}_N. \tag{5}$$

The outer product of two first order tensors $\mathbf{x} \in \mathbb{R}^p$ and $\mathbf{y} \in \mathbb{R}^n$, here denoted as vectors, is a second order tensor, which could also be represented as a matrix

$$\mathbf{x} \circ \mathbf{y} = \begin{pmatrix} x_1 \\ \vdots \\ x_p \end{pmatrix} \begin{pmatrix} y_1 \ldots y_n \end{pmatrix} = \begin{pmatrix} x_1 y_1 & \ldots & x_1 y_n \\ x_2 y_1 & \ldots & x_2 y_n \\ \vdots & \vdots & \vdots \\ x_p y_1 & \ldots & x_p y_n \end{pmatrix} \in \mathbb{R}^{p \times n}. \tag{6}$$

An n-dimensional tensor $\mathsf{X} \in \mathbb{R}^{I_1 \times I_2 \times \cdots \times I_n}$ is a *rank-1 tensor*, if it can be represented by the outer product of n vectors

$$\mathsf{X} = \bigcirc_{i=1}^{n} \mathbf{x}_i \tag{7}$$

with $\mathbf{x}_i \in \mathbb{R}^{I_i \times 1}$.

A CP (canonical polyadic) tensor, [18],

$$\mathsf{K} = [\boldsymbol{\lambda}, \mathbf{X}_1, \mathbf{X}_2, \ldots, \mathbf{X}_n] \in \mathbb{R}^{I_1 \cdots \times I_n} \tag{8}$$

is a tensor of dimension $I_1 \times I_2 \cdots \times I_n$, where the elements are the sum of the outer product of the column vectors of the so called *factor matrices* $\mathbf{X}_i \in \mathbb{R}^{I_i \times I_r}$. The weighting is done by the elements of the so called *weighting vector* $\boldsymbol{\lambda}$. The term CP decomposition is constructed by the names CANDECOMP and PARAFAC and is based on Kiers [20]. An element of the tensor K can be calculated by

$$k(i, j, \ldots, p) = \sum_{k=1}^{r} \lambda(k) x_1(i, k) x_2(j, k) \ldots x_n(p, k). \tag{9}$$

A CP tensor with weighting vector $\boldsymbol{\lambda}$ and factor matrices $\mathbf{X}_i \in \mathbb{R}^{I_i \times r}$ can be represented as the sum of the outer products

$$[\boldsymbol{\lambda}, \mathbf{X}_1, \mathbf{X}_2, \ldots, \mathbf{X}_n]$$
$$= \sum_{k=1}^{r} \lambda(k) \mathbf{x}_1(:,k) \circ \mathbf{x}_2(:,k) \circ \cdots \circ \mathbf{x}_n(:,k) = \sum_{k=1}^{r} \lambda(k) \mathop{\bigcirc}_{i=1}^{n} \mathbf{x}_i(:,k), \qquad (10)$$

see Fig. 2.

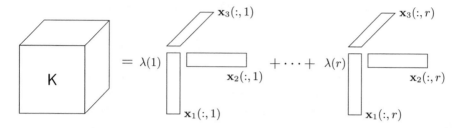

Fig. 2. CP tensor decomposition.

The rank of a CP tensor is defined as the minimal number r of rank 1 tensors, which are summed up as in (10), [21]. Using the weighting factor $\lambda(k)$ the column vectors $\mathbf{x}_i(:,k)$ of the factor matrices can be normalized. The CP decomposition of a rank r tensor, with factor matrices having r columns, can represent this tensor exactly. If the rank is not chosen sufficiently high for the decomposition to represent the original tensor exactly, an approximation of the original tensor is constructed. There is no simple algorithm to determine the rank of a tensor, since this problem is NP-hard, [1]. There exist different approaches to estimate the rank of a tensor by e.g. successively increasing the rank until a satisfactory approximation is found, [1].

The CP decomposition of a tensor $\mathsf{M} \in \mathbb{R}^{I_1 \times \cdots \times I_n}$ is calculated by solving the optimization problem

$$\min_{\boldsymbol{\lambda}, \mathbf{X}_1, \mathbf{X}_2, \ldots, \mathbf{X}_n} \frac{1}{2} \left\| \mathsf{M} - [\boldsymbol{\lambda}, \mathbf{X}_1, \mathbf{X}_2, \ldots, \mathbf{X}_n] \right\|^2, \qquad (11)$$

where $\|\ldots\|$ denotes the Frobenius norm, [22]. The choice of the rank is a trade off between accuracy of the approximation and the number of parameters to be stored. For the application examples in this contribution no exact representations are looked for but approximations are used. To decompose an incomplete tensor, i.e. a tensor where some elements are unknown, the optimization problem (11) is extended by a binary observability tensor $\mathsf{G} \in \mathbb{R}^{I_1 \times I_2 \times \cdots \times I_n}$

$$\min_{\boldsymbol{\lambda}, \mathbf{X}_1, \mathbf{X}_2, \ldots, \mathbf{X}_n} \frac{1}{2} \left\| \mathsf{G} \circledast (\mathsf{M} - [\boldsymbol{\lambda}, \mathbf{X}_1, \mathbf{X}_2, \ldots, \mathbf{X}_n]) \right\|^2, \qquad (12)$$

where $\circledast$ denotes the Hadamard product, i.e. the element-wise multiplication of the corresponding entries. The entries in G are set to be 1 if the corresponding value in M exists and to 0 otherwise, [22]. Using this extension the unknown values are not considered in the optimization problem. The optimization problem defined in (11) can be extended with the condition, that the entries in the factor matrices are only allowed to be non negative

$$\min_{\lambda, \mathbf{X}_1, \mathbf{X}_2, \ldots, \mathbf{X}_n} \frac{1}{2} ||M - [\lambda, \mathbf{X}_1, \mathbf{X}_2, \ldots, \mathbf{X}_n]||^2, \tag{13}$$

$$\textit{with} \quad x_i(j, k) > 0, \quad j = 1, \ldots, I_i, \quad i = 1, \ldots, n, \quad k = 1, \ldots, r$$

for applications, where the entries in the tensor are non-negative.

3   Signal-Based Fault Detection for Heating Systems

In this section specific properties as dependencies on the time of the day, day of the week or other measurements are used to split measurements up. The resulting measurements are stored in a tensor and used for signal-based fault detection.

3.1   Fault Detection Based on Factor Matrices

The fault-free measurements of an energy meter, sensor or measurement point are split into s sections and stored in a tensor

$$\mathsf{X} \in \mathbb{R}^{I_1 \times I_2 \times \cdots \times I_n}. \tag{14}$$

The number of sections s is the multiplication of the dimensions $s = \prod_{i=1}^{n} I_i$. For each dimension i a property, e.g. the time of the day is split in I_i intervals - not necessarily equidistant but chosen on the basis of expert knowledge. The number of properties and the number of intervals in each property are chosen such that a complete tensor is constructed, i.e. each element $x(i_1, i_2, \ldots, i_n)$ has a value. This might not always be possible, i.e. for some elements $x(i_1, i_2, \ldots, i_n)$ of the tensor X multiple or even no elements are in the measurement data under investigation. In case of multiple elements for one tensor entry the mean value of the measurements is used. In case of some entries being empty the decompose-and-unfold method is used.

Decompose-and-Unfold Method: In Fig. 3 a schematic view of an incomplete third order tensor M is shown. It is assumed, that the measurements are not independent of each other. E.g. if one dimension quantizes the time of the day and a second dimension the day of the week, it is possible, that the heat demand change from morning to noon is the same on Mondays and Tuesdays although the absolute values are different because a different number of users needs to be supplied with heat. To fill up the missing entries in the tensors M the tensor-lab toolbox, [23], for MATLAB is used. It is assumed, that if by (12) a low rank approximation is found, which can represent the existing entries of the tensors, the information of all sections is contained in this tensor. By unfolding the entire tensor the previously non existing entries can be found. To decompose the tensor

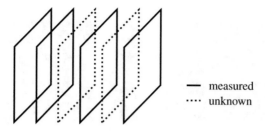

Fig. 3. Incomplete third order tensor, [14].

the canonical polyadic decomposition algorithm cpd of the tensorlab toolbox is used. All entries of the tensor, which are unknown, are defined as non existing, which is done by setting the corresponding entries to the value NaN in MATLAB. The dynamics of the system have to be represented by measurements with non-faulty values in a sufficient large number. Solely for known dynamics the tensor can be completed, an extrapolation is not possible. With the degree of freedom in the approximation, i.e. the rank r of the decomposition, the amount of information stored in the factor-matrices is increased. If the rank is chosen too high the interpretation of the factor matrices - identification of specific events - is no longer possible, because the degree of freedom is chosen too high. In this contribution the minimal rank of the decomposition is found heuristically.

The first signal-based fault detection method uses the factor matrices of a data tensor directly. The factor matrices are used to analyze the properties of measurements of a component or a system, see e.g. [7]. Nominal factor matrices are defined. If the identified factor matrices deviate from the reference for a specified time interval a fault is assumed. The second signal-based fault detection method calculates expected values from the decomposed tensor according to (9) and compares these with measured values to generate the residuals. For both approaches the chosen structure and the defined rank are very important.

3.2   Application Examples

The methods presented in Sect. 3.1 will be applied to two systems: first faulty times of operation of a water boiler are investigated and then the failure of a night-time reduction is considered.

Fault Detection for a Hot Water Supply. Using a water boiler, the tenant is supplied with warm water. At times, where no hot water is needed the hot water supply is stopped, as long as that is conform to hygiene requirements. In Fig. 4 the desired and programmed weekly operation in mean values for each hour is given. The usage time is from Monday to Friday from 6 a.m. to 11 p.m. and on Saturday from midnight to 2 p.m. In addition shortly before midnight there is an anti-blocking setting for the pumps. A third-order tensor $X \in \mathbb{R}^{I_1 \times I_2 \times I_3}$, see Fig. 1, can be defined where $I_1 = hour$, $I_2 = weekday$ and $I_3 = week$. The

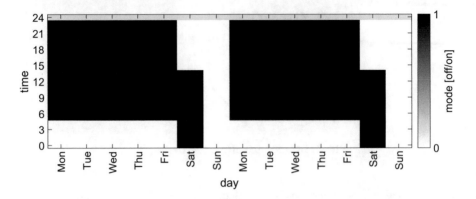

Fig. 4. Pump of the warm water supply, non-faulty operation.

reference column vectors $\mathbf{x}_i(:, k)$ of the factor martices $\mathbf{X}_i$ for the application example have been defined by (13) over a time of six fault-free weeks, see Fig. 5. The *Tensorlab Toolbox* [24] for MATLAB is used for tensor decomposition. In Fig. 5 the order of the factor matrices $\mathbf{x}_i(:, k)$ is according to the rank of the tensor, see Fig. 2. The weighting factors λ_r are factored into the column vectors such that $\lambda_r = 1$. The mean value $v(k)$ for each point in time can be calculated by $v(k) = I_1(k)I_2(k)I_3(k)$, see (9). The operation signal of the warm water supply has in case of fault free operation a complete structure, which can be represented by a third order tensor. The interpretation of the tensor is, that the first column of the factor matrices is corresponding to the operation on a weekday, the next corresponds to the operation on a Saturday and the last column corresponds to the anti-blocking setting. During the course of a week, the values are constant as expected. These factors are used as reference for the fault detection.

In Autumn of the year at hand, the data acquisition had some faults and there have been some faults in operation, see Fig. 6. For some days there are

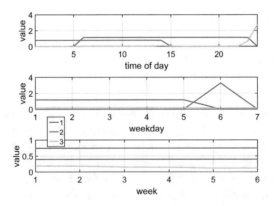

Fig. 5. Factor matrix values for the non-faulty operation.

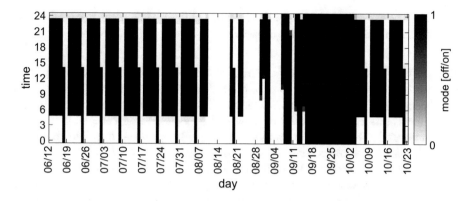

Fig. 6. Pump of the warm water supply, faulty operation.

no measurement data at all and for another period the measurement data are there but the warm water supply was constantly in operation. This occurred in the weeks 9 to 17. Figure 7 shows the factor matrices for the entire period of time. The first column of the factor matrices still stands for the operation during the week, whereas some parts are of the second columns have to be added. The second column represents a major part of the faulty non-stop operation. Thus, the first two columns correspond to the lack of measurements and the faulty operation. In the third column the representation of the anti-blocking setting is no longer visible because its impact in the cost function is smaller than impact resulting from the faulty operation. The values are no longer constant, but change over the course of a week, therefore a fault can be identified.

A disadvantage of the fault detection method is that for the detection measurement data of a period of time is needed, such that the factor matrices can be constructed reliably. Changes in the tensor can be detected if they deviate for a specified period of time.

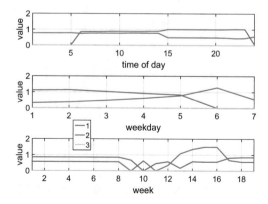

Fig. 7. Factor matrix values for the faulty operation.

This leads us to another method which will be shown on an application example involving a night-time setback.

Fault Detection for a Night-Time Setback. The room temperature is reduced in times of no usage to save energy, because of the lower transmission heat losses. For this, there exist different concepts. The reference temperature for heat circuits can be reduced or they can be switched off completely (and in case of an undershoot of a minimal temperature be switched on). The following example uses measurement data of parts of a schools heat circuit, where a night-time setback is implemented for nights and weekends. The reduced temperature in the room is realized by switching off the pumps of the corresponding heat circuits and the closing of the corresponding valve. After servicing the heating system in summer 2015 the night-time setback was deactivated.

In Fig. 8 the amount of heat transferred from the heating system to the rooms for one day over the ambient temperature is given. The data is divided into times with and without night-time setback. The behavior of the system is different in these periods of time. Without night-time setback the consumption of heat is higher, because of higher heat losses during the night. Also missing is the lower heat consumption on the weekends. Though it is not possible to define a sharp boundary between non-faulty and faulty behavior. A distinction of the behavior during the weekend in comparison to a weekday and of day and night operation does not occur. Thus, analyzing complete set of the values doesn't allow for a simple fault detection. Hence, the statistical values have to be separated in dependence of other criteria, e.g. time of the day, day of the week and ambient temperature to be able to detect this fault. The ambient temperature is subject to a continuous change and therefore needs to be quantized, i.e. put into intervals of temperatures, such that the heat losses can be put into a tensor in dependence of the ambient temperature. In this case, the values of the

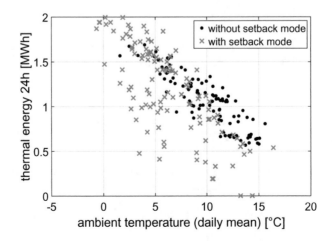

Fig. 8. Heat consumption with and without night-time setback.

ambient temperature are quantized in $2K$ intervals. An illustrative example of a quantization of continuous data is given in Sect. 4.

The heat power of the circuits is stored in a third order tensor $\mathsf{X} \in \mathbb{R}^{I_1 \times I_2 \times I_3}$, where $I_1 = time\ of\ the\ day$, $I_2 = weekday$ and $I_3 = ambient\ temperature$. The reference tensor is generated with hourly mean values over a period of 12 month. Faulty data in this period has been deleted. The incomplete tensor, constructed from the reference data, shows values that are expected from a manual data analysis, e.g. values of the factor matrix $\mathbf{X}_3$ are bigger if the ambient temperature is lower.

Nevertheless there is a variation in the values due to disturbances, which has an impact on the values estimated with the decompose-and-unfold method. For entries in the tensor, where multiple values exist in the reference data, the variance is stored in an additional tensor - with the same dimension - as basis for an evaluation. Tensor operations of the decompose-and-unfold method are performed using the *Tensorlab Toolbox* for MATLAB [24]. For fault detection an expected value of the heat power is calculated by (9) in dependence of the hour of the day, the weekday and the interval of the ambient temperature using the decomposed tensor and compared to the measured value. As threshold for detecting a fault, three times the standard deviation is chosen. If the measured value is outside this interval for three hours a fault is assumed, see Fig. 9. In the summer the night-time setback is deactivated in the heat circuit. Starting at this point in time the value of the heat power is above its expected value during the nights, see Fig. 10. Also there is a difference in the morning hours in the heat power because the rooms do not have to be heated up. Figure 11 shows the result of the fault detection, where this fault is detected. Because there is data for the entire year, the daylight-saving time is considered such that the heating up and night-time setback is always at the same time of the day.

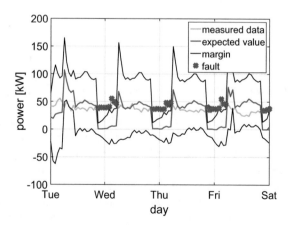

Fig. 9. Comparison of expected and measured values.

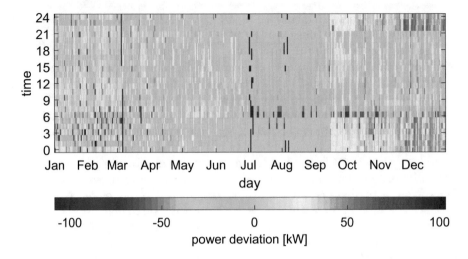

Fig. 10. Difference of measured and expected heat power.

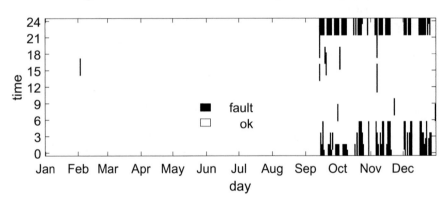

Fig. 11. Fault detection.

4 Model-Based Fault Detection for Heating Systems

Heating systems can be modeled as hybrid multilinear systems, see Sect. 1. Keeping all but one signals constant in multilinear terms and regarding just one combination of the binary signals will result in a linear behavior. Each combination of quantized and binary signals will be denoted as operating regime, see [25]. For each operating regime, a linear system can be identified around an operating point using measurement data. The measurement data is divided into parts, which will be called operational sections, if a large amount of data is available for one operating regime, see Fig. 12. The set of linear systems for all operating regimes will be denoted as parameter tensor.

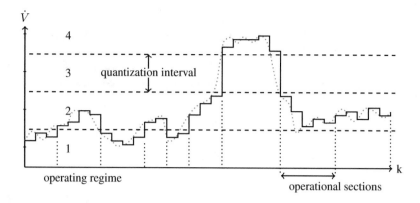

Fig. 12. Quantization.

4.1 System Definition

Continuous-valued signals are quantized using the intervals $J_1, \ldots, J_j$. Let the index of the operating regime **I** be defined by the index of the intervals of quantized signals (e.g. temperatures, flow rates) and the binary signals (e.g. on/off, opened/closed) coded 1 for `false` and 2 for `true` and let v be the number of signals - denoted as partitioning signals - used to define the operating regime. Thus, the operating regime is defined by the index vector **I**, as defined in (2). The number of intervals of the corresponding signal is $I_n, n = 1, 2, \ldots, v$. Note that $I_n = 2$ for all binary signals and that the limits of the quantization intervals are stored.

Assume the operating regime is defined by one continuous-valued signal and one binary signal, then as an example the index vector $\mathbf{I} = (i_1, i_2)$ with $i_1 \in \{1, \ldots, 7\}$ and $i_2 \in \{1, 2\}$. For each operating regime, i.e. each possible combination of the indexes in the index vector a linearization is performed.

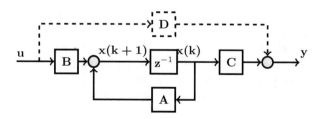

Fig. 13. Linear discrete-time state-space model.

This means with the measurement data of one operating regime a linear state-space model is estimated, Fig. 13,

$$\mathbf{x}(k+1) = \mathbf{A_I}\mathbf{x}(k) + \mathbf{B_I}\mathbf{u}(k) \tag{15}$$

$$\mathbf{y}(k) = \mathbf{C_I}\mathbf{x}(k) + \mathbf{D_I}\mathbf{u}(k) \tag{16}$$

where the time index is k, the mode of operation is $\mathbf{I}$, the state vector is $\mathbf{x} \in \mathbb{R}^n$, the input vector is $\mathbf{u} \in \mathbb{R}^m$, the output vector is $\mathbf{y} \in \mathbb{R}^p$, the system matrix is $\mathbf{A_I} \in \mathbb{R}^{n \times n}$, the input matrix is $\mathbf{B_I} \in \mathbb{R}^{n \times m}$, the output matrix is $\mathbf{C_I} \in \mathbb{R}^{p \times n}$ and the feed forward matrix is $\mathbf{D_I} \in \mathbb{R}^{p \times m}$. The state-space model defined by the matrices $\mathbf{A}$, $\mathbf{B}$, $\mathbf{C}$, and $\mathbf{D}$ must be in a canonical state-space representation.

Using the representation

$$\left(\begin{array}{c|c} \mathbf{A} & \mathbf{B} \\ \hline \mathbf{C} & \mathbf{D} \end{array} \right) = \left(\begin{array}{cccc|ccc} a_{11} & 0 & \cdots & 0 & b_{11} & \cdots\cdots & b_{1m} \\ 0 & a_{22} & \ddots & \vdots & \vdots & \ddots & \vdots \\ \vdots & \ddots & \ddots & 0 & \vdots & \ddots & \vdots \\ 0 & \cdots & 0 & a_{nn} & b_{n1} & \cdots\cdots & b_{nm} \\ \hline c_{11} & \cdots\cdots & \cdots & c_{1n} & 0 & \cdots\cdots & 0 \\ \vdots & \ddots & & \vdots & \vdots & \ddots & \vdots \\ \vdots & & \ddots & \vdots & \vdots & \ddots & \vdots \\ c_{p1} & \cdots\cdots & & c_{pn} & 0 & \cdots\cdots & 0 \end{array} \right), \qquad [14] \qquad (17)$$

the entire information of the state-space model is stored in a single matrix with dimension $\mathbb{R}^{(n+p) \times (n+m)}$. Note that in (17) the matrices $\mathbf{A}$, $\mathbf{B}$, $\mathbf{C}$, and $\mathbf{D}$ are in state-space modal representation and that the systems do not have a direct feedthrough, such that $\mathbf{D}$ is a zero matrix.

The entire multi-linear system, i.e. a linear system for each combination of the v partitioning signals, is represented using the parameter tensor M

$$\mathsf{M} \in \mathbb{R}^{(n+p) \times (n+m) \times I_1 \times I_2 \times \cdots \times I_v}, \qquad [14]. \qquad (18)$$

The operating regime $\mathbf{I}$ corresponds to the third to the $v^{th} + 2$ dimension of M. For each operating regime $\mathbf{I}$, the linear system $\left(\begin{array}{c|c} \mathbf{A} & \mathbf{B} \\ \hline \mathbf{C} & \mathbf{D} \end{array} \right)$ is contained in the tensor and accessible by keeping all but the first two dimensions constant. Despite the amount of zeros in $\mathbf{A}$ and $\mathbf{D}$, this matrix is selected to allow a generic system description.

The system identification is done for the input and output signals in specific intervals. Since a linear system is identified for each of these measurement data sets the offsets of input and output data needs to be subtracted. These offsets $\bar{\mathbf{u}}_\mathbf{I}$ and $\bar{\mathbf{y}}_\mathbf{I}$ also need to be stored, which is done by using a tensor O

$$\mathsf{O} \in \mathbb{R}^{(m+p) \times I_1 \times I_2 \times \cdots \times I_v}, \qquad [14] \qquad (19)$$

with

$$\mathsf{O}(:, \mathbf{I}) = [\bar{\mathbf{u}}_\mathbf{I} \quad \bar{\mathbf{y}}_\mathbf{I}]^T, \qquad [14]. \qquad (20)$$

4.2 Identify Parameter Tensor Slices

The identification procedure is as follows:

- Read input and output data as well as signals that determine the operating regime
- Perform some pre-processing, e.g. outlier removal, filling data gaps, depending on the data quality
- Quantize the partitioning signals, i.e. split the data into intervals
- Store the interval edges
- Sort the data by operating regime, note that measurement data from different time spans can be put together for one operating regime
- Store the operating point, i.e. the offsets on the measurement data for the input and output data in the tensor O
- Perform system identification for each operating regime and store the system matrices in the tensor M

For linear black box system identification MATLAB's N4SID algorithm of the System Identification Toolbox is used, [26], without the option of estimating a noise component. The estimated linear model is transformed into a canonical state-space model in modal representation using the MATLAB command canon with option modal. In addition the representation is normalized such that the non-zero entries of the matrix $\mathbf{C}$ for the first output are set to f_T. Since tensor algorithms are sensitive to scale of data elements [27], the parameter f_T has to be chosen such that all resulting entries have the same scale. The normalization is performed according to (21)–(25), [14].

$$\mathbf{T}_{\mathbf{norm}} = f_T \begin{pmatrix} \frac{1}{c_{11}} & 0 & \cdots & 0 \\ 0 & \frac{1}{c_{12}} & \cdots & 0 \\ \vdots & \vdots & \ddots & \vdots \\ 0 & 0 & \cdots & \frac{1}{c_{1n}} \end{pmatrix} \tag{21}$$

$$\mathbf{A}_{\mathbf{norm}} = \mathbf{T}_{\mathbf{norm}}^{-1} \mathbf{A} \mathbf{T}_{\mathbf{norm}} \tag{22}$$

$$\mathbf{B}_{\mathbf{norm}} = \mathbf{T}_{\mathbf{norm}}^{-1} \mathbf{B} \tag{23}$$

$$\mathbf{C}_{\mathbf{norm}} = \mathbf{C} \mathbf{T}_{\mathbf{norm}} \tag{24}$$

$$\mathbf{D}_{\mathbf{norm}} = \mathbf{D} \tag{25}$$

It is assumed that the matrices $\mathbf{A}$, $\mathbf{B}$, $\mathbf{C}$, and $\mathbf{D}$ stored in and restored from the parameter tensor M are the normalized ones without stating the index $norm$, for better readability.

4.3 Construct Full Parameter Tensor

For systems with many partitioning signals, it is very likely that no measurements are taken for some operating regimes. To cope with this problem, the decompose-and-unfold method introduced in Sect. 3.1 is used.

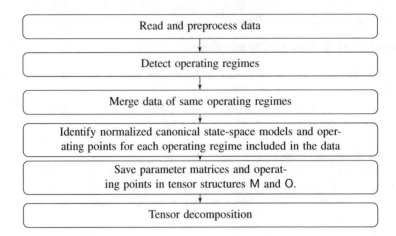

Fig. 14. Generating parameter tensor [14].

This approach is applicable if for a sufficiently large number of operating regimes a linear state-space model can be identified, i.e. if the entire system dynamic is captured by the measurement data.

Note that choosing the rank of a decomposition is not trivial. Increasing the rank is not always better. Especially when using the decompose-and-unfold method. Assume the rank of a system is three. On the one hand, a rank-two approximation can not capture all system dynamics. On the other hand, a rank-four approximation might bring in a freedom, not captured by the optimization problem (12) because in the corresponding entries the binary tensor has 0 entries, thus making the estimation of the unknown entries meaningless. In this contribution the selection of the rank is done heuristically by first choosing a high rank decomposition and then reducing rank in an iterative way until the unknown parameters are approximately the same for different initial values in the tensor decomposition. Figure 14 sums up Sects. 4.1, 4.2 and 4.3 how the parameter tensor can be created to describe the system behavior in every operating regime.

4.4 Fault Detection with Parity Equations

Parity equations are used for offline fault detection. In Fig. 15 the proposed fault detection procedure is shown.

The nominal, i.e. is non-faulty, behavior is described by the parameter tensor introduced in Sect. 4.1. A way to detect discrepancies between process and model are parity equations and observer based designs. Model and process both lead to identical or equivalent residual generators [28], once the design objectives have been selected [28]. Usually the procedure of selecting design objectives for parity equations is simpler than for observer based designs [28]. No initial value for the state vector $\mathbf{x}$ is needed, when using parity equations.

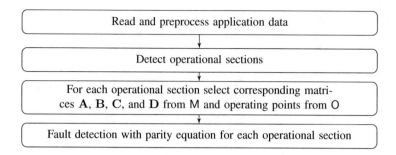

Fig. 15. Fault detection [14].

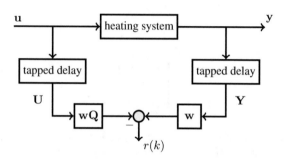

Fig. 16. Parity equation.

In Fig. 16 the input and output data are compared to the model behavior contained in $\mathbf{w}$ and $\mathbf{Q}$. The input $\mathbf{u}$ and the output $\mathbf{y}$ are delayed using tapped delay blocks by $q+1$ time steps to generate vectors $\mathbf{U}$ (26) and $\mathbf{Y}$ (27) containing all the delayed samples of the input and output respectively.

$$\mathbf{U}(k) = \begin{pmatrix} \mathbf{u}(k-q) \\ \mathbf{u}(k-q+1) \\ \mathbf{u}(k-q+2) \\ \vdots \\ \mathbf{u}(k) \end{pmatrix}, \qquad [14] \tag{26}$$

$$\mathbf{Y}(k) = \begin{pmatrix} \mathbf{y}(k-q) \\ \mathbf{y}(k-q+1) \\ \mathbf{y}(k-q+2) \\ \vdots \\ \mathbf{y}(k) \end{pmatrix}, \qquad [14] \tag{27}$$

The residual r is calculated at time k by

$$r(k) = \mathbf{w_I}\mathbf{Y}(k) - \mathbf{w_I}\mathbf{Q_I}(q)\mathbf{U}(k), \qquad [14] \tag{28}$$

for $q + 1$ time steps with

$$\mathbf{Q_I}(q) = \begin{pmatrix} \mathbf{D_I} & 0 & 0 & \ldots 0 \\ \mathbf{C_I B_I} & \mathbf{D_I} & 0 & \ldots 0 \\ \mathbf{C_I A_I B_I} & \mathbf{C_I B_I} & \mathbf{D_I} & \ldots 0 \\ \vdots & \vdots & \vdots & \ddots \vdots \\ \mathbf{C_I A_I^{q-1} B_I} & \mathbf{C_I A_I^{q-2} B_I} & \mathbf{C_I A_I^{q-3} B_I} & \ldots \mathbf{D_I} \end{pmatrix}, \qquad [14]. \qquad (29)$$

See [29] for the derivation of (28). By solving

$$\mathbf{w_I R_I} = 0, \qquad [14] \qquad (30)$$

with

$$\mathbf{R_I}(q) = \begin{pmatrix} \mathbf{C_I} \\ \mathbf{C_I A_I} \\ \mathbf{C_I A_I^2} \\ \vdots \\ \mathbf{C_I A_I^q} \end{pmatrix}, \qquad [14] \qquad (31)$$

the residual generator vector $\mathbf{w_I}$ is determined. Multiple solutions are possible for (30). Do not choose the trivial solution. Properties of fault detection are affected, by selecting $\mathbf{w_I}$.

Using the fault criterion

$$e(k) = \begin{cases} 0 & \text{if} \quad r(k) \in [b_l, b_u] \\ 1 & \text{else} \end{cases}, \qquad [14] \qquad (32)$$

the fault signal e is calculated, where 0 stands for 'no fault' and 1 for 'fault'. The residual $r(k)$ is compared to upper and lower bound, b_u and b_l for every operating section to decide if there is a fault or not. The length of a operating section is limited to $q + 1$ time steps, due to computation time and memory use. For operating sections longer than $q + 1$ time steps the data is divided and examined individually. The residual $r(k)$ is used to calculate the upper bound b_u and lower bound b_l for the data used to create $\mathbf{M}$ and are the same for every operating regime $\mathbf{I}$, which gives the decision boundary for nominal operation. Faulty operation is assumed for values of $r(k)$, which exceed the range limited by b_u and b_l.

4.5   Application to Heating System Fault Detection

The heat generation unit of the heating system of the State Office for Nature, Environment and Consumerism in Düsseldorf, Germany is investigated. The unit consists of three hydraulically balanced boilers with attached gas burners. The flow rate through the boiler can be stopped by a lid when the boiler is switched off. The heating system was already under investigation in [30, 31], where a model of the entire system is given and the controller design and implementation is described.

The heat power balance

$$c_w \rho V_b \dot{T}_s(t) = c_w \rho \dot{V}(t)(T_r(t) - T_s(t)) + P_{in}(t) \tag{33}$$

represents simplified but sufficiently the dynamical behavior of a gas burner with boiler, as illustrated in Fig. 17. The supply temperature of the boiler is denoted by T_s, the return temperature by T_r, the flow rate out of the boiler and back into it by $\dot{V}$, the input powers of the boilers by P_{in}, c_w is the specific heat capacity of water and ρ the water density,

The method explained above is performed on boiler 2, using the N4SID algorithm included in the MATLAB System Identification Toolbox for black box identification to estimate the first order state-space matrices. The state of the linear models is the supply temperature of the boiler. The burner switching signal (on/off) and the flow rate define the operating regime. No training data with the lid closed is available, as boiler 2 is the main supplier of the heating system. Therefore the lid position cannot be used to define the operating regime. The flow rate is divided in seven intervals. A model of the system is estimated for each operating regime, resulting in a tensor M of dimension $\mathbb{R}^{2\times3\times7\times2}$. The operating regime index is $I_b = (i_1, i_2)$ with $i_1 \in \{1, \dots, 7\}$ and $i_2 \in \{1, 2\}$.

Measurement data of six month is used to create the parameter tensor. In this data set thirteen of fourteen operating regimes appear. Equation (34) shows the entries of the parameter tensor corresponding to the **A** matrices – which are scalars in this case, since a first order system is under investigation – in the different operating regimes. The * marks the missing mode. The two entries in

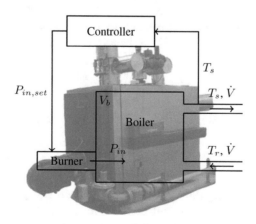

Fig. 17. Boiler and burner of a heating system [10].

the second dimension correspond to the burner switching signal and the seven entries in the first dimension correspond to the seven intervals of the flow rate.

$$\mathbf{M}(1,1,:,:) = \begin{bmatrix} 1.00 \ 0.89 \\ 0.81 \ 0.76 \\ 0.82 \ 0.77 \\ 0.68 \ 0.75 \\ 0.96 \ 0.69 \\ 0.57 \ 0.64 \\ * \quad 0.62 \end{bmatrix} \tag{34}$$

The parameters of the operation modes $I_b = (5,1)$ and $I_b = (3,2)$ in $\mathbf{A}$ and $\mathbf{B}$ respectively are not the expected values, which would be expected by looking at the other parameters. The reason for this might be found in faults in the training data or in an faulty system identification. To cope with this problem the values resulting from these operation modes are deleted (35) and the decompose-and-unfold method is used.

$$\mathbf{M}(1,1,:,:) = \begin{bmatrix} 1.00 \ 0.89 \\ 0.81 \ 0.76 \\ 0.82 \ \ * \\ 0.68 \ 0.75 \\ * \quad 0.69 \\ 0.57 \ 0.64 \\ * \quad 0.62 \end{bmatrix}, \quad [14] \tag{35}$$

For all operating regimes including the missing ones, it can be assumed, that $\mathbf{C} = 1$ and $\mathbf{D} = [0 \ 0]$, as the system can be described by (33). The result is improved, if the decompose-and-unfold method is used for $(\mathbf{A}|\mathbf{B})$ and the known, exact values for $(\mathbf{C}|\mathbf{D})$ are inserted. The resulting $\mathbf{A}$ matrices are

$$\mathbf{M}(1,1,:,:) = \begin{bmatrix} 1.00 \ 0.88 \\ 0.80 \ 0.77 \\ 0.82 \ 0.79 \\ 0.70 \ 0.73 \\ 0.63 \ 0.69 \\ 0.57 \ 0.64 \\ 0.50 \ 0.62 \end{bmatrix}, \quad [14]. \tag{36}$$

Six month of training data are used to validate the parameter tensor, see Fig. 18. Six faults are detected, all of them in operational sections where the lid of boiler 3 has been opened or closed. It is assumed, that the flow rate $\dot{V}$ of boiler 2, calculated by hydraulic parameters, the position of the boilers lids and the flow rate of the heat generation unit, in these sections is faulty due to inaccuracy in data logging.

Fault detection on real input and output data of one summer month is performed. The measured supply temperature and the resulting residual $r(k)$ for

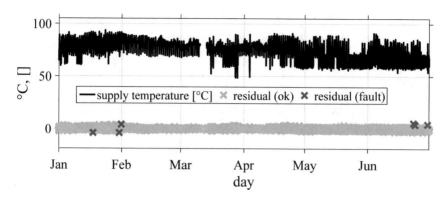

Fig. 18. Fault detection boiler 2, model validation [14].

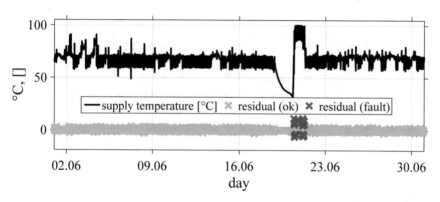

Fig. 19. Fault detection boiler 2, manual operation [14].

each operational section are shown in Fig. 19. The residual $r(k)$ is marked red in case of fault, i.e. $e(k) = 1$, and it is marked green for non-faulty sections. On June 18th, the supply temperature drops to temperatures below 35 °C after a complete system breakdown with boiler 2 being turned off in this period. As input and output data match model behavior, no fault is detected. In consequence of the lack of heat supply, on June 20th boiler 2 has been turned on in manual mode. It is ignoring controller signals and working on full power, thus heating up the system until it is stopped by an emergency switch. This does not match to expected system behavior, as the emergency switch is now acting as a two point controller. The boiler has been set back to normal operating mode on June 21st. The fault is detected correctly, as can be seen in Fig. 19.

Figure 20 shows measurement data and the fault detection result of a winter month. Another fault can be observed. Boiler 2 is not processing the controller power signal till December 14th. It does not vary the power output and in consequence is not producing the correct amount of heat. The burner switching signal (on/off) is configured correctly and the boiler is turned on at the correct periods. Without controlling the power input, the controller can not keep the

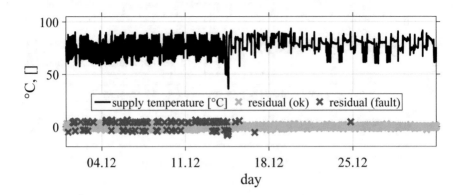

Fig. 20. Fault detection boiler 2, failure in power control [14].

system stable and the supply temperature T_s exceeds the set point temperature. A service engineer adjusted the parameters of boiler 2 on December 14th, and the system worked as expected for the remaining time of the month. The fault is detected correctly as the residual $r(k)$ has values exceeding the range limited by b_u and b_l for the faulty part of the data set. Faulty input data for calculation of $\dot{V}$ is assumed for the two errors detected after December 15th. As well as the faults detected during validation of the parameter tensor, see Fig. 18, these faults are in operational sections where the lid of boiler 3 has been opened or closed.

5   Conclusions

Different methods to detect faults in heating systems using tensor decomposition algorithms have been proposed. It has been shown that factor matrices of CP decompositions represent inherent properties of a component or (sub)-system and thus, its elements can be used directly as fault indicators in signal-based fault detection. Using a reasonable quantization, measurement data can e.g. be checked for normal operation. Moreover, tensors decompositions can be used to represent operating maps of expected values, e.g. heat consumption in dependence on time of the day, day of the week or ambient temperature and their combinations. This method is advantageous the more structured the measurement data is and the less training data is corrupted by faults. The number of false alarms and unrecognized faults strongly depends on the chosen data structure and user-defined rank of the decomposition.

This contribution also shows how tensor representations and algorithms can be useful for model-based fault detection methods. A multi-linear time-invariant model is constructed by identifying multiple linear time-invariant models for different operating regimes. Heat power balances are the basis of the models of heating systems, where for the RHS multiplications of an input like the flow rate and a state like the temperature is indicating that the system is nonlinear. By quantizing these inputs or state signals, e.g. the flow rate, operating regime and their individual linear models can be defined. Furthermore, different binary signals extend the operating regime. For each binary signal and each interval of the flow rate, a linear model is identified and all parameters of these models can be written in a tensor structure. An approximation method – the decompose-and-unfold method – is proposed, if for some operating regimes no measurement data is available. Good approximations of these operating regimes, where no measurement data is available could be found by estimating low rank approximations of the parameter tensor with specialized algorithms for incomplete tensors. An application example shows the general functionality.

Future work should be done concerning robustness of the approaches and investigating the starting values of the factor matrices. In addition, work should be done on improved estimation of system matrices on operating regimes with missing measurement data.

Acknowledgements. This work was partly supported by the project OBSERVE of the Federal Ministry for Economic Affairs and Energy, Germany (Grant-No.: 03ET1225C) and partly supported by the Free and Hanseatic City of Hamburg (Hamburg City Parliament publication 20/11568). Most of the work of the first author was done during his time at PLENUM.

References

1. Kolda, T.G., Bader, B.W.: Tensor decompositions and applications. SIAM Rev. **51**, 455–500 (2009)
2. Rehault, N., Lichtenberg, G., Schmidt, F., Harmsen, A.: Modellbasierte Qualitätssicherung des energetischen Gebäudebetriebs (ModQS). Technical report, Abschlussbericht (2013)
3. Katipamula, S., Brambley, M.: Methods for fault detection, diagnostics, and prognostics for building systems - a review, Part I. HVAC&R Res. **11**, 3–25 (2005)
4. Isermann, R., Balle, P.: Trends in the application of model-based fault detection and diagnosis of technical processes. Control. Eng. Pract. **5**, 709–719 (1997)
5. Ding, S.X.: Model-Based Fault Diagnosis Techniques. Springer, Heidelberg (2008)
6. Venkatasubramanian, V., Rengaswamy, R., Kavuri, S.N., Yin, K.: A review of process fault detection and diagnosis: Part III: process history based methods. Comput. Chem. Eng. **27**, 327–346 (2003)
7. Lichtenberg, G.: Modellbasierter Reglerentwurf komplexer Gebäudeautomationssysteme. In: OBSERVE Workshop - Nichtwohngebäude energieeffizient betreiben. (2017)
8. Lunze, J.: Automatisierungstechnik: Methoden für die Überwachung und Steuerung kontinuierlicher und ereignisdiskreter Systeme. Oldenbourg Verlag (2012)
9. Jagpal, R.: EBC Annex 34 - computer aided evaluation of HVAC system performance - technical synthesis report: computer aided evaluation of HVAC system performance (2006)
10. Pangalos, G.: Model-based controller design methods for heating systems. Dissertation, Hamburg University of Technology (2015)
11. Lichtenberg, G.: Hybrid tensor systems. Habilitation, Hamburg University of Technology (2011)

12. Pangalos, G., Eichler, A., Lichtenberg, G.: Hybrid multilinear modeling and applications. In: Obaidat, M., Koziel, S., Kacprzyk, J., Leifsson, L., Ören, T. (eds.) Simulation and Modeling Methodologies, Technologies and Applications, pp. 71–85. Springer, Cham (2015)
13. Liberzon, D.: Switching in Systems and Control. Systems & Control: Foundations & Applications. Birkhäuser, Boston (2003)
14. Sewe, E., Pangalos, G., Lichtenberg, G.: Fault detection for heating systems using tensor decompositions of multi-linear models. In: 7th International Conference on Simulation and Modeling Methodologies, Technologies and Applications, Madrid (2017)
15. Van Overschee, P., De Moor, B.: Subspace Identification for Linear Systems: Theory - Implementation - Applications. Springer, Boston (2012)
16. Mulders, A.V., Schoukens, J., Volckaert, M., Diehl, M.: Two nonlinear optimization methods for black box identification compared. IFAC Proc. Vol. **42**, 1086–1091 (2009)
17. Paduart, J., Lauwers, L., Swevers, J., Smolders, K., Schoukens, J., Pintelon, R.: Identification of nonlinear systems using polynomial nonlinear state space models. Automatica **46**, 647–656 (2010)
18. Cichocki, A., Zdunek, R., Phan, A., Amari, S.: Nonnegative Matrix and Tensor Factorizations. Wiley, Chichester (2009)
19. Müller, T., Kruppa, K., Lichtenberg, G., Réhault, N.: Fault detection with qualitative models reduced by tensor decomposition methods. IFAC-PapersOnLine **48**, 416–421 (2015)
20. Kiers, H.A.: Towards a standardized notation and terminology in multiway analysis. J. Chemom. **14**, 105–122 (2000)
21. Cichocki, A., Mandic, D., De Lathauwer, L., Zhou, G., Zhao, Q., Caiafa, C., Phan, H.A.: Tensor decompositions for signal processing applications: from two-way to multiway component analysis. IEEE Signal Process. Mag. **32**, 145–163 (2015)
22. Vervliet, N., Debals, O., Sorber, L., Lathauwer, L.D.: Breaking the curse of dimensionality using decompositions of incomplete tensors: tensor-based scientific computing in big data analysis. IEEE Signal Process. Mag. **31**, 71–79 (2014)
23. Vervliet, N., Debals, O., Sorber, L., Van Barel, M., De Lathauwer, L.: Tensorlab user guide (2016)
24. Vervliet, N., Debals, O., Sorber, L., Van Barel, M., De Lathauwer, L.: Tensorlab 3.0 (2016). www.tensorlab.net
25. Murray-Smith, R., Johansen, T.A.: Multiple Model Approaches to Modelling and Control. Taylor and Francis, London (1997)
26. Ljung, L.: System identification toolbox: user's guide (2016)
27. Fanaee-T, H., Gama, J.: Tensor-based anomaly detection: an interdisciplinary survey. Knowl. Based Syst. **98**, 130–147 (2016)
28. Gertler, J.: Analytical redundancy methods in fault detection and isolation. In: Preprints of IFAC/IMACS Symposium on Fault Detection, Supervision and Safety for Technical Processes, SAFEPROCESS 1991, pp. 9–21 (1991)
29. Isermann, R.: Fault-Diagnosis Systems: An Introduction from Fault Detection to Fault Tolerance. Springer, Heidelberg (2006)
30. Sewe, E., Harmsen, A., Pangalos, G., Lichtenberg, G.: Umsetzung eines neuen Konzepts zur Mehrkesselregelung mit Durchflusssensoren. HLH Lüftung/Klima - Heizung/Sanitär - Gebäudetechnik **1**, 37–42 (2012)
31. Pangalos, G., Lichtenberg, G.: Approach to boolean controller design by algebraic relaxation for heating systems. IFAC Proc. Vol. **45**, 210–215 (2012)

Describing Resource Allocation to Dynamically Formed Groups with Grammars

Suna Bensch[1]([⊠]) and Sigrid Ewert[2]

[1] Department of Computing Science, Umeå University, Umeå, Sweden
`suna@cs.umu.se`
[2] School of Computer Science and Applied Mathematics,
University of the Witwatersrand, Johannesburg, Johannesburg, South Africa
`Sigrid.Ewert@wits.ac.za`

Abstract. In this paper we model dynamic group formation and resource allocation with grammars in order to gain a deeper understanding into the involved processes. Modelling with grammars allows us to describe resource allocation and group formation as generative processes that provide, at any given time, information about at what stage the process of group formation and resource allocation is. We divide our model into four phases: (1) resource supply, (2) candidate group formation, (3) final group formation, and (4) resource distribution. In particular, we show that we can use permitting random context grammars to describe the first two phases. For the third phase we introduce an algorithm that determines based on a resource allocation strategy the final group to which resources are distributed. The last phase is described with random context grammars under a specific leftmost derivation mode. Our model shows that if information about the available resource and candidate group formation is distributed and kept separate, then the synchronisation of this information at a later stage (i.e. resource distribution phase) needs a more powerful grammar model.

Keywords: Resource allocation · Dynamic group formation
Grammars · Regulated rewriting
Random permitting context grammar · Random context grammar

1 Introduction

Resource allocation is an important task in many applications including (cloud) computing, wireless networks, and project management. Resource allocation is domain dependent but generally it is a process involving a decision of how an available resource should be distributed to entities or subprocesses in order to achieve a specified goal. Resources may be technology, labour, equipment or processes. In this paper we use grammars to describe how resources are allocated to dynamically formed groups consisting of individual entities. By resource

© Springer Nature Switzerland AG 2019
M. S. Obaidat et al. (Eds.): SIMULTECH 2017, AISC 873, pp. 153–176, 2019.
https://doi.org/10.1007/978-3-030-01470-4_9

allocation we refer to the distribution of certain available assets to a group of entities, taking into consideration

- the resource requests or capabilities of the individual entities, and
- a resource allocation strategy that either promotes close associations among the entities or a wide-span resource distribution.

In [2] we developed a grammar based model that described specifically how enterprises dynamically form a group (i.e. a virtual buying cooperative) in order to purchase a quantity of available goods. The actual goods allocation, or more specifically, which enterprises were chosen to purchase goods was simply determined outside the model. This paper extends the approach described in [2] to capture resource allocation to groups in general and introduces an algorithm that determines a group of entities to which the resources are allocated. Our model assumes a given network graph of connected individual entities and an available resource. From the network graph a candidate group of connected entities is formed to which resources can potentially be allocated. Then, given the chosen resource allocation strategy, a final group of entities is formed to which the resources are eventually allocated.

We model the resource allocation to dynamically formed groups with grammars and strings that these grammars generate. Grammars or rewriting systems are generative devices that generate infinitely many strings in a so-called language but are themselves finite representations. Grammars are a suitable tool for modelling because they help us discover fundamental properties of resource allocation to dynamically formed groups. With the help of grammars we can describe group formation and resource allocation as successive generative processes and since there exists a large body of knowledge regarding the formal properties of the grammars we are using, we gain an understanding of how information is distributed throughout the entire process. As grammars are easily implementable, our model and its analysis guides also implementation choices for a software that would, for example, given a network of connected entities and an available resource execute the best resource allocation.

This paper is organised as follows. In Sect. 2 we give an overview of the related work relevant to our paper. In Sect. 3 we explain our model of resource allocation to dynamically formed groups and in Sect. 4 we give the necessary definitions of the grammars we are using. In Sect. 5 we formalise our model. In Sect. 6 we give a model that we would obtain if we would use grammars of weaker generative capacity and we end our paper with some conclusions in Sect. 7.

2   Background

Effective and efficient coalition (or group) formation and resource allocation is an essential process in many applications. For example, coalition formation, from an multi-agent perspective, is modelled in [10] using Petri Nets. Resource allocation strategies are also studied in cloud computing, where resource distribution is often guided by demand and has to be solved effectively [11]. In business process management, *process modelling grammars* provide a thorough understanding of an organisation's business processes [13].

Modelling with grammars has a long tradition and started in the 1950's with Chomsky's seminal work [3] and grammars describing the syntax of natural languages. Over the years, grammars have been widely used in natural language processing [1,9]. An area within grammar research is *regulated rewriting grammars* where grammar formalisms, that have a controlled way of working, are studied. Various regulated rewriting grammars and their languages and the motivations thereof are extensively described in [4]. A random context grammar [14] (abbreviated rcg) and variants thereof are a regulated rewriting formalism. Random context grammars are equipped with a *context* that guides their steps. Context is either permitting or forbidding. Permitting context enables the grammar to work in the presence of certain context conditions and forbidding context forbids the grammar to work in the presence of certain context conditions. The language generated by an rcg is denoted rcgl. A random context grammar using permitting context only, is called a random permitting context grammar (rPcg) and a random context grammar using forbidding context only, is called random forbidding context grammar (rFcg). The language generated by an rPcg or rFcg are denoted rPcl or rFcl, respectively. In [4] it is shown that rcgl without erasing rules lie strictly between the context-free and context-sensitive languages. Rcgls generated by rcgs with erasing rules are equivalent to recursively enumerable languages. As shown in [15] rPcg with and without erasing rules have the same generative capacity. An open question is, whether rFcgs without erasing rules are equivalent to rFcgs with erasing rules. Random context grammars have been used to generate pictures [6,7] and thus are also suitable to generate visual password systems [12]. In [2] rPcg and rcg are used to describe how a group of several small business enterprises that are geographically apart, form a coalition in order to purchase available goods.

3   Resource Allocation and Dynamic Group Formation

Our model is divided into the following successive four phases:

1. Resource supply.
2. Candidate group formation.
3. Final group formation.
4. Resource distribution.

In the following we explain each of the phases and introduce a running example that we will refer to for the rest of the paper.

3.1   Resource Supply

The *resource supply* phase reflects a resource becoming available. We consider a resource to be the sum of abstract resource units that are of the same type. In [2] a resource unit represented a purchasable commodity (e.g. a package of washing powder). Our model describes the total quantity of an available resource as the sum of its abstract resource items. For example, an available resource of a type X will be represented as m distinct objects $X_1 X_2 \ldots X_m$.

Example: For our running example, we assume 5 available resource items of type X, that is, $X_1 X_2 X_3 X_4 X_5$.

3.2 Candidate Group Formation

The phase *candidate group formation* dynamically forms a group of entities based on the entities' resource requests (or resource handling capacities) to which a resource could potentially be allocated. Our model assumes a given network N of connected individual entities $A_1, \ldots, A_n$. Figure 1 illustrates an example of such a network, where each node label corresponds to one entity and an edge connecting two nodes represents that two entities are direct associates.

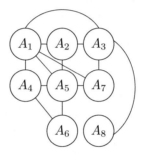

Fig. 1. The graph represents a network of individual entities. Each node label corresponds to one such entity and an edge connecting a node n_1 with a node n_2 represents that the entities at n_1 and n_2 are direct associates.

Given a network N, our model forms a candidate group C consisting of individual entities in N. The candidate group C corresponds to a subgraph in N and consists of a set of entities to which a resource could potentially be allocated. The candidate group C is formed by taking into account the resource item requests (or the resource handling capacities) of the candidate entities. In [2] a candidate group was a set of enterprises that wanted to purchase goods as one entity. The subgraph C is formed as follows. There is one entity in N that is the so-called *starter entity* and it *activates* or *calls* other entities directly connected to it to join the candidate group C. How the calling of entities exactly is interpreted, is domain-specific. For instance, the sequence of entities calling other connected entities, can represent an execution or production line during which a calling entity has to execute a certain task (a number of times) before other entities. We refer to the linear sequencing of entities calling each other as *calling sequence*. In [2] where entities were business enterprises, the calling sequence was the sequence of invitations (i.e. one enterprise inviting its business associates).

Example: For our running example, we assume the following calling sequence: A_4 (directly connected to A_1, A_5, A_6) is the start entity and calls A_5. The entity A_5 (directly connected to A_1, A_2, A_4, A_6, A_7) calls then the entities A_1 and A_2, and A_2 (directly connected to A_1, A_3, A_5) calls entity A_1. Note that the entity A_1 is represented two times in this calling sequence, which, in our model, would mean that the entity A_1 is allocated a resource two times. In order to prevent this situation, we have an algorithm (given below) that, given a calling sequence, constructs a tree θ in which every node is distinct. The tree θ contains the labels of all entities in the candidate group formation C. In our running example, the subgraph C consists of the nodes A_1, A_2, A_4 and A_5.

Then our model describes the quantity requests (or capacities) of the individual candidate entities in C, which is expressed as a number.

Example: We assume that the numbers for A_1, A_2, A_4 and A_5 are 4, 2, 3, 2 respectively. One can, for instance, interpret this as entity A_1 requesting 4 resource items, entity A_2 requesting 2 resource items, etc.

3.3   Final Group Formation

The phase *final group formation* determines the set F of entities to which the resource is then eventually allocated based on one of the strategies that we proposed in [2]:

- (Early strategy or E-strategy). This resource allocation strategy promotes close network associations by allocating resources to entities in a group that are closely connected in a network.
- (Maximal strategy or M-strategy) This resource allocation promotes a wide-span distribution of resources and favors resource allocation to the maximal number of entities.

For determining the group F, we introduce an algorithm that constructs a tree Θ in which all possible distributions of the resource to the candidate group C are displayed. The final group F is then selected according to the E-strategy or the M-strategy.

Example: The total quantity of the available resource is $X_1 X_2 X_3 X_4 X_5$. Given, the group C and the resource requests of the individual entities as above (i.e. A_1, A_2, A_4, A_5 request 4, 2, 3, 2 items, respectively), we have the following possible resource distributions:

1. Entity A_1 is allocated 4 resource items and one resource item X remains.
2. Entity A_2 is allocated 2 resource items and entity A_4 is allocated 3 items and 0 resource items X remain.
3. Entity A_2 is allocated 2 resource items and entity A_5 is allocated 2 items and one resource item X remains.
4. Entity A_4 is allocated 3 items and A_5 is allocated 2 items and 0 resource items X remain.

We are interested in all possible resource allocations disregarding the ordering of the entities (i.e. a resource allocation to A_2 and A_4 is the same as to A_4 and A_2). Both E-strategy and M-strategy favour a resource distribution in which as few as possible resource items X remain (ideally 0).

Example: The M-strategy would determine a resource distribution either to the entities A_2 and A_4 or to the entities A_4 and A_5. The E-strategy would take into account the tree θ and determine the resource distribution to entities A_4 and A_5 (because in the calling sequence the start entity A_4 calls A_5 and after that other entities are called).

3.4   Resource Distribution

The resource distribution phase allocates the resource items to the group F which consists of entities that are determined according to a chosen resource allocation strategy. The details are explained below.

4   Definitions and Preliminaries

We assume the reader to have some familiarity with the theory of formal languages (e.g. context-free grammars and languages) as in [8]. Let $\mathbb{N}$ denote the integers, and $\mathbb{N}_+ = \{1, 2, \ldots\}$. Let Σ be a set of symbols, called the alphabet. A word is a sequence of symbols and a language L is a set of words over an alphabet Σ.

Definition 1 *[5]. A random context grammar (rcg) $G = (V_N, V_T, P, S)$ is a quadruple, where*

1. *V_N is a finite set of nonterminals,*
2. *V_T is a finite set of terminals,*
3. *P is a finite set of productions of the form*

$$A \to x \ (\mathcal{P}; \mathcal{F}),$$

   where $A \in V_N$, $x \in (V_N \cup V_T)^*$ and $\mathcal{P}, \mathcal{F} \subseteq V_N$, and
4. *$S \in V_N$ is the start symbol.*

Let V denote $V_N \cup V_T$. For two strings y_1 and y_2, $y_1, y_2 \in V^*$ and a production $A \to x \ (\mathcal{P}; \mathcal{F})$ in P, we may write $y_1 A y_2 \implies y_1 x y_2$ if every $B \in \mathcal{P}$ is in the string $y_1 y_2$ and no $B \in \mathcal{F}$ is in the string $y_1 y_2$. We refer to $y_1 A y_2 \implies y_1 x y_2$ as a derivation step and to the strings $y_1 A y_2$, $y_1 x y_2$ as sentential forms. The reflexive and transitive closure of $\implies$ is denoted by $\overset{*}{\implies}$. The language generated by a grammar G is defined as

$$L(G) = \{w \mid w \in V_T^* \text{ and } S \overset{*}{\implies} w\}.$$

A random permitting context grammar (rPcg) is a random context grammar $G = (V_N, V_T, P, S)$, where for each production $A \to x \ (\mathcal{P}; \mathcal{F}) \in P$, $\mathcal{F} = \emptyset$.

A random forbidding context grammar (rFcg) is a random context grammar $G = (V_N, V_T, P, S)$, where for each production $A \to x \ (\mathcal{P}; \mathcal{F}) \in P$, $\mathcal{P} = \emptyset$. A context-free grammar (cfg) is a random context grammar $G = (V_N, V_T, P, S)$, where $\mathcal{P} = \mathcal{F} = \emptyset$ for each production $A \to x \ (\mathcal{P}; \mathcal{F}) \in P$. A language is context-free, random permitting context, random forbidding context, or random context if it is generated by a context-free, random permitting context, random forbidding context, or random context grammar respectively. In the remainder of this paper, if a context consists only of one element, we write A instead of $\{A\}$.

Example 1. Let $G = (\{S, A, B, C\}, \{a, b, c\}, P, S)$, where P is given by

$$\{S \to ABC,$$

$$A \to aA'(B; \emptyset),$$

$$B \to bB'(C; \emptyset),$$

$$C \to cC'(A'; \emptyset),$$

$$A' \to A(B'; \emptyset),$$

$$B' \to B(C'; \emptyset),$$

$$C' \to C(A; \emptyset),$$

$$A \to a(B; \emptyset),$$

$$B \to b(C; \emptyset),$$

$$C \to c\}.$$

The generated language is $L(G) = \{a^n b^n c^n \mid n \geq 1\}$. The derivation starts with rewriting S with the first rule and we obtain the sentential form ABC. Now the second and third are applicable, each of which introduces an A' or B' into the sentential form. After, for example, applying the second and third rule we obtain the sentential form $aA'bB'C$. Then the fourth rule is applied (since there is an A' in the sentential form) and we obtain the sentential form $aA'bB'cC'$. Note that at this point of the derivation process the fifth rule as well as the sixth rule can be applied. If we, at this point, would apply the sixth rule we would obtain the sentential form $aA'bBcC'$ and the fifth rule would not be applicable since the symbol B' does not occur in the sentential form. In fact, such a derivation would block and never complete since no rule could be applied. So, after obtaining the sentential form $aA'bB'cC'$ as described above, we apply the fifth, sixth, seventh rule and obtain $aA'bB'cC' \stackrel{*}{\Rightarrow} aAbBcC$. After this we start the first cycle again (applying second, third, fourth rule again) or terminate the derivation by using the last three rules. We can generate all strings consisting of an equal number of occurrences of a, b, and c.

5   Modelling Resource Allocation to Dynamically Formed Groups

In the following we construct successively fragments of an rPcg for the phases resource supply, candidate group formation and final group formation. We show that the resource distribution phase cannot be described by an rPcg but requires the generative capacity of an rcg with a specific leftmost restriction.

Let $k \geq 1$ be the number of entities in the candidate group C. The language that we want to generate is L_{RA} and reflects all possible resource allocations, in particular, the outcomes of each of the four phases: resource supply, candidate group formation, final group formation and resource distribution. We divide a string in L_{RA} by the designated symbol $\hat{\$}$, where to the left of the $\hat{\$}$ symbol the candidate group is represented and to the right of $\hat{\$}$ resource supply, final group formation and resource allocation. The non-context-free language L_{RA} is defined as:

$$L_{RA} = \{a_1^{n_1} a_2^{n_2} a_3^{n_3} \ldots a_k^{n_k} \hat{\$} a_1^{m_1} a_2^{m_2} a_3^{m_3} \ldots a_k^{m_k} x^l \mid$$

$$1 \leq i \leq k, n_i \geq 1, m_i = n_i \text{ or } 0, l \geq 0\}.$$

A string in L_{RA} reflects one resource allocation. In particular, the symbols $a_1, a_2, \ldots, a_k$ represent different entities and the number of the resource request items (or capacities) (n_1 times for entity a_1, n_2 times for entity a_2, etc.). To the right of the $\hat{\$}$ symbol we represent the final group of entities to which the resource will be allocated. Note that to the right of the $\hat{\$}$ symbol there may be fewer entities than on the left of the $\hat{\$}$ symbol (indicating that the final group F is smaller than the candidate group C). If a string in L_{RA} contains symbols x it means that not all resource items were allocated.

Figure 2 illustrates the information that we obtain after each phase. The information is represented in the sentential forms that our grammar generates. In Fig. 2 the sentential forms are generalised to the following two parts: left of the $\$$ symbol and right of the $\$$ symbol.

In what follows, we describe how a random permitting context grammar G is constructed to describe the phases up to the resource distribution phase. For the resource distribution phase we will have to construct an rcg since an rPcg will not have the necessary capacity needed to model resource allocation.

For the following subsections we assume our running example.

5.1   Modelling Resource Supply

We start with modelling the first phase resource supply. This reflects the situation in which a resource becomes available. We construct a fragment of our rPcg $G = (V_N, V_T, P, S)$ as described subsequently. We construct rules of the following form:

$$S \rightarrow A' A_{call} X'(\emptyset; \emptyset),$$

$$X' \rightarrow X' X(\emptyset; \emptyset),$$

$$X' \rightarrow \$(\emptyset; \emptyset),$$

where $A', A_{\text{call}}, X', X, \$ \in V_N$. The first rule rewrites the start symbol S and is applied once (the nonterminals A' and A_{call} are needed for the subsequent derivation steps. The second rule is applied as many times as the total number of resource items and the third rule is applied once, indicating that the resource supply phases ended. The symbol $\$$ is also used to divide the sentential forms.

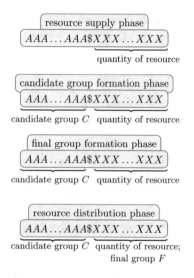

Fig. 2. The figure ([2], with permission of the author) shows which information is obtained after each phase represented as generalised sentential forms. The sentential forms are divided into left of $\$$ and right of $\$$ and are depicted as $AAA \ldots AAA\$XXX \ldots XXX$. Beneath each left and right part we explain what kind of information is conveyed.

Example: In our running example we assume to have five resource items. After applying the three initial rules above we have a sentential form of the following form (representing the result of the resource supply phase):

$$A' A_{\text{call}} \$ X X X X X.$$

5.2   Modelling Candidate Group Formation

Given the network graph N in Fig. 1, our model forms a candidate group C (a subgraph of N) for resource allocation. We assume that a starter entity begins by calling other entities directly connected to it. In our example, the calling sequence is as follows: A_4 calls A_5, A_5 calls A_1 and A_2, A_2 calls A_1. The Algorithm 1 constructs a tree θ from a given network graph N according to some calling sequence.

Algorithm 1: ([2]) Input: A network graph N and a calling sequence. Output: A tree θ displaying the candidate group C. Method: Construct a tree θ as follows:

- *Let entity A be the starter entity and label the root of θ with A and denote it tree layer L_0.*
- *For each tree layer L_i, $0 \leq i \leq n$ and for each symbol $A_1, A_2, \ldots, A_k$ in L_i do*
  - *for A_j, $1 \leq j \leq k$ calling entities $B_1, B_2, \ldots, B_{l_j}$ do*
  - *for all B_l, $1 \leq l \leq l_j$ not occurring in any tree layer L_g, $0 \leq g \leq i$ do*
  - *let $B_1, B_2, \ldots, B_{l_j}$, $1 \leq l \leq l_j$ be the children of A_j.*
- *Denote the resulting layer L_{i+1}.*

Example: The tree θ would be as illustrated in Fig. 3.

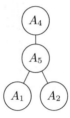

Fig. 3. A calling sequence tree θ representing that A_4 is the starter entity and calls entity A_5. Entity A_5 calls entities A_1 and A_2. No node occurs more than once.

Given a calling sequence tree θ we construct the rules for G. The rules are of the form $A \rightarrow \hat{\hat{A}} B_A C_A \ldots G_A B_{\text{call}} \ldots G_{\text{call}}$. This rule reflects that entity A called entities $B, C, \ldots, G$ by the subscript A on the nonterminals. The nonterminals with the subscript "call" give the respective entities the opportunity to call other entities. The Algorithm 2 constructs the rules with permitting context.

Algorithm 2: ([2]) Input: A calling sequence tree θ. Output: Rules with permitting context. Method: Construct the rules as follows:

- *If the root of θ is labelled by A and $A_1, A_2, \ldots, A_k$ occur in layer L_1 as children nodes of A do*
  - *add the rule*

$$A_{\text{call}} \rightarrow \hat{\hat{A}} A_{1_A} A_{2_A} \ldots A_{k_A} A_{1_{\text{call}}} A_{2_{\text{call}}} \ldots A_{k_{\text{call}}} (\$; \emptyset)$$

  to P.
- *For all tree layers L_i, $1 \leq i \leq n$ and all symbols A_j, $1 \leq j \leq l$ occurring in L_i do*
  - *for $j = 1$ add a rule*

$$A_{1_{\text{call}}} \rightarrow \hat{A}_1 B_{1_{A_1}} B_{2_{A_1}} \ldots B_{p_{A_1}} B_{1_{\text{call}}} B_{2_{\text{call}}} \ldots B_{p_{\text{call}}} (\hat{\hat{A}}; \emptyset)$$

  to P, where B_f, $1 \leq f \leq p$, are the children of node A_1 in L_{i+1} and $\hat{\hat{A}}$ is (the rightmost) A_l occurring in layer L_{i-1}.

- *for $j = 2, \ldots, l$ add a rule*

$$A_{j_{\text{call}}} \rightarrow \hat{A}_j C_{1_{A_j}} C_{2_{A_j}} \ldots C_{r_{A_j}} C_{1_{\text{call}}} C_{2_{\text{call}}} \ldots C_{r_{\text{call}}} (\hat{A}_{j-1}; \emptyset)$$

  to P, where C_g, $1 \leq g \leq r$, are the children of node A_j in L_{i+1}.
- *If a node A has no children we add the rule*

$$A_{\text{call}} \rightarrow \hat{A}(\hat{Y}, \emptyset)$$

  to P, where $\hat{Y}$ is either (the rightmost) A_l occurring in the previous layer L_{i-1} or $\hat{A}_{j-1}$.
- *For the rightmost Z in the last layer L_n that has no children add the rule*

$$Z_{\text{call}} \rightarrow \hat{Z} \# (\hat{Y}, \emptyset)$$

  to P, where $\hat{Y}$ is either (the rightmost) A_l occurring in the previous layer L_{i-1} or $\hat{A}_{j-1}$.

The nonterminal symbols of the form $\hat{A}$ are introduced in order to keep track of the order of the callings. Note how the permitting context of $\hat{A}$ nonterminals is simply carried over to the next layer in a tree or to a sister node to the right in a layer. The symbol $\#$ indicates that the calling phase ended.

Example: The following rules are constructed:

$$A_{\text{call}} \rightarrow \hat{A}_4 A_{5_{A_4}} A_{5_{\text{call}}} (\$; \emptyset),$$

$$A_{5_{\text{call}}} \rightarrow \hat{A}_5 A_{1_{A_5}} A_{2_{A_5}} A_{1_{\text{call}}} A_{2_{\text{call}}} (\hat{A}; \emptyset),$$

$$A_{1_{\text{call}}} \rightarrow \hat{A}_1 (\hat{A}_5; \emptyset),$$

$$A_{2_{\text{call}}} \rightarrow \hat{A}_2 \# (\hat{A}_1; \emptyset).$$

Given the sentential form $A' A_{\text{call}} \$ XXXXX$ and applying the rules in our example we obtain the sentential form:

$$A' \hat{A}_4 A_{5_{A_4}} \hat{A}_5 A_{1_{A_5}} A_{2_{A_5}} \hat{A}_1 \hat{A}_2 \# \$ XXXXX.$$

Now all entities that are called will either state how much of the resource they request (or can handle) or choose not to join the candidate group C. For this, we introduce for all nonterminal symbols with a double subscript and for A', simple recursive rules that are applied as many times as the number of the resource items that the respective entity requests (or can handle). For A' we introduce $A' \rightarrow A' A(\#; \emptyset)$ and $A' \rightarrow A(\#; \emptyset)$ (for the starter entity A) and for a nonterminal symbol $A_{i_{A_j}}$ we introduce rules of the form $A_{i_{A_j}} \rightarrow A_{i_{A_j}} A_{i_{A_j}}$. For an entity $A_{i_{A_j}}$ that will not join the candidate group, we introduce a rule of the form $A_{i_{A_j}} \to A_{\text{ex}i_{A_j}}$ where "ex" stands for "exit". The rules (except for the first two) are context-free.

Then we introduce rules that delete all nonterminals A in the sentential form that are marked with $\hat{A}$ or have the addition ex. Note that this is a simple design choice that can be made as desired. We make this design choice to simplify the formalisation and for better readability of the paper. In case terminal symbols a are marked with $\hat{a}$ or the addition "ex", the language L_{RA} must be changed accordingly. We introduce for all nonterminals of the form $\hat{A}$ and $A_{\text{ex}i_{A_j}}$ the rules $\hat{A} \to \lambda(\#; \emptyset)$ and $A_{\text{ex}i_{A_j}} \to \lambda(\#; \emptyset)$, respectively.

Example: Entities A_1, A_2, A_4, A_5 request 4, 2, 3, 2 resource items, respectively. No entity decides not to join the candidate group C. Applying the rules, that we construct as explained above, we obtain the following sentential form:

$$A_4 A_4 A_4 A_{5_{A_4}} A_{5_{A_4}} A_{1_{A_5}} A_{1_{A_5}} A_{1_{A_5}} A_{1_{A_5}} A_{2_{A_5}} A_{2_{A_5}} \#\$XXXXX.$$

The sentential form represents that entity A_4 requests 3 resource items, entity A_5 requests 2 resource items, etc. Furthermore, we see that A_5 was called by A_4, A_1 by A_5, etc. To make the sentential form more readable, we do not list the double subscript and obtain the following sentential form:

$$A_4 A_4 A_4 A_5 A_5 A_1 A_1 A_1 A_1 A_2 A_2 \#\$XXXXX,$$

which shows the candidate group C to the left of the $\$$ symbol consisting of the entities A_4, A_5, A_1, A_2 (and the quantity of their requested resource items) and to the right of the $\$$ symbol the available resource items.

5.3   Modelling Final Group Formation

A generalised sentential form after the candidate group formation looks as follows:

$$A_1^{n_1} A_2^{n_2} \ldots A_k^{n_k} \#\$X_1 \ldots X_m$$

illustrating the quantity of resource items that each of the k entities in C requests (or can handle) and the total number m of resource items available. In this phase the final group F to which the resources will be allocated is determined. For this we introduce an algorithm that constructs a tree that displays all possible resource allocations to the entities in C. The algorithm constructs successively a tree Θ in which the nodes are labelled with the entities in C and the edges are labelled with the quantity of available resource items at a given time step. For example, let a parent node A be connected with edges $e_1, e_2, \ldots, e_g$ to its children $A_1, A_2, \ldots, A_g$. Let the edges $e_1, e_2, \ldots, e_g$ be labelled with a number n. Such a subtree represents the situation that, after resource allocation to A, there are n resource items left to be allocated to $A_1, A_2, \ldots, A_g$. Before we give the formal algorithm, we explain the algorithm by our running example.

There are five resource items $X_1 X_2 X_3 X_4 X_5$ to be allocated and C consists of the four entities A_1, A_2, A_4 and A_5 that want 4, 2, 3 and 2 resource items respectively. As illustrated in Fig. 4, the algorithm begins constructing the tree Θ by labelling the root λ and the children of λ are all k entities in C. The edges connecting the root with the children are labelled by the number 5 (indicating that 5 resource items are available to be allocated).

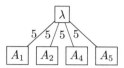

Fig. 4. Example of resource allocation using Algorithm 3 (first step).

As illustrated in Fig. 5 the leftmost node A_1 is expanded, such that the new children of A_1 are labelled with A_2, A_4, A_5 (i.e. the sisters of node A_1). The edges connecting A_1 with its children are labelled by 1, which indicates that if A_1 would have been allocated 4 resource items, 1 resource item would be available to be distributed to the rest of the entities A_2, A_4 and A_5.

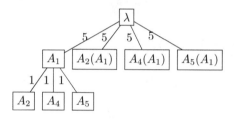

Fig. 5. Example of resource allocation using Algorithm 3 (second step).

Whenever a leftmost (depth-first) node A is expanded, an additional symbol (A) is added to its sister nodes. This indicates that if these sister nodes are expanded later in time they cannot have a child labelled A.

Then the leftmost and depth-first node A_2 should be expanded with its sister nodes A_4 and A_5, but the quantity of 1 resource item is less than the entity A_2 requests (i.e. 2). We illustrate this in Fig. 6 by letting the path end with a terminal symbol "A_1r1" where r1 stands for "rest 1". The path from the root to the leaf labelled "A_1r1" represent one possible resource allocation of 5 resource items to the entities A_1 and A_2 but in which only entity A_1 is allocated 4 items (and 1 resource item remains). Also note that A_2 becomes an additional symbol (A_2) for its sister nodes.

As illustrated in Fig. 7 the nodes $A_4(A_2)$ and $A_5(A_2)$ are expanded in a similar way as A_2. As illustrated in Fig. 8 the node $A_2(A_1)$ expands with children labelled with the labels of the sister nodes A_4 and A_5 ((A_1) indicates that A_1 cannot be a child label of A_2). The edge between $A_2(A_1)$ and its children A_4 and A_5 is labelled 3 indicating that if A_2 would have been allocated 2 resource items, 3 resource items would remain for A_4 and A_5. Note also that A_2 becomes an additional node for $A_4(A_1)$ and $A_5(A_1)$. As illustrated in Fig. 9, now the

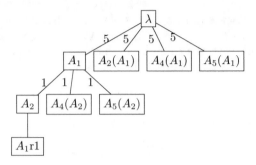

Fig. 6. Example of resource allocation using Algorithm 3 (third step).

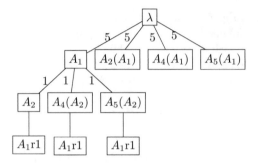

Fig. 7. Example of resource allocation using Algorithm 3 (Steps 4 and 5).

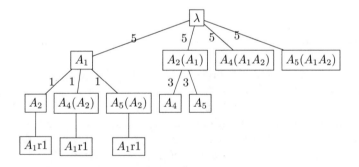

Fig. 8. Example of resource allocation using Algorithm 3 (Step 6).

leftmost and depth-first nonterminal A_4 gets a child A_5 and the edge is labelled 0 (indicating that A_4 would have taken 3 resource items).

Then A_5 would be expanded with a child labelled "A_2A_4r0" to indicate that if 5 resource items would be allocated to A_2, A_4, A_5, then A_2 would get 2, A_4 would get 3 and there would be 0 resource items for A_5.

The node $A_5(A_4)$ cannot have a child with label A_4 because of the symbol (A_4). A_5's child is labelled "A_2A_5r1" and the edge connecting the two nodes is labelled by 1.

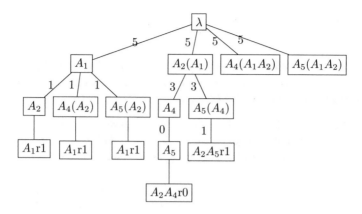

Fig. 9. Example of resource allocation using Algorithm 3 (Steps 7–9).

As illustrated in Fig. 10, the child of node $A_4(A_1A_2)$ is labelled A_5 and the connecting edge is labelled 2. Then the node A_5 is expanded with an edge labelled 0 and a leaf labelled "A_4A_5r0". The last node $A_5(A_1A_2A_4)$ has an edge labelled 3 connected to a child labelled "A_5r3".

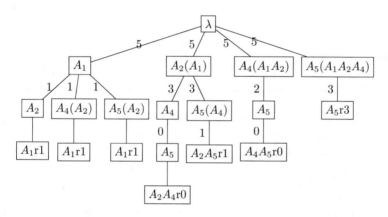

Fig. 10. Example of the output tree of Algorithm 3 displaying all possible resource allocations.

Let us understand what the tree Θ in Fig. 10 shows. We have four entities and 5 resource items and the following resource allocation possibilities:

1. If A_1 is allocated 4 resource items, 1 resource items remains, which cannot be allocated to other entities.
2. If A_2 and A_4 are allocated 2 and 3 resource items respectively, then there are 0 remaining resource items.
3. If A_2 and A_5 are allocated 2 resource items each, then there is 1 resource item left, which cannot be allocated to other entities.

4. If A_4 and A_5 are allocated 3 and 2 resource items respectively, then there are 0 remaining resource items.

We pointed out before that the E-strategy would determine resource allocation 4 and the M-strategy would determine resource allocation 2 or 4. All these possibilities are displayed in the tree Θ:

– Given the M-strategy, search in Θ for the longest path whose nodes are the largest subset of entities. If there are several equally long paths: select the one with the largest sum of resource items (i.e. edges) not exceeding the total of X.
– Given the E-strategy, search in Θ for the longest path whose nodes are the largest subset of entities. If there are several equally long paths: select the leftmost path in the tree Θ with the largest sum of items (i.e. edges) not exceeding the total of X.

Algorithm 3: **Input:** The total quantity m of the resource X. The entities $A_1, A_2, \ldots, A_k$ together with the requested resource items. **Output:** Allocation tree Θ with labelled edges. **Method:** Construct the tree as follows:

1. Let the root of Θ be labelled by λ and denote the tree layer Λ_i.
2. Let the children of λ be labelled by $A_1, A_2, \ldots, A_k$ and denote the tree layer Λ_{i+1}.
3. Label the edges connecting Λ_i with Λ_{i+1} with the total quantity m of the available resource.
4. Take the leftmost node label A_{left} in Λ_{i+1} and let its children be labelled by the sister nodes $A_{S_1}, A_{S_2}, \ldots, A_{S_j}$ of A_{left} at Λ_{i+1}. Denote the newly created layer Λ_{i+2}.
   If, however, a sister node A_{S_i}, $1 \leq i \leq j$, has an additional label A_A, then A_A cannot be a child of A_{S_i}.
5. If f is the quantity that A_{left} wanted and if $m - f \geq g$ and g is the quantity that the leftmost node A_{S_i} on Λ_{i+2} wanted, then
   – let Λ_{i+2} be Λ_{i+1} and go back to Step 4.
6. Termination of the path:
   (a) If f is the quantity that A_{left} wanted and if $m - f < g$ or $m - f = 0$, then
      – let the child of A_{left} be a leaf labelled by all nonterminal symbols on the path from λ up to A_{left}, followed by the symbol "r" (standing for "rest"), followed by the number on the edge connecting A_{left} with its parent.
   (b) If $m - f < 0$ then let the child of A_{left} be a leaf node labelled by all nonterminals symbols on the path from the root up to the parent of A_{left}, followed by the symbol "r" and the number of the edge connecting the parent of A_{left} with its parent.
7. If a path has terminated in a leaf, introduce the additional labels to the sister nodes as follows:
   – For each layer Λ let the label of the leftmost node (that is expanded) be an additional symbol to the sister nodes on Λ.

8. *Now take the deepest layer that has nonterminal symbols and take the leftmost node and go back to Step 4 and expand the path until no longer possible.*

Now given a chosen strategy a final group F is determined. We assume in our example, M-strategy and allocate to the entities A_2 and A_4 (0 resource items remaining).

5.4   Modelling Resource Distribution

Given a final group F and a sentential form

$$A_1^{n_1} A_2^{n_2} \dots A_k^{n_k} \# \$ X_1 \dots X_m,$$

we can see to the left of the symbol $\$$ the resource quantity of each entity requested (or capable to do) from the candidate group C, and to the right of the symbol $\$$ the total quantity of the resource items.

Note that in order to obtain a string in the language L_{RA}, we would first have to rewrite one A_1 on the left hand side of the symbol $\$$ and then we would have to rewrite the leftmost nonterminal X on the right side of the symbol $\$$. We basically would need to walk across the sentential form back and forth, rewrite one A_1 and then rewrite the leftmost X, until all A_1's are rewritten. Then we rewrite one A_2 and then rewrite the leftmost X, until all A_2's are rewritten, etc.

In the following we argue that such a mode of derivation cannot be encoded in the permitting context. We furthermore show that also adding a forbidding context does not suffice. What we need to model the resource distribution is an rcgs with a special kind of leftmost restriction is investigated in [4] (leftmost restriction II, p. 54). In this leftmost restriction at each step of a derivation the leftmost occurrence of a nonterminal in a sentential form which can be rewritten has to be rewritten. It is shown that the generative capacity of rcgs with this kind of leftmost restriction is considerably higher than that of rcgs (they generate recursively-enumerable languages) (see Theorem 1.4.4 in [4]). If no deleting rules are used, rcgs with this leftmost restriction generate context-sensitive languages. A simpler form of leftmost restriction in which at each step of a derivation the leftmost occurrence of a nonterminal has to be rewritten, decreases the generative capacity to context-free.

Lemma 1. *Given a sentential form*

$$w = A_1^{n_1} A_2^{n_2} \dots A_k^{n_k} \# \$ X \dots X$$

there is no rPcg that generates L_{RA}.

Proof. Observe that an rPcg must alternate between rewriting the nonterminal symbols on the left side of $\$$ and the nonterminals X on the right hand side of $\$$. First all A_1's and their leftmost X's have to be rewritten, then all A_2's and their leftmost X's have to be rewritten, etc.

Let the number of nonterminal symbols in G be n. Out of these n nonterminal symbols let $k + 3$ be the nonterminal symbols occurring in the sentential form

$w = A_1^{n_1} A_2^{n_2} \ldots A_k^{n_k} \# \$ X \ldots X$. Let the remaining m nonterminal symbols (i.e. $m = n - (k + 3)$), be special nonterminal symbols that mark each of the $k + 3$ nonterminal symbols Y with a $\hat{Y}$ symbol. That is, $\hat{A}_i$, $1 \leq i \leq k$, $\hat{\#}$, $\hat{\$}$, and $\hat{X}$. Given the sentential form w as above we observe that first one A_1 has to be rewritten introducing an $a_1 \in V_T$ into w. For this we have to have a rule of the form $A_1 \to \alpha a_1 \beta(\chi; \emptyset)$, where $\alpha, \beta \in (V_T \cup V_N)^*$, $\chi \subseteq V_N$. We observe that χ should contain one X, since the symbols on the left of the $\$$ symbol should be rewritten if there exists symbols X on the right of the $\$$ symbol. Moreover, we note that the rule $r_1 : A_1 \to \alpha a_1 \beta(\chi; \emptyset)$ must only be applied once, then a corresponding X should be rewritten before a second A_1 is rewritten, etc.

Thus, in order to prevent the rule r_1 being applicable more than once we have to introduce into its permitting context a symbol $\hat{A}_1$ such that $r_1 : A_1 \to \alpha a_1 \beta(\{\hat{A}_1, ..\}; \emptyset)$ (once applied) cannot be applied again before $\hat{A}_1$ is introduced into the sentential form w. The symbol $\hat{A}_1$ can be introduced into w by a rule r_2 that rewrites one of the nonterminal symbols in w, that is either A_i, $2 \leq i \leq k$, $\#$, $\$$, or X (but not A_1, otherwise it would be possible to rewrite more than one A_1).

For our argumentation, we assume r_2 rewrites X (but it could be any of the nonterminal symbols, except A_1) and introduces the symbol $\hat{A}_1$. Then r_2 is of the following general form: $X \to \alpha a_1 \hat{A}_1 \beta(\chi; \emptyset)$, where $\alpha, \beta \in (V_T \cup V_N)^*$. Now we argue again that, in order to prevent the rule r_2 being applicable more than once we have to introduce into the permitting context of r_2 a symbol $\hat{X}$ such that $r_2 : X \to \alpha a_1 \hat{A}_1 \beta(\{\hat{X}, ...\}; \emptyset)$ (once applied) cannot be applied again before $\hat{X}$ is introduced into the sentential form w.

The symbol $\hat{X}$ can be introduced by a rule r_3 that rewrites one of the nonterminal symbols in w, that is, either A_i, $2 \leq i \leq k$, $\#$, $\$$ (not A_i or X). But, continuing in this fashion all nonterminal symbols $\hat{Y}$ in V_N would be consumed and w could not be rewritten. This shows that there is no rPcg whose permitting contexts manage to rewrite w in the alternating fashion we descibed above. □

In the following we sketch that also allowing a forbidding context is not sufficient to model resource distribution.

Lemma 2. *Given a sentential form*

$$w = A_1^{n_1} A_2^{n_2} \ldots A_k^{n_k} \# \$ X \ldots X$$

there is no rcg that generates L_{RA}.

Sketch of proof. We can construct an rcg that rewrites in w first one nonterminal symbol A_i and then one corresponding nonterminal X on the right side of the $\$$ symbol. A sketch of such an rcg is as follows. First we construct a rule of the form $r_1 : A_i \to a_i \hat{A}_i(X; \{\hat{A}_i, \hat{X}\}$ to prevent the rule from being applicable twice in a row. Then we construct a rule $r_2 : X \to a_i \hat{X}(\hat{A}_i; \hat{X})$ to rewrite one symbol X. In order to delete the nonterminal symbols $\hat{A}_i$ and $\hat{X}$ we introduce the following two rules r_3 and r_4: $r_3 : \hat{A}_i \to \lambda(\hat{X}; \emptyset)$ and $r_4 : \hat{X} \to \lambda(\emptyset; \hat{A}_i)$. The rules would be applied successively r_1, r_2, r_3, r_4. Similar rules can be constructed for the other nonterminals. What is crucial here is, that it is not possible to encode in the permitting and forbidding contexts to rewrite the leftmost symbols X. The rules above would pick one X (not necessarily the leftmost) to be replaced, which would lead to strings not in L_{RA}. To rewrite the nonterminal symbols X in a leftmost fashion, the nonterminals would have to be numbered or distinct. One way to achieve this is to simply number the nonterminals X in the resource supply phase by rules of the following form: $r_1 : X \to X_1 X$, $r_2 : X \to X_2 X$, ..., $r_m : X \to X_m X$. But this means that m is bounded.

We conclude the following from the two lemma above.

Corollary 1. *An rcg is not sufficient to model the resource distribution phase.*

However, if we impose the leftmost restriction II described above on the rcg we can model the resource distribution phase. The Algorithm 4 constructs such an rcg.

Algorithm 4: ([2]) **Input:** The final group F chosen according to a resource allocation strategy. **Output:** Resource distribution rules with permitting and forbidding context. **Method:** Construct the rules as follows:

– *Let $A_1, A_2, \ldots, A_k$ be the entities in F that have been chosen by the E-strategy or M-strategy.*
– *For every A_i, $1 \le i \le k$ do*
– *Construct for A_1 the following rules:*

$$A_1 \to a_1 \hat{A}_1 (X; \{\hat{A}_1, \hat{X}_{A_1}\})$$
$$X \to a_1 \hat{X}_{A_1} (\hat{A}_1; \hat{X}_{A_1})$$
$$\hat{A}_1 \to \lambda(\hat{X}_{A_1}; \emptyset)$$
$$\hat{X}_{A_1} \to \lambda(\emptyset; \hat{A}_1)$$

where $\hat{A}_1$ and $\hat{X}_{A_1}$ are newly introduced nonterminal symbols in V_N.
– *Construct for all other A_i, $j = 2, \ldots, k$, the following rules*

$$A_i \to a_i \hat{A}_i (X; \{A_{i-1}, \hat{A}_{i-1}, \hat{X}_{A_{i-1}}, \hat{A}_i\})$$
$$X \to a_i \hat{X}_{A_i} (\hat{A}_i; \hat{X}_{A_i})$$
$$\hat{A}_i \to \lambda(\hat{X}_{A_i}; \emptyset)$$
$$\hat{X}_{A_i} \to \lambda(\emptyset; \hat{A}_i)$$

where $\hat{A}_i$ and $\hat{X}_{A_{i-1}}$ are newly introduced nonterminal symbols in V_N.
– *For all other nonterminals B on the left hand side of $\$$ representing entities that were not chosen by a purchasing strategy create the following rules:*

$$B \to b.$$

– *Add $X \to x(\emptyset; \gamma)$ to P to finally replace possible nonterminals X, where $\gamma = \{A_1, A_2, \ldots, A_k, \hat{A}_1, \hat{A}_2, \ldots, \hat{A}_k, \hat{X}_{A_1}, \hat{X}_{A_2}, \ldots, \hat{X}_{A_k}\}$.*

– *Add the rules* $\# \to \lambda(\$; \emptyset)$ *and* $\$ \to \hat{\$}(\emptyset; \#)$ *to* P, $\hat{\$} \in V_T$.

Example: In our example, we chose the M-strategy and the final group consists of A_2 and A_4. Thus, we construct the following rules:

$$A_2 \to a_2 \hat{A}_2(X; \{\hat{A}_2, \hat{X}_{A_2}\}),$$

$$X \to a_2 \hat{X}_{A_2}(\hat{A}_2; \hat{X}_{A_2}),$$

$$\hat{A}_2 \to \lambda(\hat{X}_{A_2}; \emptyset),$$

$$\hat{X}_{A_2} \to \lambda(\emptyset; \hat{A}_2),$$

$$A_4 \to a_4 \hat{A}_4(X; \{A_2, \hat{A}_2, \hat{X}_{A_2}, A_4\}),$$

$$X \to a_4 \hat{X}_{A_4}(\hat{A}_4; \hat{X}_{A_4}),$$

$$\hat{A}_4 \to \lambda(\hat{X}_{A_4}; \emptyset),$$

$$\hat{X}_{A_4} \to \lambda(\emptyset; \hat{A}_4),$$

$$A_1 \to a_1, A_5 \to a_5, \# \to \lambda(\$; \emptyset), \$ \to \hat{\$}(\emptyset; \#).$$

Given a sentential form w we would derive the terminal string $w' \in L_{RA}$:

$$w = A_1 A_1 A_1 A_1 A_2 A_2 A_4 A_4 A_4 A_5 A_5 \# \$ X X X X X \overset{*}{\Rightarrow}$$

$$v = a_1 a_1 a_1 a_1 a_2 a_2 a_4 a_4 a_4 a_5 a_5 \hat{\$} a_2 a_2 a_4 a_4 a_4.$$

Corollary 2. *An rcg modelling resource allocation to dynamically formed groups has to work in leftmost restriction II.*

6   Modelling Resource Allocation with only rPcg

Our previous discussions showed that the occurrences of the nonterminals X cannot be matched against the occurrences of the nonterminals to the left of the symbol $\$$ using only rPcg. The permitting context is not sufficient to enable walking back and forth a sentential form and rewriting the nonterminals in a leftmost fashion. Thus, using only rPcg to model resource allocation will necessarily change the model we discussed in the previous sections. In this section, we propose an alternative model of resource allocation using only rPcg but generating L_{RA}.

Given a network graph N. The alternative model begins, without a resource supply phase, with a starter entity $A \in N$ declaring how many resource items it can handle or requests. This is modelled by introducing to the left and to the right hand side of the symbol $\$$ the quantity that the starter entity requests or could handle. Then the starter entity calls for other entities in N, beginning to create a calling sequence. Now the called entities declare each, to the right and left of the symbol $\$$ how many resource items they would request (or handle).

Then the called entities call other entities, etc. This process yields the following generalised sentential form:

$$v = A^{n_1} B^{n_2} \dots Z^{n_p} \$ X_A^{n_1} X_B^{n_2} \dots X_Z^{n_p},$$

in which on the left of the \$ symbol there are $|Z|$ candidate entities declaring n_i, $1 \leq i \leq p$ resource items each and on the right hand side of \$ the entities pre-allocate the corresponding number of resource items X.

After this, we assume that the total quantity of a resource would become available. Given the available resource $X_1 \dots X_m$, and the $|Z|$ candidate entities, we can use Algorithm 3 to determine all possible resource allocations and then choose either E-strategy or M-strategy to determine the final group F for the actual resource allocation. The actual resource allocation would then rewrite the nonterminal symbols in the sentential form v correspondingly.

Algorithm 5 generates the fragments of the rPcg for the candidate formation phase in which all entities request or declare resource items and call other entities.

Algorithm 5: ([2]) **Input:** A calling sequence tree θ consisting of the candidate entities in C. **Output:** Production rules for the candidate formation phase in the alternative model. **Method:** Construct the rules as follows:

- *Let $S \rightarrow A\$X_A$ be a rule.*
- *For each tree layer L_i in θ, $0 \leq i \leq n$ and for each symbol $A_1, A_2, \dots, A_k$ in L_i do*
  - *for A_j, $1 \leq j \leq k$ construct declaration rules as follows*

$$A_j \rightarrow A_j A_j'(X_{A_j}; \emptyset)$$

$$X_{A_j} \rightarrow X_{A_j} X_{A_j}'(A_j'; \emptyset)$$

$$A_j' \rightarrow A_j(X_{A_j}'; \emptyset)$$

$$X_{A_j}' \rightarrow X_{A_j}(A_j; \emptyset)$$

  - *for A_j, $1 \leq j \leq k$ inviting associates $B_1, B_2, \dots, B_{l_j}$ construct the following rules*

$$A_j \rightarrow B_1 B_2 \dots B_{l_j}$$

$$X_{A_j} \rightarrow X_{B_1} X_{B_2} \dots X_{B_{l_j}}(\{B_1, B_2, \dots, B_{l_j}\}; \emptyset)$$

  - *for the leaf nodes A in θ construct the following rules*

$$A \rightarrow \lambda$$

Note that the rules have to be applied in the order they are constructed in Algorithm 5, since otherwise the derivation blocks. After applying rules constructed in Algorithm 5 we obtain a sentential form of the generalised form:

$$v = A^{n_1} B^{n_2} \dots \$ X_A^{n_1} X_B^{n_2} \dots$$

Once the final group and a resource allocation strategy have been decided context-free rules can be generated as outlined in Algorithm 6.

Algorithm 6: Input: A sentential form v resulting from the candidate formation phase in the alternative model. Output: Production rules for terminating the derivation. Method: Construct the rules as follows:

- Let $v = A^{n_1} B^{n_2} \ldots Z^{n_p} \$ X_A^{n_1} X_B^{n_2} \ldots X_Z^{n_p}$.
- Construct for all nonterminals A_j occurring to the left of $\$$ rules of the form

$$A_j \rightarrow a_j$$

- Construct for all entities B_j in the final group F, rules of the form

$$X_{B_j} \rightarrow b_j$$

- Construct for all other nonterminals X occurring on the right hand side of $\$$ the following rules:

$$X \rightarrow \lambda$$

Some of the design choices can be made differently, but the essence is that in a resource allocation model using only random permitting context, information about available resources and candidate entities cannot be distributed and coordinated later. Information has to be introduced rather locally, so that a walking across the sentential form to match nonterminal symbols against each other is avoided.

7   Conclusions and Future Work

We used grammars as a modelling tool in order to gain a deeper understanding into the processes of resource allocation and dynamic group formation. We modelled the individual phases of group formation and resource distribution as generative processes. We showed that one can use random permitting context grammars to describe the phases of *resource supply* and *candidate group formation*. For the *final group formation* phase, we introduced an algorithm that lists all possible resource allocations to candidate entities. Then a resource allocation strategy is chosen and the final group of entities can be determined. For the last phase *resource distribution* we showed that random context grammars with a specific kind of leftmost restriction describe the resource distribution by walking across the sentential form and model the allocation of resources. With our grammar model we can take a snapshot in time and see *where* or *how far* the process of group formation or the resource allocation is. For example, a sentential form can provide information how many resource items have already been allocated to an entity from the final group.

A grammar based model might be too simplistic to be directly applicable but provides valuable insight into the fundamental properties of the process being modelled. Another benefit with modelling resource allocation with grammars is that one can draw many implications from the rich mathematical theory behind regulated rewriting. For example, a fundamental property of our model that we could see using grammars is that, if information about the available resource and candidate entities is distributed among the phases *resource supply* and *candidate group formation* (and kept separate), then the synchronisation of information at a later stage (i.e. resource distribution phase) needs more powerful grammar models (i.e. rcgs with a specific leftmost restriction). We also showed that, if we would want to have less powerful grammars, that is, random permitting context grammars, we would have to adjust the resource allocation model, such that the information about the available resource and candidate entities is not kept separate but rather generated simultaneously as pre-allocation of resource items while forming the candidate group. A disadvantage with the alternative adjusted model is that a sentential form at a given time does not necessarily show where or how far the process of group formation or the resource allocation is.

Future work will focus on implementations of the formal theories on resource allocation and group formation and on evaluations with respect to efficiency and adequacy of the models. An interesting research question is, what aspects of resource allocation and group formation forbidding random context grammars can describe. Another interesting question concerns parsing. A parsing algorithm decides first if an input string w is in a certain formal language L and then assigns a structural representation to w (i.e. a derivation tree). A parsing algorithm can be useful if one wants to know whether several entities $a_1, \ldots a_l$ can be allocated certain available resource items according to a chosen resource allocation strategy. The entities and the resource distribution could be represented as a string $w = a_1^{n_1} \ldots a_l^{n_l} \$ a_1^{n_1} \ldots a_k^{n_k}$, $k \leq l$ and a parsing algorithm could decide if $w \in L$. This represents the question whether it is possible for a certain group of entities to be allocated a specific available resource.

References

1. Bensch, S., Drewes, F., Jürgensen, H., van der Merwe, B.: Graph transformation for incremental natural language analysis. Theor. Comput. Sci. **531**, 1–25 (2014)
2. Bensch, S., Ewert, S., Raborife, M.: Modelling the formation of virtual buying cooperatives with grammars of regulated rewriting. In Proceedings of the 7th International Conference on Simulation and Modeling Methodologies, Technologies and Applications, SIMULTECH 2017, pp. 45–55 (2017)
3. Chomsky, N.: Three models for the description of language. IRE Trans. Inf. Theory **2**(3), 113–124 (1956)
4. Dassow, J., Păun, G.: Regulated Rewriting in Formal Language Theory, vol. 18. EATCS Monographs on Theoretical Computer Science. Springer, Heidelberg (1989)
5. Ewert, S., Van der Walt, A.: A pumping lemma for random permitting context languages. Theor. Comput. Sci. **270**, 959–967 (2002)
6. Ewert, S.: Random context picture grammars: the state of the art. In: Drewes, F., Habel, A., Hoffmann, B., Plump, D. (eds.) Manipulation of Graphs, Algebras and Pictures, pp. 135–147. Hohnholt, Bremen (2009)
7. Ewert, S., van der Walt, A.: Random context picture grammars. Publicationes Mathematicae (Debrecen), **54**(Supp), 763–786 (1999)

8. Hopcroft, J., Ullman, J.: Introduction to Automata Theory, Languages, and Computation. Addison-Wesley, Reading (1979)
9. Joshi, A.K.: Tree adjoining grammars: how much context-sensitivity is required to provide reasonable structural descriptions? In: Dowty, D.R., Karttunen, L., Zwicky, A.M. (eds.) Natural Language Parsing. Psychological, Computational, and Theoretical Perspectives, pp. 206–250. Cambridge University Press, New York (1985)
10. Mashkov, V., Barilla, J., Simr, P., Bicanek, J.: Modeling and simulation of coalition formation. In: Proceedings of the 5th International Conference on Simulation and Modeling Methodologies, Technologies and Applications (SIMULTECH 2015), pp. 329–336 (2015)
11. Mohan, N.R.R., Raj, E.B.: Resource allocation techniques in cloud computing – research challenges for applications. In: Fourth International Conference on Computational Intelligence and Communication Networks (CICN), pp. 556–560 (2012)
12. Okundaye, B.: A tree grammar-based visual password scheme. Ph.D. thesis, University of the Witwatersrand, Johannesburg (2015)
13. Recker, J.: Evaluations of Process Modeling Grammars. Springer, Heidelberg (2011)
14. van der Walt, A.P.J.: Random context languages. Inf. Process. **71**, 66–68 (1972)
15. Zetzsche, G.: On erasing productions in random context grammars. In: 37th International Colloquium on Automata, Languages and Programming, ICALP 2010, Proceedings, Part II, pp. 175–186 (2010)

Parallel Simulation of Virtual Testbed Applications

Arthur Wahl$^{(\boxtimes)}$, Ralf Waspe, Michael Schluse, and Juergen Rossmann

Institute for Man-Machine Interaction, RWTH Aachen University,
Ahornstraße 55, Aachen, Germany
{wahl,waspe,schluse,rossmann}@mmi.rwth-aachen.de
https://www.mmi.rwth-aachen.de

Abstract. To enable the systematic evaluation of complex technical systems by engineers from various disciplines advanced 3D simulation environments that model all relevant aspects are used. In these Virtual Testbeds real-time capabilities are very hard to achieve without applying multi-threading strategies, due to the high complexity of the simulation. We present a novel simulation architecture that facilitates a modular approach to perform parallel simulations of arbitrary environments, where no explicit knowledge of the underlying simulation algorithms or model partitioning is needed by the engineer. Simulation components such as rigid-body dynamics, kinematics, renderer, controllers etc. can be distributed among different threads without concern about the specific technical realization. We achieve this by managing self-contained independent copies of the simulation model, each bound to one thread.

Keywords: Parallel simulation · Virtual Testbeds · Synchronization
Scheduling

1 Introduction

Virtual Testbeds (VT) greatly enhance the scope of the "classical simulation approach". While typical simulation applications examine only specific aspects of an application, a VT as shown in Fig. 1, enables development engineers to examine the entire system in its environment and provides a holistic view of the dynamic overall system including internal inter-dependencies in a user-defined granularity. In contrast to the classical "bottom-up"-strategy this can be seen as a "top-down"-approach. For classical fields of application of robotics, e.g. in production plants with a well-defined environment, a "bottom-up"-strategy in the development of simulation models is the approved method, because it allows very detailed insights into the analyzed subsystems. On the other hand, unpredictable effects of the interaction of multiple subsystems may easily be overseen. In particular, when it comes to building larger applications (like e.g. a space robot mission, planetary landing or exploration, autonomous working machines, advanced vehicle assistant systems, forestry application etc. as presented in the

© Springer Nature Switzerland AG 2019
M. S. Obaidat et al. (Eds.): SIMULTECH 2017, AISC 873, pp. 177–199, 2019.
https://doi.org/10.1007/978-3-030-01470-4_10

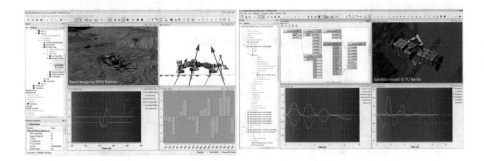

Fig. 1. Virtual Robotic Testbed for space environments [35].

Virtual Robotics Testbed [35]), they are hard to describe in an analytical way and are therefore harder to integrate into an analytical simulation model.

VTs offer advanced 3D simulation environments that realistically model all important aspects of such applications, including all relevant components and inter-dependencies. They provide (interactive) methods for reproducible tests and for the verification and validation of such systems under arbitrary boundary conditions. For this, they integrate 3D models, simulation algorithms, and real world devices.

However, such testbeds pose new challenges in the field of parallel simulation compared to the "classical simulation approach". They demand the development of a parallel simulation architecture that encompasses arbitrary application fields of VTs. Therefore, the concurrent execution of multiple simulations algorithms with different degrees of granularity must be supported. This will enable the simulation to incorporate various aspects of an application with different degrees of detail at the same time. Furthermore, the partitioning scheme provided to distribute the simulation among different threads must enable an engineer to generate partitions that are scalable and adaptable during the development or creation phase of a VT application e.g. an engineer can add new functionality or components to an existing simulation model to investigate additional aspects of an application. A partition must provide the possibility to include these new extensions. As a consequence, the integration of new simulation algorithms must also be supported. Regarding the amount of different simulation algorithms provided by a VT, an explicit parallelization of each simulation algorithm would require specific knowledge about the algorithms and their internal dependencies, provide no scalability and therefore present an inappropriate approach.

We present a novel parallel simulation architecture for arbitrary VT applications that fulfills the aforementioned requirements. The architecture generates low synchronization overhead during parallel execution and facilitates a modular distribution of a simulation among different threads without further effort. We follow a "top-down"-approach like the VT approach itself and propose the application of a functional partitioning scheme that distributes simulation algorithms among different threads. As shown in Fig. 2, simulation algorithms or

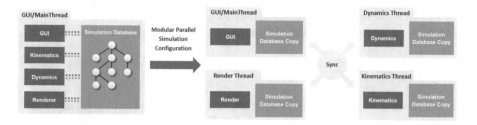

Fig. 2. Example distribution of simulation components [45].

components of a VT (such as rigid-body dynamics, kinematics, renderer, controllers etc.) can simply be assigned by an engineer to a specific thread and executed concurrently. Hereby, no explicit knowledge of the underlying algorithms is required or model partitioning is needed to perform parallel executions of a simulation. The modular distribution provides scalability since parallel execution scales with the number of components provided by a simulation model and partitions can be extended during the development phase by newly included components.

Execution of the simulation components is handled by a scheduler. Each thread owns a scheduler, where components can register and be executed correspondingly to their execution rate. The execution rate can be chosen individually for each component, similar to the approach presented in [11] for the simulation of autonomous mobile-robots teams. Components can be executed safely in parallel among different threads without interfering with each other. We achieve this by managing (partial) copies of the state data underlying the simulation models. The state data is organized in an active object oriented graph database and represents the current state of the simulation model at a given time-step. Each copy of the database acts as a self-contained, independent entity and is bound to one thread. All components assigned to one thread, work on the same copy of the database and are executed sequentially. Components cannot access or manipulate (no read/write access) the database of different threads and therefore can be executed safely in parallel without interfering with each other. Interaction between components among different threads is handled during a synchronization step. Our approach introduces a conservative synchronization algorithm with variable lookahead. Each scheduler can operate under a different lookahead (synchronization interval), depending on the execution rate of its slowest component. A change detection mechanism keeps track of changes that have occurred during a time-step. These changes represent the progress of each component and are used to update all databases to attain a consistent global simulation state. Due to the utilization of an active simulation database with an efficient change detection mechanism, our approach generates low synchronization overhead and parallel execution time is mainly dependent on the execution time of the slowest thread.

In this extended version of our article [45], we present our current results and elaborate further on the underlying simulation system and the time scheduler.

2   Related Work

In the context of simulation systems, replication has been exploited in order to evaluate simulation outcomes or to find optimal parameter settings to improve the accuracy of simulation results [9, 15, 16, 26, 41]. The approach is to run multiple independent copies of the same simulation program in parallel with different input parameters. At the end of the runs the results are averaged. This approach has been named Multiple Replication in Parallel (MRIP) [31]. Furthermore, MRIP has been applied in stochastic simulations. Stochastic simulations require the execution of multiple replicated simulation runs in order to build confidence intervals for their results [30]. MRIP has been applied to accelerate the building process [1, 5, 7]. This type of concurrent replication is not aimed at increasing the execution speed of each single run, but is aimed at efficiently providing a set of output samples by the concurrent execution of multiple independent differently parametrized sequential simulation runs. Bononi et al. introduced an approach that combines the concurrent execution of replicas with a parallel discrete event simulation by merging the execution tasks of more than one parallel simulation replica [2]. Hereby, blocking (idle CPU time) can be avoided during the synchronization barrier phase of a conservative parallel simulation (processes wait on the completion of more computational intensive processes) by switching to the execution of other replicas which already completed their synchronization phase. Parallel simulation cloning, another approach aimed at introducing replication in the context of parallel simulation was first employed by Hybinette and Fujimoto as a concurrent evaluation mechanism for alternate simulation scenarios [19, 20]. The aim of this approach is to allow fast exploration of multiple execution paths, due to sharing of portions of the computation on different paths. Parts of the simulation (logical processes) can be cloned during the simulation at predefined decision points. From then on, the original process and the clones execute along different execution paths in parallel, hence execution paths before cloning are shared. In addition, the application of virtual logical process avoids repeating unnecessary computation among clones and aims at further sharing computations after the decision point.

Opposed to replication and cloning, our approach uses multiple independent copies of the state data (organized in an active object oriented graph database) underlying the simulation model that all participate in the parallel execution of one simulation run. Through the utilization of state data copies, simulation components can be executed safely in parallel among different threads without interfering with each other. OpenSG (an open source project for portable scenegraph systems) uses a similar multi buffer approach, where partial copies of the scenegraph structure are used to introduce multi-threading [38, 44]. Besides information about the logical and spatial arrangement of a 3D scenes, as given by a scenegraph, our database approach also incorporates functionality in form of integrated simulation components, flexibly adopts to multiple data schemes, is independent from the type of simulation (continuous or discrete) and provides all the data needed for simulating VT applications [37]. In addition, each database copy offers an efficient change detection mechanism that collects all the changes that have happened to a database during one simulation time-step. Synchronization (restoring a global consistent simulation state among all copies), thereby is reduced to the distribution of such changes among all database copies.

The database represents the core of our simulation architecture. The function of the core is to store the structure and state of the simulation upon which integrated simulation components will act on. This type of architecture is described by the concept of aspect-based engines as presented by Revie [33]. An aspect-based engine is composed of multiple independent modules, called aspects. Each aspect supplies functionality to the core via aggregation. Aggregation is realized through a common interface allowing aspects to interact with the core and access shared data describing the current simulation state. Interactions between the aspects is realized entirely through manipulating the core. Compared to aspect-based engine concept by Revie our architecture adopts the aggregation-based approach by introducing an inheritance hierarchy to the proposed generic node graph structure. A generic graph node does not impose specific meaning upon its subgraphs. The structure of the simulation is defined purely via data. This information is stored within the nodes of the graph and must be interpreted by the aspects by parsing the whole graph. By extending the core to an object oriented graph we further increase flexibility and avoid the need for graph interpretation. Aspects or in our case simulation components are integrated as subgraph extensions to the graph database, realizing the interface themselves by providing their own fixed list of properties. In addition, data access is handled more efficiently by supporting introspection, since the graph structure does not have to be traversed anymore e.g. to find instances of a specific type. The highly modular nature of such an architecture allows for the accommodation of multiple projects with varying requirements and presents the ideal basis for the development of arbitrary parallel VT applications.

2.1   Multi-threading in Modern 3D Simulators

Explicit parallel programming constructs or multi-threading mechanisms are provided by most modern 3D simulators to accelerate the execution of simulation models. Several simulators introduce multi-threading only within their integrated physics engine or the mechanisms for parallel execution are restricted to small aspects of the whole simulation model: Webots [47] is a commercial development environment used to model, program and simulate complex robotic setups. Webots uses a multi-threading version of the ODE physics engine. Gazebo [14] is a 3D simulator for robotic applications that offers different sensors (cameras, distance sensors, GPS) and several types of robots (wheeled, legged robots, helicopters). Gazebo uses an "island thread" strategy to parallelize the integrated physics engine. Hereby, groups of interacting entities that are mathematically decoupled from each other are clustered into "islands" and simulated concurrently. Besides automatic parallel execution, explicit parallel programming can be used in both simulators to further introduce parallelism and increase performance, but this requires explicit knowledge of the underlying simulation algorithms plus their inter-dependencies by a development engineer. Simscape [39] is a multibody simulation environment for 3D mechanical systems that builds upon a block diagram environment for Model-Based Design (Simulink [25]). Highly detailed and specialized aspects of an application can be simulated with Simscape. Simulation models are partitioned accordingly to the sample time of each block diagram to realize parallelism. These partitions are then mapped to tasks that run on multicore processors, which results in highly specialized parallel simulation applications. V-Rep is a virtual robot experimentation platform based on a distributed control architecture. V-REP [40] uses threads to mimic the behavior of coroutines. It offers the possibility to run threaded child scripts. Each scene object in the simulation model can be associated with a child script that represents a program handling a particular function in the simulation (e.g. control a robot). Accordingly, only small aspects of a simulation model are executed in parallel.

We introduce a modular approach in the context of VT applications that does not restrict parallel execution to one component or specific aspects of a simulation model but offers a development engineer the possibility create parallel simulations for VT applications that incorporate the entire simulation model without the need of having specific technical knowledge about the underlying simulation algorithms.

3   Active Real-Time Simulation Database

The key component of our simulation architecture is an active real-time simulation database, which is an object-oriented, self-reflecting graph database, as introduced in [37]. To reach the goal outlined above, the following requirements have to be implement by the database:

- The database must support the integration of data (e.g. 3D simulation data) and algorithms in one single - now active simulation - database, supporting interface definition and providing means for state oriented as well as event based communication.
- For the (real-time) simulation performance it is important how the data can be accessed and manipulated by the simulation algorithms (e. g. to implement even complex controllers). Ideally, database management itself should be time efficient, thus leaving computing power available to the simulation routines.
- In addition to this, it must be possible to easily add new simulation algorithms or enhance existing methods, while guaranteeing stability and performance of the overall system.
- The database must provide replication mechanisms to create full or partial copies of the database (e.g. horizontal fragmentation to create subsets of the original database containing only instances regarding one aspect of the simulation like GUI, rendering, dynamics, etc.)

- A database copy is assignable to only one thread and must act as a unique and independent entity without interfering with other copies while being manipulated by the assigned thread.
- In addition to this, it must be possible to distribute changes among all databases and update a database by applying these changes.

We developed a new architecture for 3D simulation systems, which is based on a small (micro-) kernel which provides a sustainable basis for various and diverse simulation applications, eliminates unnecessary dependencies and fulfills the requirements mentioned above. This kernel is the *Versatile Simulation Database* (VSD), a real-time database containing all the data and algorithms needed for simulation applications. Fully implemented in C++, it provides the central building blocks for data management, meta-information, communication, persistence and user interface. The design of the simulation system as shown on the left side of Fig. 3 is inspired by the Object Management Group (OMG) meta model hierarchy [22].

3.1   Meta Information

The uppermost layer (labeled M2 in Fig. 3) is the meta-information system, essential for the developer flexibility, as well as the and end user friendliness of the database and the simulation system. It is the basis for persistence, user interface, parallel and distributed simulation, scripting and communication.

It mainly consists of the following classes:

- **MetaTypeVal:** Describes all data types that can be used as values (e. g. int, double, string, simple structs, enumerations, flags).
- **MetaProperty:** Describes a property (see Sect. 3.2) with its getter and setter functions, its data type and a number of additional flags. These flags describe the behavior of the property as exposed to the user (editable, savable, etc.) as well as the properties ability to be used in parallel and distributed simulation.
- **MetaMethod:** Describes a method (member function) of an instance or a type.
- **MetaInstance:** Describes an instance including its class hierarchy. Each non-abstract meta-instance is able to create corresponding instances. Each meta-instance holds a list of the corresponding meta-methods and meta-properties.

3.2   Instances and Properties

The middle layer (labeled M1 in Fig. 3) describes the data model of the simulation. In order to be able to retain semantic information and integrate data and algorithms into one single database, the VSD data model is an object oriented graph database [18], whose structure is detailed in this section. A simplified class hierarchy of the VSD core is shown on the right side of Fig. 3. All nodes in the graph database, the database itself and even the simulation environment are derived from a single base class called "Instance". This base class

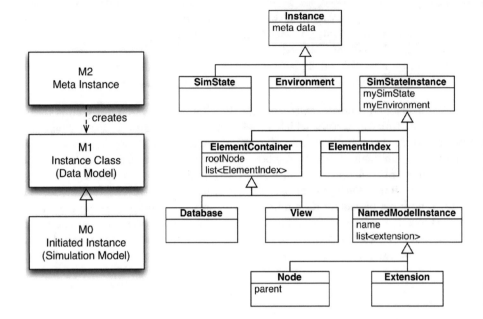

Fig. 3. The Meta Model Hierarchy (left side), The core database class hierarchy (right side).

provides mechanisms for inter-instance communication, as well as access to the meta-information system, which allows introspection of class hierarchy, properties and methods (see Sect. 3.1). The simulation model (labeled M0 in Fig. 3) is an instantiation of the data model.

Properties are standardized getter and setter functions that encapsulate the data itself. All data in the simulation system is stored as properties. Properties can encapsulate any single value or value containers (lists, vectors, etc.), whose data types are known to the meta-information system. Properties can also hold references or lists of references. References come in two different varieties, composite aggregation (with reference counting as described in [23]), and shared aggregation. All parent child relations within the database are implemented as composite aggregation references. Shared aggregation references do not change the reference counter of the instance, but are informed if the instance gets deleted.

3.3 Active Database

As mentioned above the VSD is not only a static data container, but also contains the simulation algorithms themselves. The environment, as well as all containers and element indexes actively inform interested listeners about new instance creation or deletion, as well as property modifications. Furthermore each instance sends a signal when one of its properties has been changed. Thus interested parts of the simulation systems can be informed about state changes in the simulation, eliminating the need to continuously poll the database content. With the active messaging system and the availability of element index lists we have minimized the need for data polling and tree traversal. By creating derived classes (like "Actuator", "Sensor", "Robot", etc.) from the Node or Extension base class, the simulation algorithms (actuator control, sensor simulation, robot controller, etc.) itself are integrated into VSD. Simulation algorithms that need a complete overview over the simulation state (like rigid body simulations) are integrated on the database level, but still manage their data on the node and extensions level as illustrated before.

That's why we call the VSD *active*. To achieve a complete decoupling of the different system components with well defined interfaces (introspectable using the meta-information system), methods are provided for event as well as state based communication. These methods can be used to let the components exchange information as defined by the algorithm developer or the simulation expert.

4 Scheduling

A comprehensive overview of the evolution of simulation is given in [27], while [12] specializes on parallel and distributed simulation. Relevant to the scope of this paper is the notion of time as proposed by [32]:

- *physical time*: "the time in the physical system that is being modeled"
- *simulation time*: "representation of the physical time for the purposes of simulation"
- *wall time*: "elapsed real time during execution of the simulation, as measured by a hardware clock"

An overview of multi-domain simulation is given in [8]. For the applications shown later on in this paper the work [28] is of interest, where the challenges of combining a simulation system with a virtual reality environment are described. Simulation is mostly described as a series of events, meaning points in time at which the state of simulation changes. For this paper it is assumed that the current simulation state is held within a database and that the simulation is advanced by executing time ordered tasks, which in turn may trigger events that change the simulation state.

4.1 Task Scheduler and Event Queue

The term scheduler is most often used in handling of computer processes with a known maximum duration, whose execution has to be finished at a specified wall time. One of the earliest works is by [24]. The scheduling of parallel execution of this tasks is described in many publications, such as [21] or [17]. Due to the fact that the simulation time is not necessarily tied to the wall time and that events that happen at the same simulation time do not have to be executed at the same wall time, an event list or event queue is most often employed in simulations. A basic event is depicted on the left side in Fig. 4. An event has a name (this may be a name, a number or just a pointer) and a next execution (simulation) time. Periodically reoccurring events may also have a step time.

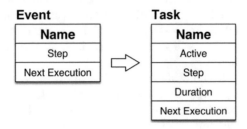

Fig. 4. A task extends the concept of an event and can also have a duration.

The basic methods for event queues have been well established by [6,43] or [10]. In an event queue all pending events are listed in the order in which they have to be executed. Figure 5 shows the execution of an event queue up to a given target time. If the execution time of the first event is smaller then the targeted simulation time the event is executed and removed from the list. The next execution time for that event is then calculated and it is reinserted into the list. This process is repeated until the execution time of the first event in the queue exceeds the target time.

Various methods have been proposed to minimize the insertion time of the event into the queue such as the use of calenders [3,29] or CPU caching [46]. The use of event lists in distributed simulation has also been examined, for example by [4]. However none of these proposed methods work for both real-time task scheduling (for hardware in the loop scenarios) and event list handling.

Our proposed method is based on the concept of tasks. A task will be executed at its designated simulation time, which in its basic implementation is not linked

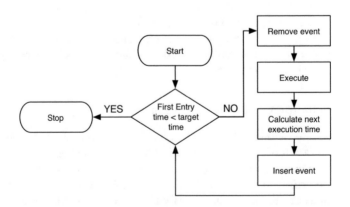

Fig. 5. An event list executed up to a given target time.

to the wall time and therefore not in itself suitable for real-time applications. A task (as shown in Fig. 4) has a name, an optional step time and a next execution time. The next execution time can be calculated via step time or set manually by the task itself. The task duration is given in simulation time and is a tool for maintaining data cohesiveness in parallel simulation. If no duration is given it is assumed that the duration equals the step time. The task scheduler of the simulation system does not enforce the wall time duration of tasks. If a task is supposed to be real time capable, it is up to the developer or modeler to ensure that a task is executable within its wall time duration. Tasks are collected in task lists, which execute the contained tasks up to a given simulation time like the event lists shown in Fig. 5.

For selective task execution both task lists and tasks can have contexts, which can be built-in or user defined. A task list will only execute tasks that share at least one common context. A special context for tasks is *"always"*, which forces a task list to execute a task even if no common context exists.

4.2   Time Controller

The simulation time of a task list is advanced in time slices. As shown in Fig. 6, the rate (in both simulation and wall time) at which the simulation is advanced is dictated by a time controller, which can be adapted to the application domain. A scheduler consists of a time controller and a task list. For a time sliced simulation the time controller advances the simulation in fixed intervals called *granularity*. For real time simulation the time to execute all tasks within a slice is measured. The time controller keeps track of how the simulation-time progresses compared to the wall-time. If the wall time for the execution is less then the simulation time the system can wait until the wall time for the next execution is reached. If the wall time duration exceeds simulation time, the controller can advice its associated task list to skip tasks that are not critical for the simulation (for example rendering may be skipped).

Fig. 6. The execution of tasks within a list is triggered by a time controller [45].

For most 3D simulations a granularity equal to the render rate is sufficient. However for real-time scenarios, such as hardware-in-the-loop or robot control a much smaller granularity can be chosen. These systems can theoretically have any granularity supported by the underlying operation system. As show in [34], a granularity of under four milliseconds on a headless real time operating system is feasible. For simulations with a fixed granularity step a dedicated task can be used to automatically track and increase the simulation time. For dedicated discreet event simulation (DES) applications a special time controller can be implemented that does not advance the simulation time in fixed slices. Instead it always instructs the task list to advance the simulation to the execution time of its first task. DES tasks can compute their next execution time without any given step time.

5   Modular Parallel Simulation of Virtual Testbeds

We introduce a parallel simulation approach that meets the challenges presented by VT architectures. The approach must be applicable to arbitrary application fields of VT, facilitate the concurrent execution of multiple simulation algorithms with different degrees of granularity and allow for scalability as well as the integration of new algorithms. VTs are composed of a variety of different integrated simulation components such as rigid-body dynamics, kinematics, sensors, actuators, renderer etc. which can be regarded as logical, independent processes. We propose a partitioning scheme that utilizes the modular VT structure and partition applications along their simulation components. We follow the "top-down"-approach and distribute simulation components among simulation threads for parallel execution.

5.1   Simulation Thread

A simulation thread owns a scheduler, with a time controller and a task list, a (partial) copy of the main VSD and a unique context id e.g. "simulation thread $1 - n$". Components can be assigned by a development engineer to a simulation thread by choosing a thread specific context id and hereby activating their execution through the schedulers corresponding task list, see Fig. 7. A scheduler executes only components with a matching context id. The execution rate can be chosen individually for each component. Components assigned to one simulation thread are executed sequentially. The execution order is given by the execution rate of the individual components. In consequence, the execution order of a set generated by sequential simulation run is also maintained by a simulation thread during parallel execution. A distinction must be made between sequences of components that have to be executed in a predefined order and sequences not dependent on the execution order. Sequences that have to be executed in a predefined order can only be assigned to one thread, to preserve that order. Our partitioning scheme facilitates the possibility to spawn an individual simulation thread for each sequence, thus providing the scalability to exploit the maximum number of cores on a CPU. A time controller of a simulation thread provides thread safe functionality to start, stop and pause a scheduler.

5.2   Simulation Architecture

As shown in Fig. 8, four different basic types of controllers can be assigned to a simulation thread: a time controller for multi-thread execution, a time controller for rendering, a time controller for the GUI update of the simulation system and a time controller for sequential execution. Each time controller controls the

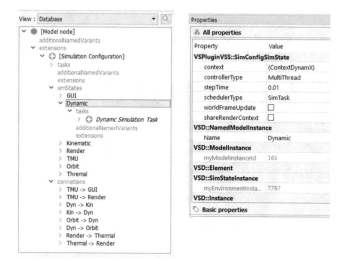

Fig. 7. GUI: Dynamics simulation component is assigned to the simulation thread "Dynamic" by choosing a thread specific context id "ContextDynamX". Time controller as well as scheduler type can be set via the property widget.

execution of a corresponding scheduler. Both time controller types for parallel and sequential execution are derived from abstract classes that provide functionality to start, stop and pause a scheduler, as well as functionality to monitor how simulation-time progresses compared to wall-time. A simulation task scheduler can either be controlled by a corresponding time controller during parallel or sequential execution. Tasks with rendering content are executed by a render scheduler. A render scheduler determines the render update rate and is responsible for updating all transformation matrices of the applications geometry before rendering the scene. The GUI update of the simulation system is handled by the GUI scheduler. Responsiveness of the application can be achieved by solely handling GUI event updates in the main thread.

All scheduler classes are also derived from an abstract base class. The abstract base class provides functionality to manage the execution of simulation tasks, handle synchronization and holds a reference to the corresponding time controller. The architecture presented here is capable of concurrently executing distributed render and non-render simulation components among multiple simulation threads.

5.3   Simulation Database Copies

Each simulation thread owns a full or partial copy of the main threads VSD. The copying process starts by first creating all instances of the VSD. Then the graph structure of the main threads VSD is traversed. This is done by recursively invoking all reference properties of an instance, starting at the root node and subsequently creating the referenced instances. Property references represent the

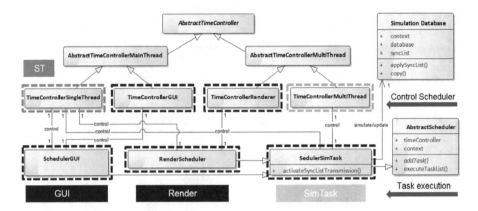

Fig. 8. Simulation architecture [45].

branches of the graph structure. Traversing the graph by following all branches and copying the referenced instances results in the generation of the whole graph structure for the copied VSD. During the next stage of the copying process the data values of all instances represented by their properties are copied. The last stage of the copying process handles all initialization processes, evoked by the generation of instances. The copying process inherently includes the copying of simulation components as well as the initialization of their properties. Therefore, a copy of the VSD acts as a complete, self-contained, independent entity that provides all the necessary data and functionality to execute a simulation run for the assigned components. Memory saving mechanism can be applied during the copying process e.g. by using filters (black/white lists) to exclude instances of a specific type from being copied or to enable data sharing.

5.4 Configuration of Simulation Threads

The creation of multiple independent simulation databases, that each facilitates all the data and functionality needed for simulation, represents the core idea of our parallel simulation approach. Each simulation thread has the ability to execute an independent, autonomous part of the whole simulation. All components assigned to one thread, work on the same copy of the VSD. This ensures that access to the VSD of a simulation thread only happens successively by components that share the same context. Typical problems that come with multithreading like simultaneous data access, race conditions etc. are avoided, because components from different threads do not share the same VSD, but access their own copy.

The approach presented here, provides the ability to create complex configurations of simulation threads. An example configuration for parallel simulations of on-orbit satellite servicing applications, as presented in the Virtual Space Robotics Testbed [36] is shown in Fig. 11.

6   Synchronization

To achieve correctness of parallel executed simulation runs, synchronization across the processors is required. The goal is to ensure that a parallel execution of a simulation produces the same result as an equivalent sequential execution. Therefore, dependencies have to be preserved among partitions and events. Intermediate results have to be processed in the right order during the computation of the simulation state across the processors to attain a global, consistent simulation state. Assuming that the simulation consists of a collection of logical processes that resulted from a partition scheme and that the logical processes communicate by exchanging timestamped messages or events, this can be achieved during synchronization by preserving the local causality constraint [13]. The local causality constraint states that in order to achieve identical simulation results, each logical process or partition must process events in non-decreasing time-stamp order. As a consequence results are also repeatable.

Our approach introduces a conservative synchronization architecture with variable lookahead (synchronization interval). Each scheduler can operate under a different lookahead, depending on the execution of its slowest component. A change detection mechanism keeps track of changes that have occurred during a time-step. These changes represent the progress of each component and are used to update all VSDs.

6.1   Synchronization Architecture

The change detection mechanism of the VSD stores changes inside a list (sync list). The sync list represents the difference between the actual and the last simulation state. It can be used to update other VSDs. Synchronization is realized through the exchange and application of sync lists. A VSD can connect to another VSD, it is interested in receiving updates from, by enabling a synchronization connection. Therefore in Fig. 10, we introduce the following communication mechanism:

A synchronization connection is a sender/receiver-relation between two VDSs. Communication is done via synchronization tasks that function as containers for the sync lists to be sent. These task are transmitted and executed during the synchronization phase of the scheduling main-routine, before the next simulation time-step starts. After establishing the synchronization connections, each sender scheduler possesses a list of synchronization tasks, containing all receiver references and the necessary information to transmit sync lists to them. The list is processed during each scheduling-cycle of the sender scheduler. Hereby, the current sync list of the sender VSD together with the current time-step of the sender scheduler are passed to the receiver scheduler. This is done by creating a new synchronization task containing the aforementioned information and inserting the task into the corresponding task-list of the receiver scheduler. That way, the receiver does not have to halt immediately when receiving a synchronization task. The execution of simulation components is not influenced by the broadcast of synchronization tasks. The receiver continues the execution of simulation components and only processes received synchronization tasks when the scheduling-routine arrives at that processing step. In return, a sender with a lower scheduling time-step than the receiver can perform unimpeded, multiple scheduling cycles and send multiple sync lists to the receiver while the receiver may still be processing simulation components until both reach the same time-step. The received synchronization tasks are applied in time-stamp order and the simulation state of the receiver VSD is updated by applying all sync lists. Pursuing this strategy allows an independent execution of sender and receiver schedulers with different time-steps until both schedulers reach the same time-step and the synchronization algorithm is applied to ensure that the causality constraint is not violated (Fig. 9).

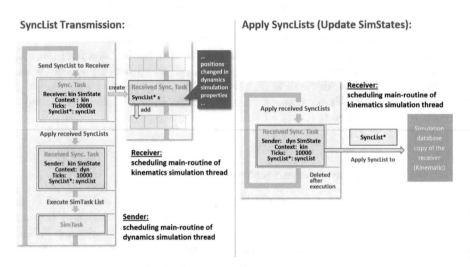

Fig. 9. Synchronization architecture [45].

6.2   Synchronization Algorithm

Simulation threads can be paused during the synchronization phase of the scheduling main-routine. Sender and Receiver schedulers have to wait either for the dispatch or application of synchronization tasks if they progress faster than other schedulers that participate in the synchronization process. Because a sync list keeps value copies of properties changed in the VSD, a sender scheduler has not to wait until all connected receivers with a smaller time-step have finished applying the sync list. We propose the following synchronization sequence, see Fig. 10.

A scheduler starts the synchronization process and the main-routine by sending the current sync list containing all changes from the last scheduling time-step ($t_{i-1}^{scheduler}$) to its receivers. Hereby, receivers are woken up that have waited on receiving that sync list. Next the scheduler processes the sync lists received so

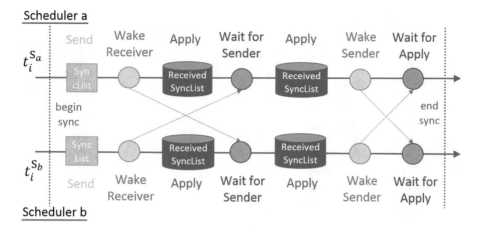

Fig. 10. Synchronization sequence [45].

far in time-stamp order ($t_i^{synclist} < t_i^{scheduler}$). After the first processing step, the scheduler checks if it has progressed faster than the remaining senders. If the scheduler has progressed faster, it has to wait until all senders have advanced sufficiently and transmitted their sync lists up to the current time-step of that scheduler. The scheduler is immediately woken up as soon as a sync list arrives and starts processing it. In the last step of synchronization process the scheduler itself waits on the application of all transmitted sync lists and is woken up as soon as the last receiver has applied the transmitted sync lists.

Processing synchronization tasks in the sequence presented here, guarantees that synchronization tasks are not processed out of time-stamp order, that synchronization tasks do not contain future VSD changes and that no synchronization tasks are missed during a scheduling-cycle. All changes contained by the received sync lists are applied during one time-step in exact the same order to the VSD of a receiver as they happened in the original VSDs of the senders. Hereby, the causality constraint is preserved and repeatable as well as identical results are produced during the parallel execution.

7   Results/Validation

The modular parallel simulation architecture presented here is constantly being developed further through the integration of new simulation components and more VT applications. The results are very promising regarding the performance. So far, we tested several scenarios of the Virtual Space Robotics Testbed with different simulation component distribution configurations. In all cases the average performance during parallel execution met the desired outcome predicted by our partitioning scheme. The average performance was mainly dependent on the total execution time of the slowest distributed component. Our approach is capable of achieving performance gains that scale with the number of threads, as shown for the following complex simulation scenario.

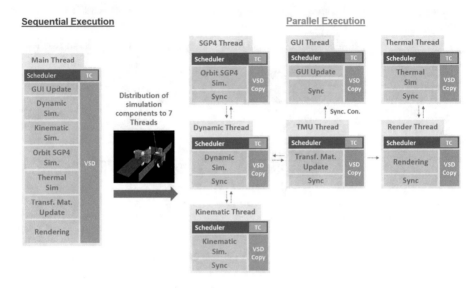

Fig. 11. Modular parallel simulation of a satellite reconfiguration scenario.

For the parallel simulation of a modular satellite reconfiguration scenario where a chaser satellite is docked to a satellite composed of individual building blocks [48] and utilizes robotic manipulators to exchange malfunctioning components, we choose a configuration of seven simulation threads (see Fig. 11), each connected to each other by synchronization connections.

- One simulation thread is running the multi-body dynamics simulation component which is handling all dynamically modeled components like, motors, grippers, robot manipulators, the docking interfaces between the building blocks etc.
- A second simulation thread is executing the kinematics simulation as well as the robot control components that realize the movement of the robotic manipulators.
- A third thread is running the orbit simulation using a SGP4 [42] propagation model.
- A fourth thread is executing a thermal simulating component developed by our project partners at TUB, which is using the incident solar radiance to compute the temperatures of the satellites building blocks.
- A fifth thread is updating all the transformation matrices and propagating them to the render thread which is using the updated geometry information to re-render the whole scene during each simulation step.
- A sixth thread is rendering the scene.
- The main thread is solely responsible for updating the GUI, by running the GUI event loop, in doing so the GUI remains responsive throughout the whole parallel simulation.

During the parallel simulation of this scenario, we achieved a constant speed-up factor of 4.16 (see Figs. 12 and 13). The combined average execution time (actual execution time of the simulated component plus synchronization time) of the slowest simulation thread during parallel execution was 3.02 ms. The total average execution time of all components during sequential simulation was 12.57 ms. Dividing both values delivers the aforementioned speed-up factor. Synchronization overhead was very low, on the average around 210 μs for the render thread, which was the component with longest execution time and therefore had no wait time during synchronization. In addition the render thread also had the highest synchronization overhead, because it had to apply all transformation matrix updates. Despite the harsh execution times, in the range of a few milliseconds, our approach was capable of achieving a stable and constant speed-up during each simulation time-step, resulting in a linear progression of simulation time. Memory usage increased by 67 MB per VSD copy from initially 649 MB (consisting of: 321 MB for the simulation model and 328 MB for the simulation architecture including all components of the Virtual Space Robotics Testbed) to 1051 MB after six additional VSD copies were generated for the configuration presented above and the simulation was started. Model geometries were shared among the simulation threads. Textures were only generate for the render thread. Due to the processing of synchronization tasks in time-stamp order, the parallel simulation of this scenario produced repeatable and identical results compared to the equivalent sequential execution. The results were generated on a Intel(R) Core(TM) i9-7900X multi-core CPU with 10 physical cores at 4.0 GHz and 32 GB of RAM.

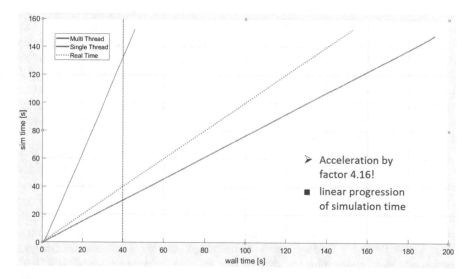

Fig. 12. Performance measurements and execution times of the satellite reconfiguration scenario.

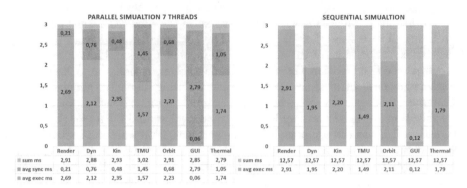

Fig. 13. Average execution and synchronization times (application of vsd changes and wait time) of the simulation components of the satellite reconfiguration scenario during parallel and sequential simulation.

8   Conclusion/Future Work

We introduced a parallel simulation approach that addresses the challenges posed by VTs and developed a parallel simulation architecture that presents an ideal platform for further research regarding the parallel execution of arbitrary VT applications. Scalability is provided by the functional partitioning scheme that allows a modular distribution of various simulation components among threads. The scheduler based simulation architecture facilitates the concurrent execution of multiple distributed simulations algorithms with different degrees of granularity. Distributed components are executed safely in parallel without interfering with each other, due to the utilization and assignment of independent VSD copies to simulation threads. Each VSD copy facilitates all the data and functionality needed for simulation. The change detection mechanisms provides an efficient distribution method of VSD updates among the simulation threads. In addition, the synchronization architecture allows an independent execution of sender and receiver schedulers. As a result, our approach generates low synchronization overhead and provides speed-ups that are scalable with the number of utilized threads during parallel execution.

Regarding the scalability during parallel execution, our modular approach is limited by the number of simulation components provided by an VT application and the synchronization overhead. In future, we would like to investigate further partitioning schemes and synchronization algorithms for VT applications to increase the scalability of our approach. Our functional partitioning scheme could be extended by a spatial partitioning to generate more complex configurations of distributed VT components. For example, an application simulating an automobile production plant with multiple robotic production units, each realizing one step in the production process could be partitioned spatially along its production areas. Following the spatial partitioning scheme a robot unit could be assigned to one simulation thread which provides an instance of the dynamic and kinematic components and simulated concurrently along with all the other production units. Our synchronization algorithm could be enhanced by transmitting data copies instead of references. Hereby, simulations threads would not have to wait on the application of their transmitted VSD changes by their receivers. Furthermore we would like to investigate the application of computational load balancing strategies to realize an automatic distribution of components among all simulation threads during parallel execution. Computational load could be identified by measuring the execution time of each scheduler during a simulation time-step. An autonomic distribution and execution of components, according to a load balancing strategy, would intercept possible computational load changes during the parallel execution of a simulation run and restore performance by redistributing components among threads with less computational load.

Acknowledgments. Parts of this work were developed in the context of the research project ViTOS. Supported by the German Aerospace Center (DLR) with funds of the German Federal Ministry of Economics and Technology (BMWi), support code 50 RA 1304.

References

1. Ballarini, P., Forlin, M., Mazza, T., Prandi, D.: Efficient parallel statistical model checking of biochemical networks. arXiv preprint arXiv:0912.2551 (2009)
2. Bononi, L., Bracuto, M., D'Angelo, G., Donatiello, L.: Concurrent replication of parallel and distributed simulations. In: Workshop on Principles of Advanced and Distributed Simulation, PADS 2005, pp. 234–243. IEEE (2005)
3. Brown, R.: Calendar queues: a fast 0 (1) priority queue implementation for the simulation event set problem. Commun. ACM **31**(10), 1220–1227 (1988)
4. Dahl, J., Chetlur, M., Wilsey, P.: Event list management in distributed simulation. In: Euro-Par 2001 Parallel Processing (2001)
5. Eickhoff, M.: Sequential analysis of quantiles and probability distributions by replicated simulations. Ph.D. thesis, University of Canterbury (2007)
6. van Emde Boas, P., Kaas, R., Zijlstra, E.: Design and implementation of an efficient priority queue. Math. Syst. Theory **127**, 99–127 (1976)
7. Ewald, R., Leye, S., Uhrmacher, A.M.: An efficient and adaptive mechanism for parallel simulation replication. In: ACM/IEEE/SCS 23rd Workshop on Principles of Advanced and Distributed Simulation, PADS 2009, pp. 104–113. IEEE (2009)
8. Faruque, M.O., Dinavahi, V., Steurer, M., Monti, A., Strunz, K., Martinez, J.a., Chang, G.W., Jatskevich, J., Iravani, R., Davoudi, A.: Interfacing issues in multi-domain simulation tools. IEEE Trans. Power Deliv. **27**(1), 439–448 (2012)
9. Forlin, M., Mazza, T., Prandi, D.: Predicting the effects of parameters changes in stochastic models through parallel synthetic experiments and multivariate analysis. In: 2010 Ninth International Workshop on Parallel and Distributed Methods in Verification, and Second International Workshop on High Performance Computational Systems Biology, pp. 105–115. IEEE (2010)
10. Franta, W.R., Maly, K.: An efficient data structure for the simulation event set. Commun. ACM **20**(8), 596–602 (1977)
11. Friedmann, M., Petersen, K., von Stryk, O.: Simulation of multi-robot teams with flexible level of detail. In: International Conference on Simulation, Modeling, and Programming for Autonomous Robots, pp. 29–40. Springer (2008)

12. Fujimoto, R.M.: Parallel and distributed simulation. In: Farrington, P.A., Nembhard, H.B. (eds.) Proceedings of the 1999 Winter Simulation Conference, pp. 122–131 (1999)
13. Fujimoto, R.M.: Parallel simulation: parallel and distributed simulation systems. In: Proceedings of the 33rd Conference on Winter Simulation, pp. 147–157. IEEE Computer Society (2001)
14. GAZEBO: Open source robotics foundation (2017). http://www.gazebosim.org
15. Glasserman, P., Heidelberger, P., Shahabuddin, P., Zajic, T.: Splitting for rare event simulation: analysis of simple cases. In: Proceedings of the 28th Conference on Winter Simulation, WSC 1996, pp. 302–308. IEEE Computer Society, Washington, DC (1996). https://doi.org/10.1145/256562.256635
16. Glynn, P.W., Heidelberger, P.: Analysis of parallel replicated simulations under a completion time constraint. ACM Trans. Model. Comput. Simul. 1(1), 3–23 (1991). https://doi.org/10.1145/102810.102811
17. Graham, R.: Bounds for certain multiprocessing anomalies. Bell Syst. Tech. J. 45(9), 1563–1581 (1966)
18. Gyssens, M., Paredaens, J., van den Bussche, J., van Gucht, D.: A graph-oriented object database model. IEEE Trans. Knowl. Data Eng. 6, 572–586 (1994)
19. Hybinette, M., Fujimoto, R.: Cloning: a novel method for interactive parallel simulation. In: Proceedings of the 29th Conference on Winter Simulation, pp. 444–451. IEEE Computer Society (1997)
20. Hybinette, M., Fujimoto, R.M.: Cloning parallel simulations. ACM Trans. Model. Comput. Simul. (TOMACS) 11(4), 378–407 (2001)
21. Ka, A., Mok, L.: Fundamental design problems of distributed systems for the hard-real-time environment, vol. 1. MIT thesis, May 1983
22. Kurtev, I., van den Berg, K.: MISTRAL: a language for model transformations in the MOF meta-modeling architecture. In: European MDA Workshops: Foundations and Applications, MDAFA 2003 and MDAFA 2004, Twente, The Netherlands, 26–27 June 2003, and Linköping, Sweden, 10–11 June 2004. Revised Selected Papers (2005)
23. Levanoni, Y., Petrank, E.: An on-the-fly reference-counting garbage collector for java. ACM Trans. Program. Lang. Syst. 28, 1–69 (2006)
24. Liu, C., Layland, J.: Scheduling algorithms for multiprogramming in a hard-real-time environment. J. ACM (JACM) 1, 46–61 (1973)
25. Matlab Simulink: The mathworks, inc. (2017). http://www.mathworks.de/
26. Mota, E., Wolisz, A., Pawlikowski, K.: A perspective of batching methods in a simulation environment of multiple replications in parallel. In: Proceedings of the 32nd Conference on Winter Simulation, WSC 2000, pp. 761–766. Society for Computer Simulation International, San Diego (2000). http://dl.acm.org/citation.cfm?id=510378.510487
27. Nance, R.E., Sargent, R.G.: Perspectives on the evolution of simulation. Oper. Res. 50(1), 161–172 (2002)
28. Nielebock, S., Ortmeier, F., Schumann, M., Winge, A.: From discrete event simulation to virtual reality environments. In: Ortmeier, F., Daniel, P. (eds.) Computer Safety, Reliability, and Security, pp. 508–516. Springer, Heidelberg (2012)
29. Oh, S.H., Ahn, J.S.: Dynamic calendar queue. In: Proceedings of the 32nd Annual Simulation Symposium, pp. 20–25. IEEE (1999)
30. Passerat-Palmbach, J., Caux, J., Siregar, P., Mazel, C., Hill, D.: Warp-level parallelism: enabling multiple replications in parallel on GPU. arXiv preprint arXiv:1501.01405 (2015)

31. Pawlikowski, K., Yau, V.W., McNickle, D.: Distributed stochastic discrete-event simulation in parallel time streams. In: Proceedings of the 26th Conference on Winter Simulation, pp. 723–730. Society for Computer Simulation International (1994)

32. Perumalla, K.S.: Parallel and distributed simulation: traditional techniques and recent advances. In: Proceedings of the 38th Conference on Winter Simulation, pp. 84–95. Winter Simulation Conference (2006)

33. Revie, D.: An aspect-based engine architecture. In: GPU Pro 4: Advanced Rendering Techniques, pp. 267–284. AK Peters/CRC Press (2013)

34. Rossmann, J., Dimartino, M., Priggemeyer, M., Waspe, R.: Practical applications of simulation-based control. In: Proceedings of the IEEE International Conference on Advanced Intelligent Mechatronics (AIM2016), Banff, Alberta, Canada, 12–16 July 2016, pp. 1376–1381. IEEE (2016). Electronic ISBN 978-1-5090-2065-2, USB ISBN 978-1-5090-2064-5, Print on Demand (PoD) ISBN 978-1-5090-2066-9

35. Rossmann, J., Schluse, M.: Virtual robotic testbeds: a foundation for e-robotics in space, in industry – and in the woods. In: Developments in E-systems Engineering (DeSE), pp. 496–501. IEEE (2011)

36. Rossmann, J., Schluse, M., Rast, M., Atorf, L.: eRobotics combining electronic media and simulation technology to develop (not only) robotics applications. In: Kadry, S., El Hami, A. (eds.) E-Systems for the 21st Century – Concept, Developments, and Applications, Chap. 10, vol. 2. Apple Academic Press (2016). ISBN 978-1-77188-255-2

37. Roßmann, J., Schluse, M., Waspe, R.: Combining supervisory control, object-oriented petri-nets and 3D simulation for hybrid simulation systems using a flexible meta data approach. In: SIMULTECH, pp. 15–23 (2013)

38. Roth, M., Voss, G., Reiners, D.: Multi-threading and clustering for scene graph systems. Comput. Graph. **28**(1), 63–66 (2004)

39. Simscape Multibody: The mathworks, inc. (2017). http://www.mathworks.de/

40. V-REP: Coppelia robotics gmbh (2017). http://www.coppeliarobotics.com

41. Vakili, P.: Massively parallel and distributed simulation of a class of discrete event systems: a different perspective. ACM Trans. Model. Comput. Simul. **2**(3), 214–238 (1992). https://doi.org/10.1145/146382.146389

42. Vallado, D., Crawford, P., Hujsak, R., Kelso, T.: Revisiting spacetrack report# 3. In: AIAA/AAS Astrodynamics Specialist Conference and Exhibit, p. 6753 (2006)

43. Vaucher, J., Duval, P.: A comparison of simulation event list algorithms. Commun. ACM **18**(4), 223–230 (1975)

44. Voß, G., Behr, J., Reiners, D., Roth, M.: A multi-thread safe foundation for scene graphs and its extension to clusters. In: EGPGV 2002, pp. 33–37 (2002)

45. Wahl, A., Waspe, R., Schluse, M., Rossmann, J.: A modular approach for parallel simulation of virtual testbed applications. In: De Rango, F., Ören, T., Obaidat, M.S. (eds.) Proceedings of the 7th International Conference on Simulation and Modeling Methodologies, Technologies and Applications (SIMULTECH 2017), Madrid, Spain, 26–28 July 2017, vol. 1, pp. 255–262. SCITEPRESS (2017). ISBN 978-989-758-265-3

46. Wang, W., Yang, Y.: The speedup of discrete event simulations by utilizing CPU caching (2012)

47. Webots: CYBERBOTICS Ltd. (2017). http://www.cyberbotics.com/webots

48. Weise, J., Briess, K., Adomeit, A., Reimerdes, H.G., Göller, M., Dillmann, R.: An intelligent building blocks concept for on-orbit-satellite servicing. In: Proceedings of the International Symposium on Artificial Intelligence Robotics and Automation in Space (iSAIRAS), Turin, Italy (2012)

Complete Lyapunov Functions: Computation and Applications

Carlos Argáez[1]([⊠]), Peter Giesl[2], and Sigurdur Hafstein[1]

[1] Science Institute, University of Iceland, 107, Reykjavík, Iceland
{carlos,shafstein}@hi.is
[2] Department of Mathematics, University of Sussex,Falmer BN1 9QH, UK
P.A.Giesl@sussex.ac.uk
http://www.hi.is/~carlos
http://users.sussex.ac.uk/~pag20/
http://www.hi.is/~shafstein

Abstract. Many phenomena in disciplines such as engineering, physics and biology can be represented as dynamical systems given by ordinary differential equations (ODEs). For their analysis as well as for modelling purposes it is desirable to obtain a complete description of a dynamical system. Complete Lyapunov functions, or quasi-potentials, describe the dynamical behaviour without solving the ODE for many initial conditions. In this paper, we use mesh-free numerical approximation to compute a complete Lyapunov function and to determine the chain-recurrent set, containing the attractors and repellers of the system. We use a homogeneous evaluation grid for the iterative construction, and thus improve a previous method. Finally, we apply our methodology to several examples, including one to compute an epigenetic landscape, modelling a bistable network of two genes. This illustrates the capability of our method to solve interdisciplinary problems.

Keywords: Dynamical system · Complete Lyapunov Function
Quasi-potential · Mesh-free collocation · Radial basis functions

1 Introduction

Let us consider a general autonomous ordinary differential equation (ODE) $\dot{\mathbf{x}} = \mathbf{f}(\mathbf{x})$, where $\mathbf{x} \in \mathbb{R}^n$. A classical (strict) Lyapunov function [29] is a scalar-valued function that can be used to analyze the basin of attraction of one attractor such as an equilibrium or a periodic orbit. It attains its minimum at the attractor, and is otherwise strictly decreasing along solutions of the ODE.

A generalization of this idea is the notion of a complete Lyapunov function [10,20,28,35], which characterizes the complete behaviour of the dynamical system. It is a scalar-valued function $V : \mathbb{R}^n \to \mathbb{R}$, defined on the whole phase space, not just in a neighbourhood of one particular attractor. It is non-increasing along solutions of the ODE. The phase space can be divided into the area where the

© Springer Nature Switzerland AG 2019
M. S. Obaidat et al. (Eds.): SIMULTECH 2017, AISC 873, pp. 200–221, 2019.
https://doi.org/10.1007/978-3-030-01470-4_11

complete Lyapunov function strictly decreases along solution trajectories and the one where it is constant. For the first case, the complete Lyapunov function characterizes the gradient-like flow. There solutions pass through and the larger this area is, the more information is obtained from the complete Lyapunov function. Note that, by definition, the complete Lyapunov function needs to be constant along solution trajectories on each transitive component of the chain-recurrent set.

Furthermore, there are other methods to analyze the general behaviour of dynamical systems such as the direct simulation of solutions with many different initial conditions. This, however, is costly and can only give limited information about the general behaviour of the system, unless estimates are available, e.g. when shadowing solutions. More sophisticated methods include the computation of invariant manifolds, forming the boundaries of basins of attraction of attractors [37]; here, additional analysis of the parts with gradient-like flow is necessary. The cell mapping approach [18] or set oriented methods [12] divide the phase space into cells and compute the dynamics between these cells, see also for example [30]; these ideas have also been used to compute complete Lyapunov functions [9].

Since our first method was proposed in [31] to compute complete Lyapunov functions using mesh-free collocation and thus to divide the phase space into the chain-recurrent set and the gradient-like flow, several improvements have been proposed. We will shortly review such improvements and explain the novelties of this current work.

The general idea [31] is to compute a complete Lyapunov function v by approximating the solution of $V'(\mathbf{x}) = -1$, where $V'(\mathbf{x}) = \nabla V(\mathbf{x}) \cdot \mathbf{f}(\mathbf{x})$ denotes the orbital derivative, the derivative along solutions of the ODE. We use mesh-free collocation with Radial Basis Functions (RBF) to approximately solve this partial differential equation (PDE): we choose a finite set of collocation points X and compute an approximation v to V which solves the PDE at all collocation points. Note, however, that the PDE cannot be fulfilled at all points of the chain-recurrent set, such as an equilibrium or periodic orbit. For that reason, we used the failure of the method to approximate in certain areas to classify them [31]: we separate the collocation points X into a set X^0, where the approximation fails, and X^-, where it works well.

In general, a complete Lyapunov function should be constant in X^0. For that reason, in a subsequent step we solved the PDE $V'(\mathbf{x}) = 0$ for all $\mathbf{x} \in X^0$ and $V'(\mathbf{x}) = -1$ for all $\mathbf{x} \in X^-$, which is a more accurate approximation of a complete Lyapunov function. This procedure is then iterated, either for a specified number of iterations or until no more points are added to X^0. The computed function v gives us the following information about the ODE under consideration: the set X^0, where $v'(\mathbf{x}) \approx 0$, approximates the chain-recurrent set, including equilibria, periodic orbits and homoclinic orbits, while the set X^-, where $v'(\mathbf{x}) \approx -1$, approximates the area with gradient-like flow, where solutions pass through. The function v, through its level sets, gives additional information about the stability and attraction properties: minima of v correspond to attractors, while maxima represent repellers.

Let us give more details on how we determine whether the approximation near a collocation point is poor (or fails), resulting in placing this collocation point into the set X^0, or whether the approximation is good, and the collocation point is placed into X^-. By construction, at the collocation points we have $v'(\mathbf{x}_j) = V'(\mathbf{x}_j)$, which thus will always be a perfect approximation. We thus evaluate $v'(\mathbf{x})$ for all $\mathbf{x}$ of an evaluation grid $Y_{\mathbf{x}_j}$ of points near the collocation point $\mathbf{x}_j$. We fix a critical value $\gamma \leq 0$ and place the collocation point $\mathbf{x}_j$ into X^0 if $v'(\mathbf{x}) > \gamma$ holds for at least one $\mathbf{x} \in Y_{\mathbf{x}_j}$; otherwise $\mathbf{x}_j$ is placed into X^-.

While the basic method is already capable of classifying the chain-recurrent set in many examples, there were several shortcomings which were addressed in subsequent improvements: (a) in examples, where the velocity $\|\mathbf{f}(\mathbf{x})\|$ varies considerably, the chain-recurrent was either over- or under-estimated and (b) the method does not produce a complete Lyapunov function with a continuous derivative. Here, $\|\cdot\|$ denotes the Euclidean norm in $\mathbb{R}^n$.

To address (a), we have introduced a method [32] to analyze systems that have a large change in their velocity, namely where $\|\mathbf{f}(\mathbf{x})\|$ varies considerably over the phase space. We proposed to normalize the system by "almost" the norm. In particular, this is done by replacing the original system $\dot{\mathbf{x}} = \mathbf{f}(\mathbf{x})$ by the system

$$\dot{\mathbf{x}} = \hat{\mathbf{f}}(\mathbf{x}), \quad \text{where } \hat{\mathbf{f}}(\mathbf{x}) = \frac{\mathbf{f}(\mathbf{x})}{\sqrt{\delta^2 + \|\mathbf{f}(\mathbf{x})\|^2}} \tag{1}$$

with small parameter $\delta > 0$. The dynamical system defined by (1) with $\hat{\mathbf{f}}(\mathbf{x})$ has the same solution trajectories as $\dot{\mathbf{x}} = \mathbf{f}(\mathbf{x})$, but these are traversed at a more uniform speed, namely $\|\hat{\mathbf{f}}(\mathbf{x})\| = \frac{\|\mathbf{f}(\mathbf{x})\|}{\sqrt{\delta^2 + \|\mathbf{f}(\mathbf{x})\|^2}} \approx 1$. The smaller δ is, the closer the speed is to 1. While the normalized method generally produces better results, it comes at a higher computational cost due to the evaluation of $\hat{\mathbf{f}}$.

To address problem (b), we have proposed different improvements. After solving $V'(\mathbf{x}) = -1$, we again split the collocation points into X^0 and X^-. However, instead of solving $V'(\mathbf{x}) = r(\mathbf{x})$ with $r(\mathbf{x}) = -1$ for $\mathbf{x} \in X^-$ and $r(\mathbf{x}) = 0$ for $\mathbf{x} \in X^0$, which results in a discontinuous function $r(\mathbf{x})$, we make the right-hand side $r(\mathbf{x})$ smooth by using the distance $\eth(\mathbf{x})$ between the point $\mathbf{x}$ and the set X^0. In detail, we solve the PDE

$$V'(\mathbf{x}) = r(\mathbf{x}) := \begin{cases} 0 & \text{if } \mathbf{x} \in X^0, \\ -\exp\left(-\frac{1}{\xi \cdot \eth^2(\mathbf{x})}\right) & \text{if } \mathbf{x} \in X^-, \end{cases} \tag{2}$$

where $\eth(\mathbf{x}) = \min_{\mathbf{y} \in X^0} \|\mathbf{x} - \mathbf{y}\|$ is the distance between the point $\mathbf{x}$ and the set X^0 and $\xi > 0$ is a parameter [32].

Using (2) we guarantee that $r(\mathbf{x})$ and thus V is a smooth function such that for $\mathbf{x}$ with a large distance to X^0, $r(\mathbf{x})$ is close to -1 while for close distances it raises up to zero.

A different method to solve problem (b) defines $r(\mathbf{x}_j)$ to be the average value of $v'(\mathbf{y})$ over all $\mathbf{y} \in Y_{\mathbf{x}_j}$, this is done regardless of whether $\mathbf{x}_j$ lies in X^0 or X^-.

This again results in a smooth function $r(\mathbf{x})$, and the distinction between X^0 and X^- is just used to determine the chain-recurrent set [33,34].

In previous work, we have used different sets for the evaluation points $Y_{\mathbf{x}_j}$. In [31], in two dimensions, we used points on two circumferences around each collocation point. Later we introduced a new evaluation grid consisting of points along the direction of the flow, $\mathbf{f}(\mathbf{x}_j)$ at each collocation point. That allowed to expand our method from two to higher dimensions without increasing the amount of evaluation points exponentially [34].

Combining this with fixing $r(\mathbf{x}_j)$ by averaging over all $v'(\mathbf{y})$ with $\mathbf{y} \in Y_{\mathbf{x}_j}$ may result in the right-hand side converging to zero as the number of iterations goes to infinity. This would result in a constant complete Lyapunov function, which is a trivial complete Lyapunov function and does provide any information about the dynamics. To fix that, we introduced a methodology in [33] that scales the orbital derivative condition, which shows to be efficient to avoid trivial solutions. However, for few iterations that consideration is not necessary.

In this paper, we use an evaluation grid to evaluate the complete Lyapunov function that consists of a homogeneous distribution of points on two circumferences around each collocation point. We use an iterative method to construct a complete Lyapunov function by replacing the right-hand side $r(\mathbf{x}_j)$ by the average of $v'(\mathbf{y})$ over all points of the evaluation grid at the respective collocation point $\mathbf{x}_j$. Depending on the example, we either consider the original equation $\dot{\mathbf{x}} = \mathbf{f}(\mathbf{x})$ or the normalized one, see (1). Furthermore, we apply our methodology to four examples. The first three are from [32]. The final example is a biological system to prove the capabilities of our method in applications.

Let us give an overview over the paper: in Sect. 2 we discuss complete Lyapunov functions as well as mesh-free collocation to approximate solutions of a general linear PDE. In Sect. 3 we present our algorithm to compute a complete Lyapunov function. In Sect. 4 we apply the method to four examples and discuss the results in detail.

2   Preliminaries

2.1   Complete Lyapunov Functions

We will consider a general autonomous ODE

$$\dot{\mathbf{x}} = \mathbf{f}(\mathbf{x}), \text{ where } \mathbf{x} \in \mathbb{R}^n. \tag{3}$$

A complete Lyapunov function [10] is a continuous function $V \colon \mathbb{R}^n \to \mathbb{R}$ which is constant along solution trajectories on the chain-recurrent set, including local attractors and repellers, and decreasing along solution trajectories elsewhere. In contrast to classical Lyapunov functions [29], which are defined on the basin of attraction of just one attractor, a complete Lyapunov function characterizes the flow on the whole phase space and distinguishes between the chain-recurrent set and the gradient-like flow. Thus, it captures the long-term behaviour of the system.

Auslander [5] and Conley [10] proved the existence of complete Lyapunov functions for dynamical systems on a compact metric space. The idea is to consider corresponding attractor-repeller pairs and to construct a function which is 1 on the repeller, 0 on the attractor and decreasing in between. Then these functions are summed up over all attractor-repeller pairs. This was generalized to more general spaces by Hurley [19,20,28].

The smaller the part of the phase space where the complete Lyapunov function is constant, the more information is provided by a complete Lyapunov function. There exists a complete Lyapunov function which is only constant on the generalized chain-recurrent set [5], thus providing further information about the system as the generalized chain-recurrent set is a subset of the chain-recurrent set.

In [9,16,24] a computational approach to construct complete Lyapunov functions was proposed. The discrete-time system given by the time-T map was considered, the phase space was subdivided into cells and the dynamics between them were computed through an induced multivalued map using the computer package GAIO [11]. An approximate complete Lyapunov function is then computed using graph algorithms [9]. This approach requires a high number of cells even for low dimensions. We will use a different methodology, inspired by the construction of classical Lyapunov functions, which is faster and works well in higher dimensions. In [6], the approach of [9] is compared to the RBF method for equilibria (see below) for one particular example; here, the method of [9] works well only on the chain-recurrent set, while the RBF method is very efficient on the gradient-like part.

In [8], a complete Lyapunov is constructed as a continuous piecewise affine (CPA) function, affine on each simplex of a fixed simplicial complex. However, it is assumed that information about local attractors is a available, while the proposed method in this paper does not require any information about the system under consideration.

In [1] a quasi-potential, which is very similar to a complete Lyapunov function, is constructed by numerical integration along solution trajectories with different initial conditions. The landscape (plot of the complete Lyapunov function) and the level sets are used to analyze a biological system modelling two genes that inhibit each other, forming a double-negative feedback loop structure. We will analyze this system with our method in Sect. 4.4.

2.2 Mesh-Free Collocation

For classical Lyapunov functions, several numerical construction methods have recently been proposed, e.g. [2–4,7,13,17,21–23,25] see also the review [15]. Our algorithm will be based on the RBF (Radial Basis Function) method, a special case of mesh-free collocation, which approximates the solution of a linear PDE, specifying the orbital derivative.

Mesh-free methods, particularly based upon Radial Basis Functions, provide a powerful tool for solving generalized interpolation problems efficiently. We assume that the target function belongs to a Hilbert space H of continuous functions (often a Sobolev space) with norm $\| \cdot \|_H$ and with reproducing kernel $\varphi : \mathbb{R}^n \times \mathbb{R}^n \to \mathbb{R}$, given by a suitable Radial Basis Function Φ through $\varphi(\mathbf{x}, \mathbf{y}) := \Phi(\mathbf{x} - \mathbf{y})$, where $\Phi(\mathbf{x}) = \psi(\|\mathbf{x}\|)$ is a radial function. Examples for Radial Basis Functions include the Gaussians, multiquadrics and inverse multiquadrics; we, however, will use the compactly supported Wendland functions in this paper, which will be defined below.

We assume that the information $r_1, \dots, r_N \in \mathbb{R}$ of a target function $V \in H$ generated by N linearly independent functionals $\lambda_j \in H^*$ is known. The optimal reconstruction of the function V is the solution of the minimization problem $\min\{\|v\|_H : \lambda_j(v) = r_j, 1 \le j \le N\}$. It is well-known [36] that the solution can be written as $v(\mathbf{x}) = \sum_{j=1}^N \beta_j \lambda_j^{\mathbf{y}} \varphi(\mathbf{x}, \mathbf{y})$, where the coefficients β_j are determined by the interpolation conditions $\lambda_j(v) = r_j$, $1 \le j \le N$.

In our case, we consider the PDE $V'(\mathbf{x}) = r(\mathbf{x})$, where $r(\mathbf{x})$ is a given function and $V'(\mathbf{x}) = \nabla V(\mathbf{x}) \cdot \mathbf{f}(\mathbf{x})$ denotes the orbital derivative. We choose N points $\mathbf{x}_1, \dots, \mathbf{x}_N \in \mathbb{R}^n$ of the phase space and define functionals $\lambda_j(v) := (\delta_{\mathbf{x}_j} \circ L)^{\mathbf{x}} v = v'(\mathbf{x}_j) = \nabla v(\mathbf{x}_j) \cdot \mathbf{f}(\mathbf{x}_j)$, where L denotes the linear operator of the orbital derivative $LV(\mathbf{x}) = V'(\mathbf{x})$ and δ is Dirac's delta distribution. The right-hand sides are $r_j = r(\mathbf{x}_j)$ for all $1 \le j \le N$. The approximation is then

$$v(\mathbf{x}) = \sum_{j=1}^N \beta_j (\delta_{\mathbf{x}_j} \circ L)^{\mathbf{y}} \Phi(\mathbf{x} - \mathbf{y}),$$

where Φ is a positive definite Radial Basis Function [36], and the coefficients $\beta_j \in \mathbb{R}$ can be calculated by solving a system of N linear equations. A crucial ingredient is the knowledge on the behaviour of the error function $|V'(\mathbf{x}) - v'(\mathbf{x})|$ in terms of the so-called fill distance $h = \sup_{\mathbf{y} \in K} \inf_{j=1,\dots,N} \|\mathbf{y} - \mathbf{x}_j\|$ which measures how dense the points $\{\mathbf{x}_1, \dots, \mathbf{x}_N\} \subset K$ are in the compact set $K \subset \mathbb{R}^n$, since it gives information when the approximate solution has a negative orbital derivative. Such error estimates were derived, for example in [13,14], see also [26,36].

The advantage of mesh-free collocation over other methods for solving PDEs is that scattered points can be added to improve the approximation, no triangulation of the phase space is necessary, the approximating function is smooth and the method works in any dimension.

In this paper, we use Wendland functions [27] as Radial Basis Functions through $\psi(\|\mathbf{x}\|) := \psi_{l,k}(c\|\mathbf{x}\|)$, where $c > 0$, $k \in \mathbb{N}$ is a smoothness parameter and $l = \lfloor \frac{n}{2} \rfloor + k + 1$. Wendland functions are positive definite Radial Basis Functions with compact support, which are polynomials on their support; the corresponding (Reproducing Kernel) Hilbert Space is norm-equivalent to the Sobolev space $W_2^{k+(n+1)/2}(\mathbb{R}^n)$. They are defined by recursion: for $l \in \mathbb{N}, k \in \mathbb{N}_0$ we define

$$\psi_{l,0}(r) = (1 - r)_+^l$$

$$\psi_{l,k+1}(r) = \int_r^1 t \psi_{l,k}(t) \mathrm{d}t \tag{4}$$

for $r \in \mathbb{R}_0^+$, where $x_+ = x$ for $x \ge 0$ and $x_+ = 0$ for $x < 0$.

As collocation points $X \subset \mathbb{R}^n$ we use a hexagonal grid (5) with $\alpha \in \mathbb{R}^+$ constructed according to

$$\left\{ \alpha \sum_{k=1}^{n} i_k w_k : i_k \in \mathbb{Z} \right\}, \text{ where} \tag{5}$$

$$w_1 = (2e_1, 0, 0, \ldots, 0)$$
$$w_2 = (e_1, 3e_2, 0, \ldots, 0)$$
$$\vdots \quad \vdots$$
$$w_n = (e_1, e_2, e_3, \ldots, (n+1)e_n) \text{ and}$$
$$e_k = \sqrt{\frac{1}{2k(k+1)}}, \quad k \in \mathbb{N}.$$

We set $\psi_0(r) := \psi_{l,k}(cr)$ with positive constant c and define recursively $\psi_i(r) = \frac{1}{r}\frac{d\psi_{i-1}}{dr}(r)$ for $i = 1, 2$ and $r > 0$. The explicit formulas for v and its orbital derivative are

$$v(\mathbf{x}) = \sum_{j=1}^{N} \beta_j \langle \mathbf{x}_j - \mathbf{x}, \mathbf{f}(\mathbf{x}_j) \rangle \psi_1(\|\mathbf{x} - \mathbf{x}_j\|),$$

$$v'(\mathbf{x}) = \sum_{j=1}^{N} \beta_j \Big[-\psi_1(\|\mathbf{x} - \mathbf{x}_j\|)\langle \mathbf{f}(\mathbf{x}), \mathbf{f}(\mathbf{x}_j) \rangle$$
$$+ \psi_2(\|\mathbf{x} - \mathbf{x}_j\|)\langle \mathbf{x} - \mathbf{x}_j, \mathbf{f}(\mathbf{x}) \rangle \cdot \langle \mathbf{x}_j - \mathbf{x}, \mathbf{f}(\mathbf{x}_j) \rangle \Big]$$

where $\langle \cdot, \cdot \rangle$ denotes the standard scalar product in $\mathbb{R}^n$, β is the solution to $A\beta = \mathbf{r}$, $r_j = r(\mathbf{x}_j)$ and A is the $N \times N$ matrix with entries

$$a_{ij} = \psi_2(\|\mathbf{x}_i - \mathbf{x}_j\|)\langle \mathbf{x}_i - \mathbf{x}_j, \mathbf{f}(\mathbf{x}_i) \rangle \langle \mathbf{x}_j - \mathbf{x}_i, \mathbf{f}(\mathbf{x}_j) \rangle$$
$$- \psi_1(\|\mathbf{x}_i - \mathbf{x}_j\|)\langle \mathbf{f}(\mathbf{x}_i), \mathbf{f}(\mathbf{x}_j) \rangle$$

for $i \neq j$ and

$$a_{ii} = -\psi_1(0)\|\mathbf{f}(\mathbf{x}_i)\|^2.$$

More detailed explanations on this construction are given in [13, Chap. 3].

If no collocation point $\mathbf{x}_j$ is an equilibrium for the system, i.e. $\mathbf{f}(\mathbf{x}_j) \neq \mathbf{0}$ for all j, and all collocation points are pairwise distinct, then the matrix A is positive definite and the system of equations $A\beta = \mathbf{r}$ has a unique solution. Note that this holds true independent of whether the underlying discretized PDE has a solution or not, while the error estimates are only available if the PDE has a solution.

3   Algorithm

Starting with scattered collocation points X, we solve the equation $V'(\mathbf{x}) = -1$, where $V'(\mathbf{x}) := \nabla V(\mathbf{x}) \cdot \mathbf{f}(\mathbf{x})$ denotes the orbital derivative, the derivative along solutions of (3). Note that the equation $V'(\mathbf{x}) = -1$ does not have a solution on chain-recurrent sets in general; e.g. along a periodic orbit, the orbital derivative must integrate to 0. However, as mentioned above, we can still compute a (unique) approximation by the method described in Sect. 2.2.

In the next step we check for each collocation point $\mathbf{x}_j$ in X whether the approximation was poor (then $\mathbf{x}_j \in X^0$) or good (then $\mathbf{x}_j \in X^-$). Then we approximate the solution of the new problem $V'(\mathbf{x}) = -1$ for $\mathbf{x} \in X^-$ and $V'(\mathbf{x}) = 0$ for $\mathbf{x} \in X^0$; the set X^0 indicates the (generalized) chain-recurrent set.

To determine whether the approximation was poor or good, we evaluate $v'(\mathbf{x})$ for test points $\mathbf{x}$ around each collocation point $\mathbf{x}_j$ – for a good approximation we expect $v'(\mathbf{x}) \approx -1$. In view of our goal to compute a complete Lyapunov function, we classify collocation points as poor if the orbital derivative near them is larger than 0 or a chosen critical value, i.e. $v'(\mathbf{x}) > \gamma$, for certain points $\mathbf{x}$ near the collocation point $\mathbf{x}_j$. As points to check near a collocation point $\mathbf{x}_j$ we choose points on two circumferences around each collocation point $\mathbf{x}_j$.

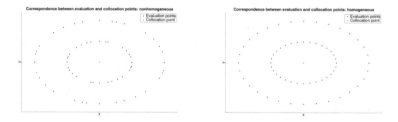

Fig. 1. Improvement over the evaluation points; both have 32 points on each circumference. Left: non-homogeneous distribution. Right: homogeneous distribution. The middle point represents the collocation point which is the centre of both circumferences.

Previously, in [31], we used points distributed on two concentric circumferences whose centre is the collocation point. However, the distribution of such points was not homogeneous along the circumferences, see Fig. 1. In this paper, however, we use points homogeneously distributed over two concentric circumferences whose centre is the collocation point. Originally in [31], the grid $Y_{\mathbf{x}_j}$ consisted of the points (6), while in this paper we use the points $Y_{\mathbf{x}_j}$ in (7).

$$
\begin{aligned}
&\mathbf{x}_j \pm r\alpha(\cos(\theta), -\sin(\theta)) \\
&\mathbf{x}_j \pm r\alpha(\sin(\theta), \cos(\theta)) \\
&\mathbf{x}_j \pm \tfrac{r}{2}\alpha(\cos(\theta), \sin(\theta)) \\
&\mathbf{x}_j \pm \tfrac{r}{2}\alpha(\cos(\theta), -\sin(\theta))
\end{aligned}
\tag{6}
$$

$$
\begin{aligned}
&\mathbf{x}_j + r\alpha\left(\cos\tfrac{2\pi k}{M}, \sin\tfrac{2\pi k}{M}\right) \\
&\mathbf{x}_j + \tfrac{r}{2}\alpha\left(\cos\tfrac{2\pi k}{M}, \sin\tfrac{2\pi k}{M}\right)
\end{aligned}
\tag{7}
$$

For (6), we have used $\theta = \{0.0°, 11.25°, 22.5°, 45°, 56.25°, 67.5°, 75°, 105°\}$ and for (7) we use $M = 32$. In both cases, r is a scaling parameter and 32 is the total amount of points we distribute in each circumference, resulting in 64 points per collocation points in total. Figure 1 shows the distributions of both evaluation grids.

Figure 1 shows an improvement on the evaluation capabilities of our method. The fact that the left-hand side figure in Fig. 1 is not homogeneously distributed is seen with 4 blank spots in each of the two circumferences. Furthermore, it also shows regions in which we can find points close to each other, while in other regions the distance between points is large. Considering that the Lyapunov function is first computed on the collocation points and then evaluated over the evaluation grid, the fact that the evaluation is non-homogeneous as shown in Fig. 1, weakens the evaluation when compared with the homogeneous case.

In [31], the aim was to obtain a description of the chain-recurrent set X^0 and the gradient-like flow X^-. That description allowed to set the orbital derivative to 0 for points in X^0 and to keep the orbital derivative to be -1 for points in X^-. This results in a non-continuous right-hand side.

In this paper we use not only the information whether points lie in X^0 or X^-, but also the average value of $v'(\mathbf{y})$ around each collocation to determine the PDE in the subsequent step. In particular, in the subsequent iteration we set the orbital derivative at each collocation point to be the average value of the Lyapunov function derivative for all evaluation points around the respective collocation points. The evaluation grid, however, is different from [33,34] in which a directional grid was used. The advantage of that particular grid is that it could be used for 3-dimensional systems without increasing its size exponentially. In our case, however, we use the averaging over the circumference values, which can also be generalized to higher dimensions.

The evaluation grid in this paper uses the contributions in all directions around each collocation point instead of just in one direction and will thus produce more accurate results. Setting $r = 0.5$ ensures that the evaluation grids for adjacent collocation points do not overlap. In our computations, as we did in [31], we use 64 evaluation points around each collocation point.

Next, we summarize our algorithm.

1. Create the collocation points X. Compute the approximate solution v_0 of $V'(\mathbf{x}) = -1$, set $i = 0$
2. For each collocation point $\mathbf{x}_j$, compute $v_i'(\mathbf{x})$: if $v_i'(\mathbf{x}) > \gamma$ for a point $\mathbf{x} \in Y_{\mathbf{x}_j}$, then $\mathbf{x}_j \in X^0$, otherwise $\mathbf{x}_j \in X^-$, where $\gamma \leq 0$ is a chosen critical value
3. Compute the approximate solution v_{i+1} of

$$V'(\mathbf{x}_j) = r(\mathbf{x}_j) = \frac{1}{2M} \left(\sum_{\mathbf{y} \in Y_{\mathbf{x}_j}} v_i'(\mathbf{y}) \right) -, \text{ where } z_- = z \text{ for } z \leq 0 \text{ and } z_- = 0$$

   otherwise
4. Set $i \to i+1$ and repeat steps 2. and 3. until no more points are added to X^0 or until a certain predefined number of iterations is reached

Note that the mesh-free collocation method only requires us to know the right-hand side $r(\mathbf{x})$ of the PDE at the collocation points, hence, it is sufficient to define r at all collocation points $\mathbf{x}_j$.

4   Examples

In the following we apply the method to four examples in two dimensions and then analyze their behaviour. Note that in all examples we use the notation $\mathbf{x} = (x, y)$. We have done 10 iterations of the method for each example. The first three systems (8), (9) and (10) have been previously studied in [31], and we compare our new method with the previous results. The final system in Sect. 4.4 models a bistable network of two genes.

For the first two examples we have used the original equation $\dot{\mathbf{x}} = \mathbf{f}(\mathbf{x})$ and have achieved very good results, while for the last two examples, we have employed the slightly slower "almost" normalized method, introduced in [32], using (1). For all examples, we have used the new evaluation grid (7) with $r = 0.5$ and $M = 32$.

4.1   Two Circular Periodic Orbits

We consider system (3) with right-hand side

$$\mathbf{f}(x, y) = \begin{pmatrix} -x(x^2 + y^2 - 1/4)(x^2 + y^2 - 1) - y \\ -y(x^2 + y^2 - 1/4)(x^2 + y^2 - 1) + x \end{pmatrix}. \tag{8}$$

This system has an asymptotically stable equilibrium at the origin, $\Omega_0 = \{(0,0)\}$ since the Jacobian at the origin is $\begin{pmatrix} -\dfrac{1}{4} & -1 \\ 1 & -\dfrac{1}{4} \end{pmatrix}$ with eigenvalues $\lambda_{1,2} = -0.25 \pm i$.

Moreover, the system has two periodic circular orbits: an asymptotically stable periodic orbit at $\Omega_1 = \{(x, y) \in \mathbb{R}^2 \mid x^2 + y^2 = 1\}$ and a repelling periodic orbit at $\Omega_2 = \{(x, y) \in \mathbb{R}^2 \mid x^2 + y^2 = 1/4\}$.

To compute the Lyapunov function with our method we used Wendland function $\psi_{5,3}$ with the parameter $c = 1$. The collocation points were set in a region $(-1.5, 1.5) \times (-1.5, 1.5) \subset \mathbb{R}^2$ and we used a hexagonal grid (5) with $\alpha = 0.018$. This setting gives a total amount of 36,668 collocation points and 2,346,752 evaluation points. We computed this example with the original system $\dot{\mathbf{x}} = \mathbf{f}(\mathbf{x})$.

Figure 2 shows the approximation v_0 of $V'(\mathbf{x}) = -1$ as well as its orbital derivative v_0'. The function v_0 (Fig. 2, left) clearly displays the stable periodic orbit and equilibrium as local minima of v_0, while the unstable periodic orbit is a local maximum. The orbital derivative v_0' (Fig. 2, right) is mostly -1 apart from the two periodic orbits and the equilibrium, where it is close to 0. The orbital derivative is clearly discontinuous.

Several improvements have been made to generate a continuous derivative [32–34] in subsequent iterations. In this paper, we use the method described above, solving $V'(\mathbf{x}) = r(\mathbf{x})$, where $r(\mathbf{x}_j)$ is given by the average value of the orbital derivative at the evaluation grid around $\mathbf{x}_j$, using the critical value $\gamma = -0.5$. Figure 3 shows v_{10} and its orbital derivative after 10 iterations.

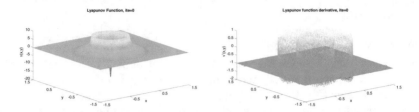

Fig. 2. Complete Lyapunov function for (8), iteration 0. The left-hand figure shows the approximated complete Lyapunov function $v_0(x, y)$. The right-hand figure shows its orbital derivative $v_0'(x, y)$. The Lyapunov function's minimum corresponds to the asymptotically stable equilibrium. The two periodic orbits with radius 1 and 1/2 are asymptotically stable (Ω_1) and unstable (Ω_2) and are local minima and maxima of v_0 respectively. The orbital derivative v_0' fails to be negative on Ω_1, Ω_2, and at the equilibrium, moreover, v_0' is not continuous.

Figure 3 shows the tenth iteration of the new method. The orbital derivative v_{10}' is now continuous, which can be seen more clearly in Figs. 4 and 5, comparing the iterations 0 and 10 on a part of the phase space. Figure 4 shows a section of the Lyapunov function derivative v_{10}' over a part of the chain-recurrent set, namely Ω_2. Figure 5 shows a section of the previous Fig. 4, which gives a clear picture of the improved continuity of the orbital derivative.

Finally, we are interested in the behaviour of the chain-recurrent set. Figure 6 shows the points **y** of the evaluation grid, where $v_i'(\mathbf{y}) > \gamma$ with $\gamma = -0.5$, which approximate the chain-recurrent set. We find the three expected connected components of the chain-recurrent set already in iteration 0; and the sets do not change much when considering iteration 10.

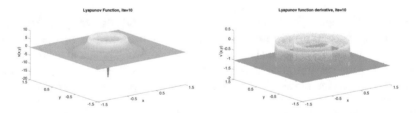

Fig. 3. Complete Lyapunov function for (8), iteration 10. The left-hand figure shows the approximated complete Lyapunov function $v_{10}(x, y)$. The right-hand figure shows its orbital derivative $v_{10}'(x, y)$; now continuous. The Lyapunov function's minimum corresponds to the asymptotically stable equilibrium. The two periodic orbits with radius 1 and 1/2 are asymptotically stable (Ω_1) and unstable (Ω_2) and are local minima and maxima of v_{10} respectively. The orbital derivative v_{10}' fails to be negative on Ω_1, Ω_2, and at the equilibrium.

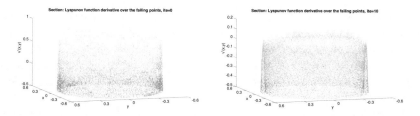

Fig. 4. Complete Lyapunov function derivative v_0' (left) and v_{10}' for (8) over Ω_2 for iterations 0 and 10, $\gamma = -0.5$.

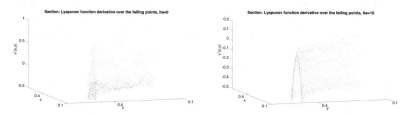

Fig. 5. Complete Lyapunov function derivative v_0' (left) and v_{10}' for (8) over part of Ω_2 for iterations 0 and 10, $\gamma = -0.5$.

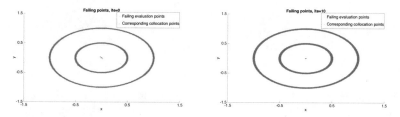

Fig. 6. The figures show the evaluation points $\mathbf{y}$ around each collocation point, where $v'(\mathbf{y}) > \gamma = -0.5$; left: iteration 0 and right: iteration 10. The three chain-recurrent sets are clearly visible in both iterations.

4.2   Van-der-Pol Oscillator System

We consider system (3) with right-hand side

$$\mathbf{f}(x, y) = \begin{pmatrix} y \\ (1 - x^2)y - x \end{pmatrix}. \tag{9}$$

System (9) is the two-dimensional form of the Van-der-Pol oscillator. This describes the behaviour of a non-conservative oscillator reacting to a non-linear damping. The origin is an unstable focus, which can be seen from its Jacobian at the origin with eigenvalues $\lambda_{1,2} = 0.5 \pm 0.866025i$.

In this case, we have used the Wendland function $\psi_{4,2}$ with parameter $c = 1$. We set our collocation points in the region $(-4, 4) \times (-4, 4) \subset \mathbb{R}^2$, using a hexagonal grid (5) with $\alpha = 0.046$. We have a total amount of $36,668$ collocation points and $2,346,751$ evaluation points. We have used the original system, namely $\mathbf{f}(\mathbf{x})$.

Figure 7 shows the approximation v_0 of $V'(\mathbf{x}) = -1$ as well as its orbital derivative v_0'. The function v_0 (Fig. 7, left) clearly displays the stable periodic orbit as minimum and the unstable equilibrium at the origin as maximum of v_0. The orbital derivative v_0' (Fig. 7, right) is mostly -1 apart from the periodic orbits and the equilibrium, where it is close to 0. The orbital derivative is clearly discontinuous.

Figure 8 shows the tenth iteration of the new method. The orbital derivative v_{10}' is now continuous, which can be seen more clearly in Figs. 9 and 10, comparing the iterations 0 and 10 on a part of the phase space. Figure 9 shows a section of the Lyapunov function derivative v_{10}' over a part of the chain-recurrent set, namely the periodic orbit. Figure 10 shows a section of the previous Fig. 9, which gives a clear picture of the improved continuity of the orbital derivative.

Figure 11 shows the points $\mathbf{y}$ of the evaluation grid, where $v_i'(\mathbf{y}) > \gamma$ with $\gamma = -0.5$, which approximate the chain-recurrent set. We find the two expected connected components of the chain-recurrent set already in iteration 0; and the sets do not change much when considering iteration 10.

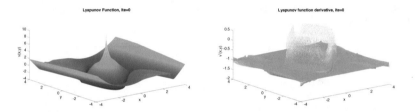

Fig. 7. Complete Lyapunov function for (9), iteration 0. The left-hand figure shows the approximated complete Lyapunov function $v_0(x, y)$. The right-hand figure shows its orbital derivative $v_0'(x, y)$. The Lyapunov function's maximum corresponds to the unstable equilibrium at the origin. The stable periodic orbit is a local minimum of v_0. The orbital derivative v_0' fails to be negative on the periodic orbit and at the equilibrium, moreover, v_0' is not continuous.

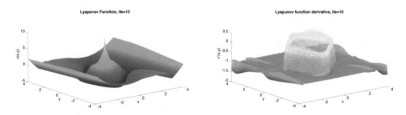

Fig. 8. Complete Lyapunov function for (9), iteration 10. The left-hand figure shows the approximated complete Lyapunov function $v_{10}(x, y)$. The right-hand figure shows its orbital derivative $v_{10}'(x, y)$; now continuous. The Lyapunov function's maximum corresponds to the unstable equilibrium at the origin. The stable periodic orbit is a local minimum of v_0. The orbital derivative v_0' fails to be negative on the periodic orbit and at the equilibrium.

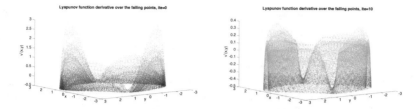

Fig. 9. Complete Lyapunov function derivative v_0' (left) and v_{10}' for (9) over the periodic orbit for iterations 0 and 10, $\gamma = -0.5$.

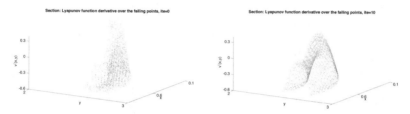

Fig. 10. Complete Lyapunov function derivative v_0' (left) and v_{10}' for (9) over part of the periodic orbit for iterations 0 and 10, $\gamma = -0.5$.

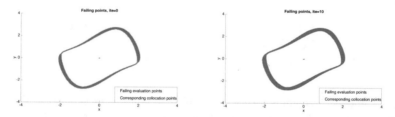

Fig. 11. The figures show the evaluation points $\mathbf{y}$ around each collocation point, where $v'(\mathbf{y}) > \gamma = -0.5$; left: iteration 0 and right: iteration 10. The chain-recurrent sets are clearly visible in both iterations.

Note that there is a considerable improvement since [31]. Let us notice that we had improved the results of [31] in [32] by using the almost-normalized method (1), at a higher computational cost. However, in Fig. 11 we use the same (original) method as in [31]. The difference is in the amount of collocation points and the new distribution of the evaluation points. In particular, even after ten iterations, there are no points apart from the periodic orbit and the equilibrium marked as failing, whereas in previous methods, more and more "noise" was added at other parts of the phase space.

4.3   Homoclinic Orbit

As in [31], we also consider here the following example with right-hand side

$$\mathbf{f}(x,y) = \begin{pmatrix} x(1 - x^2 - y^2) - y((x - 1)^2 + (x^2 + y^2 - 1)^2) \\ y(1 - x^2 - y^2) + x((x - 1)^2 + (x^2 + y^2 - 1)^2) \end{pmatrix}. \tag{10}$$

The origin is an unstable focus, which can be seen from the eigenvalues of its Jacobian at the origin, which are $\lambda_{1,2} = 1 \pm 2i$. Furthermore, the system has an asymptotically stable homoclinic orbit at a circle centred at the origin and with radius 1, connecting the equilibrium $(1, 0)$ with itself.

We have used the Wendland function $\psi_{4,2}$ with parameter $c = 1$. We set our collocation points in the region $(-1.5, 1.5) \times (-1.5, 1.5) \subset \mathbb{R}^2$ with a hexagonal grid (5) with $\alpha = 0.0125$. In this example, we have used the almost-normalized method, i.e. we have replace $\mathbf{f}$ by $\hat{\mathbf{f}}$ according to (1) with $\delta^2 = 10^{-8}$, and we have used $\gamma = -0.75$.

Figure 12 shows the approximation v_0 of $V'(\mathbf{x}) = -1$ as well as its orbital derivative v_0'. The function v_0 (Fig. 12, left) clearly displays the stable homoclinic orbit as minimum and the unstable equilibrium at the origin as maximum of v_0. The orbital derivative v_0' (Fig. 12, right) is mostly -1 apart from the homoclinic orbits and the origin. The orbital derivative is clearly discontinuous. In contrast to previous examples, the orbital derivative has large values around the equilibrium at $(1, 0)$, corresponding to the homoclinic orbit.

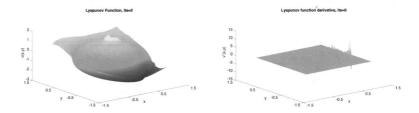

Fig. 12. Complete Lyapunov function for (10), iteration 0. The left-hand figure shows the approximated complete Lyapunov function $v_0(x, y)$. The right-hand figure shows its orbital derivative $v_0'(x, y)$. The Lyapunov function's maximum corresponds to the unstable equilibrium at the origin. The stable homoclinic orbit is a local minimum of v_0. The orbital derivative v_0' fails to be negative on the periodic orbit and at the origin, moreover, v_0' is not continuous.

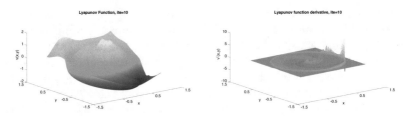

Fig. 13. Complete Lyapunov function for (10), iteration 10. The left-hand figure shows the approximated complete Lyapunov function $v_{10}(x, y)$. The right-hand figure shows its orbital derivative $v_{10}'(x, y)$; now continuous. The Lyapunov function's maximum corresponds to the unstable equilibrium at the origin. The stable homoclinic orbit is a local minimum of v_0. The orbital derivative v_0' fails to be negative on the periodic orbit and at the origin.

Figure 13 shows the tenth iteration of the new method. The orbital derivative v'_{10} is now continuous, which can be seen more clearly in Figs. 14 and 15, comparing the iterations 0 and 10 on a part of the phase space. Figures 14 shows a section of the Lyapunov function derivative v'_{10} over the chain-recurrent set. Figure 15 shows a section of the previous Fig. 14.

Figure 16 shows the points $\mathbf{y}$ of the evaluation grid, where $v'_i(\mathbf{y}) > \gamma$ with $\gamma = -0.75$, which approximate the chain-recurrent set. We find the two expected connected components of the chain-recurrent set already in iteration 0; and the sets do not change much when considering iteration 10.

We notice an important improvement when compared with [32] in which we notice that the amount of areas incorrectly marked as failing points is almost negligible.

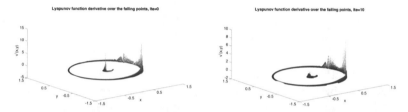

Fig. 14. Complete Lyapunov function derivative v'_0 (left) and v'_{10} for (10) over the homoclinic orbit and the origin for iterations 0 and 10, $\gamma = -0.75$.

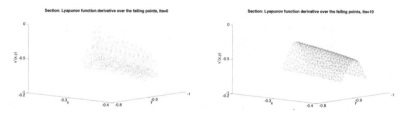

Fig. 15. Complete Lyapunov function derivative v'_0 (left) and v'_{10} for (10) over a part of the homoclinic orbit for iterations 0 and 10, $\gamma = -0.75$.

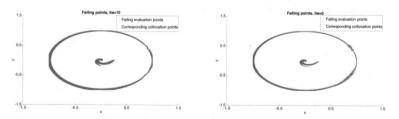

Fig. 16. The figures show the evaluation points $\mathbf{y}$ around each collocation point, where $v'(\mathbf{y}) > \gamma = -0.75$; left: iteration 0 and right: iteration 10. The chain-recurrent sets are clearly visible in both iterations.

4.4   Application to Biology

As a final example, we consider a system from [1], Eqs. (6) and (7), which models a network of two genes that suppress each other to form a double-negative feedback loop. The general model is given by

$$\mathbf{f}(x,y) = \begin{pmatrix} B_X + \frac{\text{fold}_{YX} K_{DYX}^{n_H}}{K_{DYX} + y^{n_H}} - \deg_X x \\ B_Y + \frac{\text{fold}_{XY} K_{DXY}^{n_H}}{K_{DXY} + x^{n_H}} - \deg_X y \end{pmatrix}. \tag{11}$$

where x and y represent the concentrations of the two gene products; for the meaning of the parameters, see [1]. We use the parameter values from [1] to obtain the following system

$$\mathbf{f}(x,y) = \begin{pmatrix} \frac{2}{10} + 2\left(\frac{\left(\frac{7}{10}\right)^4}{\left(\frac{7}{10}\right)^4 + y^4} \right) - x \\ \frac{2}{10} + 2\left(\frac{\left(\frac{5}{10}\right)^4}{\left(\frac{5}{10}\right)^4 + x^4} \right) - y \end{pmatrix}. \tag{12}$$

In [1], the authors compute a complete Lyapunov function, which they call quasi-potential, by numerical integration along solution trajectories with many different initial conditions. The resulting energy landscape, the plot of $v(x,y)$, is used to analyze the development and stability of cellular states.

Although this is a biological system in which only non-negative values of x and y are of biological interest, we set our collocation points in the region $(-1,6) \times (-1,6) \subset \mathbb{R}^2$ with a hexagonal grid (5) with $\alpha = 0.041$. We use the "almost" normalized method (1) with $\delta^2 = 10^{-8}$ and $\gamma = -0.5$. Our settings gave us a total amount of $35,728$ collocation points and $2,286,592$ evaluation points.

This system has three equilibria

$$\mathbf{z}_1 = (0.223344, 2.12342), \ \mathbf{z}_2 = (0.542514, 1.03822), \ \mathbf{z}_3 = (2.18526, 0.205466).$$

$\mathbf{z}_1$ and $\mathbf{z}_3$ are stable nodes; the eigenvalues of the corresponding Jacobians are $\lambda_1 = -1.23942$ and $\lambda_2 = -0.760578$, and $\lambda_1 = -1.05331$, $\lambda_2 = -0.946693$, respectively $\mathbf{z}_2$ is a saddle with corresponding eigenvalues of the Jacobian $\lambda_1 = -2.98147$, $\lambda_2 = 0.981468$.

Figure 17 shows the approximation v_0 of $V'(\mathbf{x}) = -1$ as well as its orbital derivative v_0'. The function v_0 (Fig. 17, left) clearly displays the two stable equilibria as minima and the unstable equilibrium as saddle of v_0. The orbital derivative v_0' (Fig. 17, right) is mostly -1 apart from the three equilibria; this can be seen clearer in Fig. 20. Indeed, if we take a close up to the area where the critical values are, we see that the points after the tenth iteration are clearer distinguishable, see Fig. 20. Figure 18 shows the tenth iteration of the new method, which gives a similar picture.

The chain-recurrent set, given by the failing points, is shown in Fig. 19. We can identify the three critical points $\mathbf{z}_1$, $\mathbf{z}_2$ and $\mathbf{z}_3$ easily with our method.

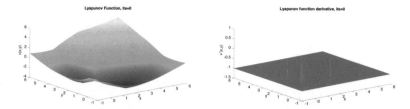

Fig. 17. Complete Lyapunov function for (12), iteration 0. The left-hand figure shows the approximated complete Lyapunov function $v_0(x, y)$. The right-hand figure shows its orbital derivative $v_0'(x, y)$. The complete Lyapunov function's saddle point corresponds to the unstable equilibrium while the two asymptotically stable equilibria are local minima of v_0. The orbital derivative v_0' fails to be negative at the three equilibria, which can be better seen in Fig. 20.

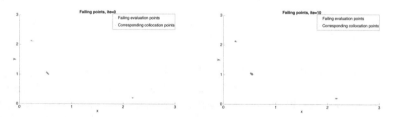

Fig. 18. Complete Lyapunov function for (12), iteration 10. The left-hand figure shows the approximated complete Lyapunov function $v_{10}(x, y)$. The right-hand figure shows its orbital derivative $v_{10}'(x, y)$; now continuous. The Lyapunov function's saddle point corresponds to the unstable equilibrium while the two asymptotically stable equilibria are local minima of v_0. The orbital derivative v_0' fails to be negative at the three equilibria.

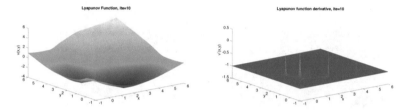

Fig. 19. The figures show the evaluation points **y** around each collocation point, where $v'(\mathbf{y}) > \gamma = -0.5$; left: iteration 0 and right: iteration 10. The three equilibria are clearly visible in both iterations.

As in [1], we analyze the contour plots of v_{10} for iteration 10, Fig. 21. They are very similar to the results obtained in [1].

We see that our numerical computation of a complete Lyapunov function of system (12) is capable of reproducing the behaviour of the biological system. Furthermore, we can see that our method reproduces the results from [1] but

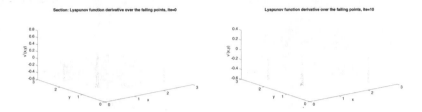

Fig. 20. Complete Lyapunov function derivative for iteration 0 (left) and iteration 10 (right) in the part of the phase space where it differs from −1 and in which the equilibria can be seen.

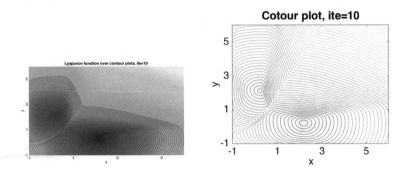

Fig. 21. Contour plots of the complete Lyapunov function v_{10} for iteration 10 and system (12).

without the need of computing numerical integration for particular solutions which shows that our method is a powerful to solve problems in other disciplines.

5   Conclusions and Outlook

In this paper we have presented several improvements to our previous methodology introduced in [31]. First of all, we used an evaluation grid which homogeneously distributes its points around the collocation points and thus avoids favouring specific regions. As seen in Fig. 1, in our previous method [31], an inhomogeneous distribution would favour the weighting at the regions in which the points are closer and hinder the regions in which the points are more distant. With a homogeneous grid, all points have the same weighting. Secondly, we also enhanced the results for (9) previously presented in [31] without using the "almost"-normalized method presented in [32]. Moreover, we are able to avoid overestimating the chain-recurrent set due to "noise".

Using the "almost"-normalized method presented in [32] with the averaging idea presented in [34] we even managed to enhance the results previously seen in [31,32] for the system (10). As can be seen in Fig. 16, after 10 iterations the amount of "noise" that harms our definition of the chain-recurrent set, is reduced. We can then conclude that averaging all around the collocation point, instead of in just one direction, has a better impact on determining the chain-recurrent sets. The reason is that close to the failing points, there are small areas that fail as well. Accordingly, if we take only one direction the failing points close a particular failing collocation point and away from that direction will not be noticed. However with a circumference they will be noticed and weighted. The directional evaluation grid has, however, another advantages. It favours the direction of the trajectories and it avoids exponential growth of the evaluation points.

Summarizing, in this paper we were capable of obtaining a good estimate for the chain-recurrent set for different systems using both the "almost"-normalized method presented in [32] as well as the common method. This shows that if the system does not present high variations of its speed the common method works as well.

In all our computations we have obtained very clear chain-recurrent sets with small "noise" or none of it. In short, we have presented an enhanced method to compute a complete Lyapunov function fulfilling the conditions for its derivative which is smooth and continuous. That allows to better determine the qualitative behaviour of a given ODE, using both the values of the complete Lyapunov function and its orbital derivative. Furthermore, we provide complete Lyapunov functions V that are constant along solutions in the chain-recurrent set, i.e. the orbital derivative is zero ($V'(\mathbf{x}) = 0$), and are strictly decreasing along solutions in the gradient-like set, i.e. $V'(\mathbf{x}) < 0$. We have shown that the method is a useful tool for applications and can determine the energy landscape of a quasi-potential.

Acknowledgements. First author in this paper is supported by the Icelandic Research Fund (Rannís) grant number 163074-052, Complete Lyapunov functions: Efficient numerical computation. Special thanks to Dr. Jean-Claude Berthet for all his good comments and advices on C++.

References

1. Bhattacharya, S., Zhang, Q., Andersen, M.E.: A deterministic map of Waddington's epigenetic landscape for cell fate specification. BMC Syst. Biol. **5**, 85 (2011)
2. Doban A.: Stability domains computation and stabilization of nonlinear systems: implications for biological systems. Ph.D. thesis. Eindhoven University of Technology (2016)
3. Doban A., Lazar M.: Computation of Lyapunov functions for nonlinear differential equations via a Yoshizawa-type construction. In: 10th IFAC Symposium on Nonlinear Control Systems NOLCOS 2016, Monterey, California, USA, 23–25 August 2016, IFAC-PapersOnLine, 2016, vol. 49, no. 18, pp. 29–34 (2016). ISSN 2405-8963
4. Anderson, J., Papachristodoulou, A.: Advances in computational Lyapunov Analysis using sum-of-squares programming. Discrete Contin. Dyn. Syst. Ser. B **20**, 2361–2381 (2015)
5. Auslander, J.: Generalized recurrence in dynamical systems. Contr. Diff. Eq. **3**, 65–74 (1964)

6. Björnsson J., Giesl P., Hafstein S.: Algorithmic verification of approximations to complete Lyapunov functions. In: Proceedings of the 21st International Symposium on Mathematical Theory of Networks and Systems (no. 0180), pp. 1181–1188 (2014)

7. Björnsson J., Giesl P., Hafstein S., Kellett C., Li H.: Computation of continuous and piecewise affine Lyapunov functions by numerical approximations of the Massera construction. In: Proceedings of the CDC 53rd IEEE Conference on Decision and Control (2014)

8. Björnsson, J., Giesl, P., Hafstein, S., Kellett, C., Li, H.: Computation of Lyapunov functions for systems with multiple attractors. Discrete Contin. Dyn. Syst. **9**, 4019–4039 (2015)

9. Ban, H., Kalies, W.: A computational approach to Conley's decomposition theorem. J. Comput. Nonlinear Dyn. **1**, 312–319 (2006)

10. Isolated Invariant Sets and the Morse Index, C. Conley, American Mathematical Society, CBMS Regional Conference Series no. 3 (1978)

11. Dellnitz, M., Froyland, G., Junge, O.: The algorithms behind GAIO - set oriented numerical methods for dynamical systems. In: Ergodic Theory Analysis and Efficient Simulation of Dynamical Systems, pp. 145–174, 805–807. Springer, Berlin (2001)

12. Dellnitz, M., Junge, O.: Set oriented numerical methods for dynamical systems. In: Handbook of Dynamical Systems, North-Holland Amsterdam, vol. 2, pp. 221–264 (2002)

13. Giesl, P.: Construction of Global Lyapunov Functions Using Radial Basis Functions. Lecture Notes in Math., vol. 1904. Springer (2007)

14. Giesl, P., Wendland, H.: Meshless collocation: error estimates with application to Dynamical Systems. SIAM J. Numer. Anal. **45**, 1723–1741 (2007)

15. Giesl, P., Hafstein, S.: Review of computational methods for Lyapunov functions. Discrete Contin. Dyn. Syst. Ser. B **20**, 2291–2331 (2015)

16. Goullet, A., Harker, S., Mischaikow, K., Kalies, W., Kasti, D.: Efficient computation of Lyapunov functions for morse decompositions. Discrete Contin. Dyn. Syst. Ser. B **20**, 2419–2451 (2015)

17. Hafstein, S.: An algorithm for constructing Lyapunov functions. Electron. J. Diff. Eqns. (2007)

18. Hsu, C.S.: Cell-to-Cell Mapping. Applied Mathematical Sciences, vol. 64, Springer, New York (1987)

19. Hurley, M.: Noncompact chain recurrence and attraction. Proc. Am. Math. Soc. **115**, 1139–1148 (1992)

20. Hurley, M.: Lyapunov functions and attractors in arbitrary metric spaces. Proc. Am. Math. Soc. **126**, 245–256 (1998)

21. Johansson, M.: Piecewise linear control systems, Ph.D. thesis: Lund University Sweden (1999)

22. Johansen, T.: Computation of Lyapunov functions for smooth nonlinear systems using convex optimization. Automatica **36**, 1617–1626 (2000)

23. Kamyar, R., Peet, M.: Polynomial optimization with applications to stability analysis and control - an alternative to sum of squares. Discrete Contin. Dyn. Syst. Ser. B **20**, 2383–2417 (2015)

24. Kalies, W., Mischaikow, K., VanderVorst, R.: An algorithmic approach to chain recurrence. Found. Comput. Math. **5**, 409–449 (2005)

25. Marinósson, S.: Lyapunov function construction for ordinary differential equations with linear programming. Dyn. Syst. Int. J. **17**, 137–150 (2002)

26. Narcowich, F.J., Ward, J.D., Wendland, H.: Sobolev bounds on functions with scattered zeros with applications to radial basis function surface fitting. Math. Comp. **74**, 743–763 (2005)
27. Wendland, H.: Error estimates for interpolation by compactly supported Radial Basis Functions of minimal degree. J. Approx. Theory **93**, 258–272 (1998)
28. Hurley, M.: Chain recurrence semiflows and gradient. J. Dyn. Diff. Eq. **7**, 437–456 (1995)
29. Lyapunov A. M.: The general problem of the stability of motion. Int. J. Control **55**, 521–790 (1992). Translated by A. T. Fuller from Édouard Davaux's French translation (1907) of the 1892 Russian original
30. Osipenko, G.: Dynamical Systems Graphs and Algorithms. Lecture Notes in Mathematics, vol. 1889. Springer, Berlin (2007)
31. Argáez, C., Giesl, P., Hafstein, S.: Analysing dynamical systems towards computing complete Lyapunov functions. In: Proceedings of the 7th International Conference on Simulation and Modeling Methodologies Technologies and Applications, SIMULTECH 2017, Madrid, Spain (2017)
32. Argáez, C., Giesl, P., Hafstein, S.: Computational approach for complete Lyapunov functions. In: Proceedings in Mathematics and Statistics. Springer (2018, accepted for publication)
33. Argáez, C., Giesl, P., Hafstein, S.: Iterative construction of complete Lyapunov functions. Submitted
34. Argáez, C., Giesl, P., Hafstein, S.: Computation of complete Lyapunov functions for three-dimensional systems. Submitted
35. Conley, C.: The gradient structure of a flow I. Ergodic Theory Dyn. Syst. **8**, 11–26 (1988)
36. Wendland, H.: Scattered Data Approximation. Cambridge Monographs on Applied and Computational Mathematics, vol. 17. Cambridge University Press, Cambridge (2005)
37. Krauskopf, B., Osinga, H., Doedel, E. J., Henderson, M., Guckenheimer, J., Vladimirsky, A.,Dellnitz, M., Junge, O.: A survey of methods for computing (un)stable manifolds of vector fields. Int. J. Bifur. Chaos Appl. Sci. Eng. **15**, 763–791 (2015)

Path Bundling in Modular Bipartite Networks

Victor Parque[⊠], Satoshi Miura, and Tomoyuki Miyashita

Department of Modern Mechanical Engineering,
Waseda University, 3-4-1 Okubo, Shinjuku-ku, Tokyo 169-8555, Japan
parque@aoni.waseda.jp, miura-s@akane.waseda.jp, tomo.miyashita@waseda.jp

Abstract. Path bundling consists in compounding multiple routes in a polygonal map to minimize connectivity in a network structure. Being closely related to the Steiner Tree Problem, yet with a different scope, path bundling aims at computing minimal trees while preserving network connectivity among origin-destination pairs to allow the joint transport of information, goods, and people. In this paper, we propose a method to tackle the path bundling problem in modular bipartite networks by using a two-layer optimization with a convex representation. Exhaustive computational experiments in diverse polygonal domains considering convex and non-convex geometry show the feasibility and the efficiency of the proposed approach, outperforming the state of the art in generating comparatively shorter trees, and improved scalability as a function of edges in bipartite networks.

Keywords: Path bundling · Optimization
Modular bipartite networks

1 Introduction

Being ubiquitous in scenarios involving transportation, communication and routing, the path bundling problem consists in compounding routes which are structurally different by computing anchoring points at intermediate joints. The goal in this regard is to minimize a distance metric, in which the anchoring points serve as coordinating foci to enable the efficient transport and communication.

Naturally, the above problem is significant in scenarios in which the means for transport and communication are scarce, and the environment is hard to navigate. Thus, it is natural to compound paths in a network in order to achieve efficient transport and communication. As such, the result of path bundling is a tree which ensures a minimal length configuration. For instance, consider the configuration of a ZigBee network in which the limited number of *coordinator nodes* is a constraint to enable many-to-many communication links [1], or consider the problem of designing the optimal network for a wire harness on mechanical systems (e.g. cars or spaceships) [2], or consider the problem of coordination among robotic systems to navigate an area [3,4]. These problems are the key motivation in this paper to design algorithms to compute minimal-length trees which optimize the communication/transport in a network structure.

© Springer Nature Switzerland AG 2019
M. S. Obaidat et al. (Eds.): SIMULTECH 2017, AISC 873, pp. 222–238, 2019.
https://doi.org/10.1007/978-3-030-01470-4_12

1.1   Related Works

The research on path bundling has its origins in the shortest-path problem [5–10], in which the main goal is to find the most optimal link between single origin-destination pairs in a polygonal domain. Indeed, well-known algorithms are Dijsktra [5] and A* [6], along with their extensions. Here, path planning in triangulated space is highly accurate and efficient. As such, for a map with polygonal obstacles having n vertices, the Delaunay Triangulation ca be computed to render a connected graph with $O(n)$ nodes, each of which represents the triangles conforming the free-space. Then, by using the Delaunay-based connected graph, path planning can be efficiently performed by graph search methods on the adjacency matrix of the Delaunay-based connected graph. In particular, If one uses the Dijkstra-based search [5,8], time complexity ca be achieved by $O(n.log(n))$, yet it is possible to use A* search [6,9] along with a funneling algorithm [7,10] to obtain a quasi-linear time path planning.

In a general scenario, path bundling has been studied in wireless networks [11–16], and network visualization [17–22]. In more particular domains, the closest developments to path bundling is related to edge bundling problem in network visualization field. In this scheme, most of the conventional works have focused on the geometry-based clustering of edges [17], the force-based edge bundling in which edges enable attraction-based bundling [17,18], the clustering and attraction to the skeleton of adjacent edges [19], and the kd-tree based optimization of the centroid points of close edges [20]. However, due to having a different scope, the above algorithms have rendered compounded networks which are aesthetically pleasing or topologically compact, but not necessarily optimal in the sense of minimizing a connectivity metric.

1.2   Contributions

In [23], we proposed an approach for path bundling of bipartite networks, in which we confirmed the feasibility and efficiency by 7,500 computational experiments in relevant benchmark scenarios. In this paper, we extend our previous approach [23] to handle modules and enable the efficient, the effective and the scalable path bundling when considering modular bipartite networks. The basic idea of our approach consists in compounding edges in modular bipartite networks by a two-layer optimization, in which we first compute the optimal roots of path bundles for each module of the bipartite network, and then compute the optimal roots of path bundles across modules. Then, by using an exhaustive set of computational experiments in diverse and representative class of polygonal domains and modular bipartite networks, we show that (1) it is possible to obtain modular bundled paths with an optimized connectivity metric, (2) the convergence is achievable by using the DIviding RECTangles algorithm, (3) the proposed approach outperforms the state of the art bundling [23] over all scenarios, rendering trees which are comparatively shorter by 7.83% and 85.32%, and (4) the maximal reduction in length occurs when the number of edges in the bipartite network tends to be large, implying the improved scalability for large-scale modular bipartite networks.

2   Computing Modular Path Bundles

In this section we describe the basic concept of our proposed approach. We first describe the basic concept of our proposal, then describe the main components involved in our algorithmic development.

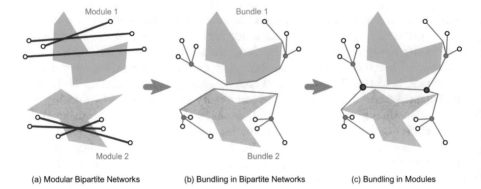

Fig. 1. Basic idea of path bundling in modular bipartite networks.

2.1   Basic Idea

The basic concept of path bundling in modular bipartite networks is depicted by Fig. 1. In our approach, the expected output is a tree structure in which paths avoid collision with obstacles as well as aim at minimizing the total distance among leaves. In our approach, there exists two roots in the tree, each of which is a coordinating point in each side of the modular network.

In order to compute the bundled path in modular bipartite networks, the following inputs are necessary:

- A modular bipartite graph $G = (V, E)$ wherein every edge $e \in E$ represents communication/transportation needs between origin-destination pairs, and edges in the set E grouped in M modules, following a cluster-like configuration such that $E = E_1 \cup E_2 \cup ...E_m... \cup E_M$, and $V = V_1 \cup V_2 \cup ...V_m... \cup V_M$. An example of a modular bipartite network is depicted by Fig. 1-(a), in which the bipartite network has 6 edges, grouped into two modules, namely Module 1 and Module 2.
- A polygonal map configuration $\rho = \{\rho_1, \ldots, \rho_s\}$, in which a-priori knowledge of obstacles $\rho \in M$ define unfeasible areas for navigation/transportation. An example of a polygonal configuration is shown by Fig. 1-(a), in which the map involves the presence of two polygons $\rho = \{\rho_1, \rho_2\}$, each of which is colored differently.

It is important to note that the number of edges and the locations of the source-destination pairs in the graph G is defined by the characteristics of the communication/transportation problem and/or is set by the network designer.

Algorithm 1. Bundling of Modular Bipartite Networks.

1: **procedure** BUNDLE(G)

2:    **Input** G ▷ Modular Bipartite Graph

3:    **Output** $(x_o, x_1, x_2, ..., x_m, ..., x_M)$ ▷ Roots of Path Bundle

4:    **for** $m \leftarrow 1$ **to** M **do**

5:        $x_m \leftarrow$ Minimize $F_m(x)$ with $x \in \mathbf{T}_m$

6:    **end for**

7:    $x_o \leftarrow$ Minimize $F_o(x)$ with $x \in \mathbf{T}_o$

8: **end procedure**

In order to compute globally-optimal path bundles in modular bipartite networks, we extend our previous work [23] to allow the existence of modules in bipartite networks. In this paper, the basic idea of computing modular path bundles is summarized by Fig. 1 and Algorithm 1, in which we first compute the roots of path bundles in each module of the bipartite network, as depicted by Fig. 1-(b) and lines 4–6 in Algorithm 1, and we then compute the roots of path bundles across modules, as depicted by Fig. 1-(c) and line 7 in Algorithm 1. In the following subsections, we describe the representation of bundled paths, as well as the approach for optimization in the above described procedures.

2.2   Bundle Representation

In this paper, we represent Path Bundles by using the roots of a tree encoded in a triangular space [23], in which each point is represented as follows:

$$P = (i, \alpha, \beta), \tag{1}$$

where P is the coordinate in triangular space, for $i \in [|T|]$, $[a] = \{1, 2, ..., a\}$ and $T = \{t_1, t_2, \ldots, t_i, \ldots, t_n\}$ is the set of Delaunay triangles of the free space of the polygonal map ρ, and $\alpha, \beta \in [0, 1]$ are arbitrary constants. The reader may note that the above representation holds a convex property due to the known boundaries of (i, α, β).

Also, the above representation can be easily mapped into a Cartesian plane. Here, the equivalent 2-dimensional cartesian coordinate can be computed as follows [22]:

$$(P_x, P_y) = (1 - \alpha)A_i + \sqrt{\alpha}(1 - \beta)B_i + \sqrt{\alpha}\beta C_i, \tag{2}$$

where P_x and P_y is the horizontal and vertical cartesian coordinate of point P, A_i, B_i, C_i are the coordinates of the vertices of the i-th triangle $t_i \in T$.

Then, based on the above representation, the path bundle of the m-th module of the bipartite network is represented by the following tuple:

$$x_m = \Big(\underbrace{(i_P^m, \alpha_P^m, \beta_P^m)}_{x_m^P}, \underbrace{(i_Q^m, \alpha_Q^m, \beta_Q^m)}_{x_m^Q} \Big), \tag{3}$$

where i_P^m is the integer number related to the i-th triangle of the m-th module of the coordinate P of bipartite network G, and $\alpha_P^m, \beta_P^m, \alpha_Q^m, \beta_Q^m \in [0, 1]$ are Real numbers.

The main motivation behind handling coordinates P and Q in the above representation is due to our aim to encode two roots, namely P and Q, each of which denotes a coordination foci in each direction of the bipartite network. Also, since we assume the presence of modules in the bipartite network, we also assume the existence of m roots, each of which has two coordinates P and Q and each root $P - Q$ is encoded by the triangular coordinate x_m.

Also, in the above notation, the following holds: $i_P^m, i_Q^m \in [|T_m|]$, in which the set T_m is computed as follows:

$$T_m = DT\Big(\mathbb{C}_{V_m} - \rho_1 \cup \rho_2 \ldots \cup \rho_s \Big), \tag{4}$$

where DT represents the Delaunay Triangulation operator, $\mathbb{C}_{V_m}$ represents the Convex Hull of the polygon resulting from the union of the vertices of the m-th module of the bipartite network G, and $\rho_1 \cup \rho_2 \ldots \cup \rho_s$ represents the union of the obstacles of the polygonal map ρ. The basic idea of the above is to compute the convex set of the free space within the region of the m-th module of the bipartite network.

Furthermore, the representation depicted by Eq. 3 has the following unique features:

- Path bundles encoded by x_m guarantee to avoid overlaps with obstacles,
- Explicit computation of point inside polygon is avoided, thus each x_m guarantees to be outside of non-navigable space.
- The nature of the search space, as well as the lower and upper bounds become clear due to the convexity property of the encoding of x_m. Concretely speaking, the encoding of the coordinate holds $x_m \in \mathbf{T}_m$ [23], in which:

$$\mathbf{T}_m = \mathbb{N}^{[|T_m|]} \times \mathbb{R}^{[0,1]} \times \mathbb{R}^{[0,1]} \times \mathbb{N}^{[|T_m|]} \times \mathbb{R}^{[0,1]} \times \mathbb{R}^{[0,1]}, \tag{5}$$

  thus, in line of the above, the lower bound of x_m is $(1, 0, 0, 1, 0, 0)$, and the upper bound of x_m is $(|T_m|, 1, 1, |T_m|, 1, 1)$ for $m \in [M]$. A-priori knowledge of these bounds is key to allow the searching through optimization heuristics.
- Compared to the Cartesian-based representation of Path Bundles, the computation of point inside polygons becomes unnecessary; which is efficient in case of optimizing path bundles in polygonal maps with large number of vertices; since computing point inside polygons takes $O(v)$ time with a ray-tracing algorithm, and takes $O(log(v))$ time with a binary search algorithm, considering a polygonal map with v vertices.

Finally, the representation of path bundle across modules is realized by the following equation:

$$x_o = \left(\underbrace{(i_P^o, \alpha_P^o, \beta_P^o)}_{x_o^P}, \underbrace{(i_Q^o, \alpha_Q^o, \beta_Q^o)}_{x_o^Q} \right), \tag{6}$$

in which the following holds: $i_P^o, i_Q^o \in [|T_o|]$, and the set T_o is computed as follows:

$$T_o = DT\left(\mathbb{C}_{x_1 \cup x_2 \cup \ldots \cup x_m} - \rho_1 \cup \rho_2 \ldots \cup \rho_s \right), \tag{7}$$

where DT represents the Delaunay Triangulation operator, $\mathbb{C}_{x_1 \cup x_2 \cup \ldots \cup x_m}$ represents the Convex Hull of the polygon resulting from the union of the roots of the path bundles of the m modules of the bipartite network G. In line of the above, the following holds: $x_o \in \mathbf{T}_o$, and

$$\mathbf{T}_o = \mathbb{N}^{[|T_o|]} \times \mathbb{R}^{[0,1]} \times \mathbb{R}^{[0,1]} \times \mathbb{N}^{[|T_o|]} \times \mathbb{R}^{[0,1]} \times \mathbb{R}^{[0,1]}, \tag{8}$$

in which the lower bound of x_o is $(1, 0, 0, 1, 0, 0)$, and the upper bound of x_o is $(|T_o|, 1, 1, |T_o|, 1, 1)$ for $m \in [M]$.

The basic idea behind the representation of Eq. 6 is to compute the convex free space which involves the convex hull of the roots of the modules in the bipartite network, and which allows defining the search space of the path bundles across modules. Furthermore, by allowing the configuration of $\mathbf{T}_m$ and $\mathbf{T}_o$, we aim at enabling the focalized search spaces, which can be used by sampling-based optimization algorithms.

2.3   Optimization Problem

The basic idea of computing path bundles is depicted by two optimization problems, one which aims at solving the path bundling problem for each module, and another which aims at solving the path bundling across modules.

Concretely speaking, in order to compute path bundles *for each module*, we aim at computing x_m for each $m \in [M]$, as follows:

$$\begin{aligned} \underset{x}{\text{Minimize}} \quad & F_m(x) \\ \text{subject to} \quad & x \in \mathbf{T}_m \end{aligned} \tag{9}$$

and, one we aim at computing the roots x_o of path bundles *across modules*, as follows:

$$\begin{aligned} \underset{x}{\text{Minimize}} \quad & F_o(x) \\ \text{subject to} \quad & x \in \mathbf{T}_o \end{aligned} \tag{10}$$

where, x is the encoding (representation) of the bundled path, $F_m(x)$ and $F_o(x)$ are the global distance metrics which evaluate the quality of the bundled paths, and $\mathbf{T}_m$ and $\mathbf{T}_o$ denote the search space of feasible bundled paths defined by Eqs. 5 and 8, respectively.

The key motivation of the above is as follows: once path bundles are defined for each module locally, it becomes possible to compute the path bundle across modules globally by using the roots of the bundles from each module. Also, the search spaces $x \in \mathbf{T}_m$ and $x \in \mathbf{T}_o$ allow for focalized sampling not only inside roots of modular bundles, but also across the roots of multiple bundles. Furthermore, by solving two optimization problems, we aim at finding the structure of higher-order (non-shallow) trees; in which, for simplicity and without loss of generality, we assume a 2-order tree. It is possible to modify Eqs. 9 and 10 in order to enable trees with more than two roots and enable higher-order trees. In future work, we aim at extending to these types of trees.

In line of the above, the objective function to compute bundles for each module is defined as follows:

$$F_m(x) = \sum_{e \in E_m} d(e_S, x_m^P) + d(x_m^P, x_m^Q) + \sum_{e \in E_m} d(e_F, x_m^Q) \tag{11}$$

And, the objective function to compute bundles across modules is as follows:

$$F_o(x) = \sum_{m \in [M]} d(x_m^P, x_o^P) + d(x_o^P, x_o^Q) + \sum_{m \in [M]} d(x_m^Q, x_o^Q) \tag{12}$$

where, $d(.,.)$ is the Euclidean obstacle-free shortest distance metric between two points, e_S is the coordinate of the *origin* node of the edge $e \in E_m$ in the m-th module, e_F is the coordinate of the *destination* node of the edge $e \in E_m$ in the m-th module, x_m^P and x_m^Q are roots of bundles defined by Eq. 3, and x_o^P and x_o^Q are roots of bundles defined by Eq. 6.

Note that the above definitions consider the coordinates x_m^P and x_m^Q, and x_o^P and x_o^Q to be in triangular space. The conversion to the cartesian coordinates is straightforward due to the definition of Eq. 2.

Furthermore, since Eqs. 11 and 12 are nonlinear and multimodal in cases polygonal maps with non-convex obstacles, the use gradient-based approaches to solve Eqs. 9 and 10 are ineffective due to the fact of biasing to local-optima.

In line of the above, we use Direct Global Optimization Algorithm (DIRECT) [24] following our previous studies in path bundling in shallow bipartite networks [23], which has shown the efficiency and effectiveness in computing roots in shallow (non-modular) trees when compared to a representative class of gradient-free optimization algorithms, as follows: Differential Evolution with Successful Parent Selection/Best1 [25], Particle Swarm Optimization with Niching Properties [26], Real-Binary Differential Evolution (RBDE) [27], and SHADE, Success History Parameter Adaptation for Differential Evolution [28].

Furthermore, DIRECT is relevant deterministic algorithm in the literature due to the fact of considering convexity and mesh partitioning, and the fact of using the DIviding RECTangles concept, which samples solutions vectors at the center of hypercubes, and then subdivides potentially optimal hypercubes recursively. Parameters for DIRECT use the same configuration of our previous study [23], with the key difference that we use the maximum number of evaluations restricted to [2000, 2500], due to the observed facts in quick convergence.

3 Computational Experiments

In order to evaluate the feasibility of our proposed approach, we used a class with convex and non-convex polygonal domains, as well as different configurations of modular bipartite networks. This section describes our experimental conditions, as well as our obtained insights.

3.1 Experimental Settings

Our computing environment was an Intel i7-4930K @ 3.4 GHz with Windows 8.1, and our algorithms were implemented in Matlab 2017b. In order to enable a meaningful evaluation of our proposed approach, we considered the following environmental settings:

- No. of Modules in the bipartite graph $= M = \{2, 4, 6, 8, 10\}$
- No. of Edges in each module of bipartite graph,
  $|E_m| = \{5, 10, 15, 20, 25\}$,
- Number of Polygonal Obstacles: $\{1, 2, 3, 4, 5\}$,
- No. Sides in each Polygonal Obstacle: $\{5, 10\}$.
- For each independent experiment, the maximum number of functions evaluations is set between 2000 and 2500.
- In each independent experiment, the initial solutions of route bundles $x_m \in \mathbf{T}_m$ and $x_o \in \mathbf{T}_o$ are initialized randomly and independently.

In order to show the kind of environments and bipartite networks used in our experiments, Figs. 2 and 3 show the polygonal domains and bipartite networks with obstacles, each of which having (a) 5 sides and (b) 10 sides, respectively. In this figure, we show a matrix-like configuration, where the x-axis denotes the number of polygons in the bipartite network, and the y-axis denotes the number of edges per each module in the bipartite network. The origin and destination pairs of edges and modules in the polygonal environments are arbitrarily generated (1) to enable the exhaustive evaluation of our proposed approach, and (2) to evaluate our approach under different and arbitrary initialization conditions. By the above considerations, bias in random luckiness is avoided.

In a side note, the key motivation behind choosing values in the number of edges up to 25, and modules up to 10 is due to our interest in evaluating the performance of path bundles in scenarios being close to the number of communication needs in indoor environments, where the complexity of the environment is controlled by (1) the number of obstacles in the polygonal map, and/or (2) the number of sides in each obstacle. In line with the above remarks, the large number of obstacles and the large number of sides in each obstacle induce in both complex polygonal environments and large number of triangulations; thus representing a challenging search space for our proposed approach. In Our future work, we aim at using environmental configurations considering large-scale scenarios being close to outdoor environments.

Also, the key rationale of using [2000, 2500] as the upper bound of the number of function evaluations is due to our interest of evaluating the effectiveness and

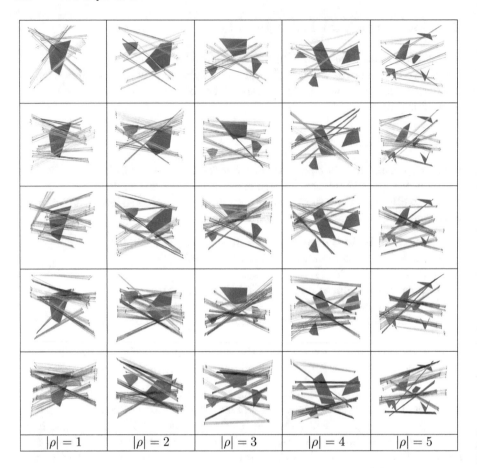

Fig. 2. Polygonal Map with **5** sides and Bipartite Networks with $|\rho|$ modules, each with **5** (top) edges, **10** edges, **15** edges, **20** edges and **25** (bottom) edges.

efficiency of our proposed approach under restrictive computational resources, which is in line of our previous observations [23]. The reader may note that the use of the number of function evaluations as a surrogate metric for efficiency is relevant to avoid bias in hardware or algorithmic implementation.

Then, in line of the above configurations, as a result, 250 experimental conditions were evaluated[1], and 5×10^5 functions evaluations were computed[2].

3.2 Results and Discussion

In order to portray the kind of tree structures obtained by our proposed approach, as well as to evaluate the efficiency in path bundling in modular bipartite

---

[1] $5 \times 5 \times 5 \times 2$.

[2] 250×2000.

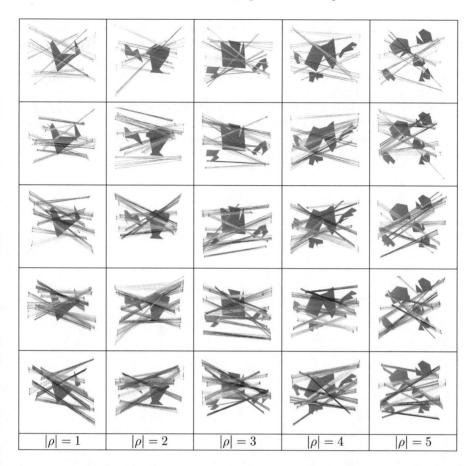

Fig. 3. Polygonal Map with **10** sides and Bipartite Networks with $|\rho|$ modules, each with **5** (top) edges, **10** edges, **15** edges, **20** edges and **25** (bottom) edges.

networks, Figs. 4 and 5 show examples of the optimized path bundles; and Figs. 6 and 7 show the convergence behaviour. Note that our results in Figs. 4 and 5 follow the same organization of Figs. 2 and 3; that is, the x-axis denotes the number of polygons in the bipartite network, and the y-axis denotes the number of edges per each module in the bipartite network.

In regards to the obtained path bundles, by observing the results being rendered in Figs. 4 and 5, we can confirm the following facts:

- Regardless of the configuration of the polygonal domain being evaluated, it is possible to generate tree structures aiming at minimizing the total tree length in modular bipartite networks.
- The location of the anchoring points of the modular bundled paths are not necessarily close to the centroid of the origin and destination pairs of the modular bipartite graph.

- The path between the anchoring points between modules is not necessarily a straight line; and, regardless of increasing the number of edges in each module of the bipartite graph, there exist sub-trees with overlapping geometry. This observation implies the need to increase the depth to compute anchoring points in each module.
- Regardless of the configuration of the bipartite network, the location of anchoring points between independent experimental instances is close to each other. This observation implies that anchoring points tend to locate at similar locations when the polygonal environment is fixed.

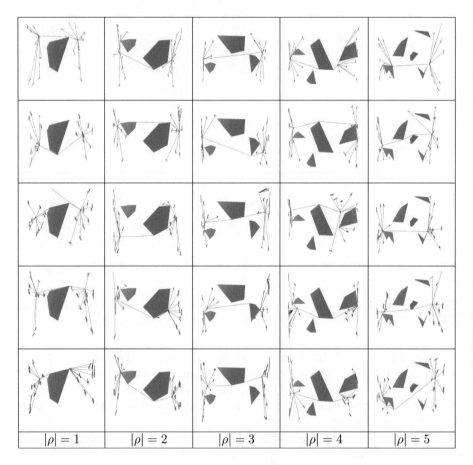

Fig. 4. Polygonal Map with **5** sides and Bipartite Networks with $|\rho|$ modules, each with **5** (top) edges, **10** edges, **15** edges, **20** edges and **25** (bottom) edges.

In regards to the convergence behaviour of the proposed algorithm, by observing Figs. 6 and 7, it is possible to confirm the following facts:

- Regardless of the configuration of the polygonal domain being evaluated, it is possible to converge to the bundled paths minimizing the global distance

metric within $[2000, 2500]$ function evaluations. This observation confirms the efficiency of our previous studies in path bundling of shallow bipartite networks [23].
- Increasing the number of edges in each module of the bipartite network has a direct effect on increasing the length of the minimal tree by some small factor smaller than 1.
- In line with the above fact, increasing the number of modules of bipartite networks has a direct effect on increasing the length of the minimal tree by some small factor larger than 1, but smaller than 3.

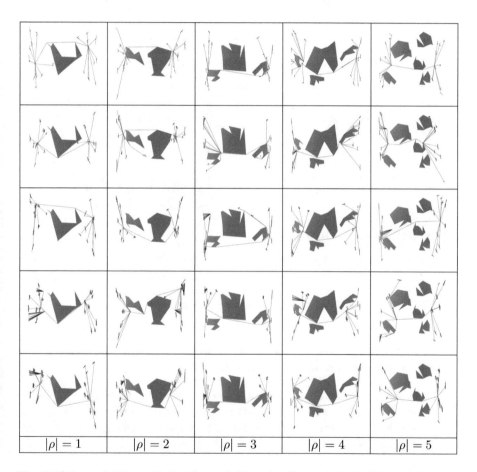

Fig. 5. Polygonal Map with **5** sides and Bipartite Networks with $|\rho|$ modules, each with **5** (top) edges, **10** edges, **15** edges, **20** edges and **25** (bottom) edges.

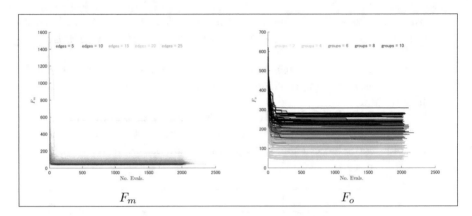

Fig. 6. Convergence in polygonal maps with 5 sides.

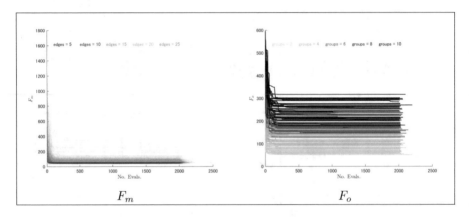

Fig. 7. Convergence in polygonal maps with 10 sides.

In regards to the comparison to the state of the art, by observing Figs. 8 and 9, it is possible to confirm the following facts:

- The proposed method outperforms the state of the art bundling of bipartite networks [23] over all evaluated environmental benchmarks, as can be observed by Figs. 8 and 9. The proposed approach obtained trees which are comparatively shorter, reducing the length of trees in the range between 7.83% and 85.32%.
- In line with the above, the maximal reduction occurs in scenarios in which the total number of edges in the bipartite network tends to be large, as can be observed by Fig. 8-(a) and 9-(a). This observation confirms the usefulness and the scalability of the proposed approach when considering large number of edges being grouped into clusters.

- There is no significant performance difference when considering polygonal obstacles between 5 and 10 sides, since similar tree lengths are obtained by path bundling, as can be observed by Figs. 8-(a) and 9-(a).

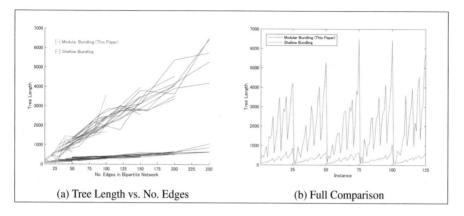

(a) Tree Length vs. No. Edges   (b) Full Comparison

Fig. 8. Comparison to the state of the art method in Polygonal Maps with 5 sides.

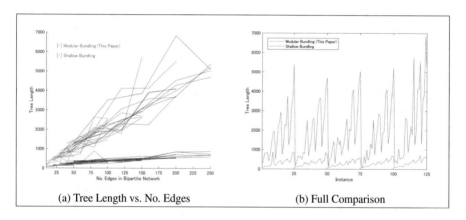

(a) Tree Length vs. No. Edges   (b) Full Comparison

Fig. 9. Comparison to the state of the art method in Polygonal Maps with 10 sides.

We believe that the above observations brings important implications to design higher-order algorithms that solve the path bundling problem effectively and efficiently. In line of the above, we provide the following propositions:

- Whenever the configuration of the number of edges, or the number of modules, or the disposition of polygons is expected to change (as a result of increasing/decreasing the number of groups and nodes at both origin-destination pairs), it may be possible to use pre-computed paths and roots as initial solutions for path bundling, since the new paths are expected to be structurally similar, and the new location of roots is expected to be near.

– Instead of computing shallow trees for path bundling, it may be possible to compute path bundling with hierarchy considerations, as the one proposed in this paper. Furthermore, a hierarchical convex search space (as the one proposed in this paper), may be key for effective and efficient sampling of the search space.

The above insights imply the feasibility and efficiency to obtain path bundles with hierarchical considerations being applicable to polygonal environments with both convex and non-convex obstacles. Although this study has focused on computing path bundles in bipartite network topologies, it may be possible to canonically sample general directed graphs [29] as well as undirected graphs [30] to enhance performance, yet the study of the above is left for future work. Also, although this study has considered modules with predefined composition, it may be possible to find the optimal modular composition of edges by k-subset sum [31], yet devising efficient algorithms are left in our future agenda. Furthermore, in future work, we aim at using large number of edges and diverse obstacle configurations reminiscent of outdoor environments.

4   Conclusion

In this paper, we have proposed an approach for designing modular path bundles based on the idea of compounding edges in modular bipartite networks on polygonal maps. The unique point of our proposed approach is to compute modular path bundles in two-layer configuration, in which we first compute the roots of path bundles in each module of the bipartite network, and then compute the roots of path bundles across modules. Exhaustive computational experiments using a diverse and representative class of polygonal domains, and modular bipartite networks, have shown that (1) it is possible to obtain modular bundled paths with an optimized tree length metric via [2000, 2500] function evaluations, (2) the convergence is efficiently achieved by using the DIviding RECTangles algorithm, (3) the proposed approach outperforms the conventional shallow bundling [23] over all evaluated environmental benchmarks, reducing the length of trees in the range between 7.83% and 85.32%, and (4) the maximal reduction in length occurs in scenarios in which the total number of edges in the bipartite network tends to be large, implying the usefulness and scalability for large-scale modular bipartite networks.

In future work, we aim at exploiting our insights in modular path bundling. Also, we aim at exploring the generalization ability in polygonal environments reminiscent of outdoor configurations, as well as dynamic configurations. We believe our approach opens new possibilities to develop effective algorithms for path bundling with hierarchical considerations being applicable to polygonal environments with both convex and non-convex obstacles

References

1. Parque, V., Miura, S., Miyashita, T.: Optimization of ZigBee networks using bundled routes. In: International Conference on Advanced Mechatronics (ICAM) (2015)
2. Hermansson, T., Bohlin, R., Carlson, J., Söderberg, R.: Automatic routing of flexible 1D components with functional and manufacturing constraints. Comput.-Aided Des. **79**, 27–35 (2016)
3. Li, W.T., Liu, Y.C.: Dynamic coverage control for mobile robot network with limited and nonidentical sensory ranges. In: 2017 IEEE International Conference on Robotics and Automation (ICRA), pp. 775–780 (2017)
4. Kantaros, Y., Zavlanos, M.M.: Global planning for multi-robot communication networks in complex environments. IEEE Trans. Robot. **32**, 1045–1061 (2016)
5. Dijkstra, E.W.: A note on two problems in connexion with graphs. Numerische Mathematik **1**, 269–271 (1959)
6. Hart, P.E., Nilsson, N.J., Raphael, B.: A formal basis for the heuristic determination of minimum cost paths. IEEE Trans. Syst. Sci. Cybern. **4**(2), 100–107 (1968)
7. Chazelle, B.: A theorem on polygon cutting with applications. In: Proceedings of the 23rd IEEE Symposium on Foundations of Computer Science, pp. 339–349 (1982)
8. Cormen, T., Leiserson, C., Rivest, R.: Introduction to Algorithms. MIT Press, Cambridge (1993)
9. Kallmann, M.: Path planning in triangulations. In: Proceedings of the Workshop on Reasoning, Representation, and Learning in Computer Games, IJCAI, pp. 49–54 (2005)
10. Lee, D.T., Preparata, F.P.: Euclidean shortest paths in the presence of rectilinear barriers. Networks **14**(3), 393–410 (1984)
11. Falud, R.: Building Wireless Sensor Networks, 4th edn. O'Reilly Media, Sebastapol (2014)
12. Wightman, P., Labardor, M.: A family of simple distributed minimum connected dominating set-based topology construction algorithms. J. Netw. Comput. Appl. **34**, 1997–2010 (2011)
13. Torkestani, J.A.: An energy-efficient topology construction algorithm for wireless sensor networks. Comput. Netw. **57**, 1714–1725 (2013)
14. Panigrahi, N., Khilar, P.M.: An evolutionary based topological optimization strategy for consensus based clock synchronization protocols in wireless sensor network. Swarm Evol. Comput. **22**, 66–85 (2015)
15. Szurley, J., Bertrand, A., Moonen, M.: An evolutionary based topological optimization strategy for consensus based clock synchronization protocols in wireless sensor network. Signal Process. **117**, 44–60 (2015)
16. Singh, S.P., Sharma, S.: A survey on cluster based routing protocols in wireless sensor networks. Procedia Comput. Sci. **45**, 687–695 (2015)
17. Cui, W., Zhou, H., Qu, H., Wong, P.C.: Geometry-based edge clustering for graph visualization. IEEE Trans. Vis. Comput. Graph. **14**, 1277–1284 (2008)
18. Selassie, D., Heller, B., Heer, J.: Divided edge bundling for directional network data. IEEE Trans. Vis. Comput. Graph. **17**(12), 2354–2363 (2011)
19. Ersoy, O., Hurther, C., Paulovich, F., Cabtareiro, G., Telea, A.: Skeleton-based edge bundlig for graph visualization. IEEE Trans. Vis. Comput. Graph. **17**(12), 2364–2373 (2011)

20. Gansner, E.R., Hu, Y., North, S., Scheidegger, C.: Multilevel agglomerative edge bundling for visualizing large graphs. In: IEEE Pacific Visualization Symposium, pp. 187–194 (2011)
21. Holten, D., van Wijk, J.J.: Force-directed edge bundling for graph visualization. In: Euro-graphics, IEEE-VGTC Symposium on Visualization (2009)
22. Osada, R., Funkhouser, T., Chazelle, B., Dobkin, D.: Shape distributions. Eurographics ACM Trans. Graph. 21(4), 807–832 (2002)
23. Parque, V., Miura, S., Miyashita, T.: Computing path bundles in bipartite networks. In: SIMULTECH 2017 - Proceedings of the 7th International Conference on Simulation and Modeling Methodologies, Technologies and Applications, pp. 422–427. SciTePress (2017)
24. Jones, D.R.: Direct Global Optimization Algorithm. Encyclopedia of Optimization. Kluwer Academic Publishers (1999)
25. Guo, S.-M., Yang, C.-C., Hsu, P.H., Tsai, J.S.H.: Improving Differential Evolution With a Successful-Parent-Selecting Framework. IEEE Trans. Evol. Comput.19, 717–730 (2015)
26. Qu, B.Y., Liang, J.J., Suganthan, P.N.: Niching particle swarm optimization with local search for multi-modal optimization. Inf. Sci. 197, 131–143 (2012)
27. Sutton, A.M., Lunacek, M., Whitley, L.D.: Differential evolution and non-separability: using selective pressure to focus search. In: Proceedings of 9th Annual Conference GECCO, July, pp. 1428–1435 (2007)
28. Tanabe, R., Fukunaga, A.: Success-history based parameter adaptation for differential evolution. In: Proceedings of IEEE Congress on Evolutionary Computation, Cancun, Mexico, pp. 71–78 (2013)
29. Parque, V., Kobayashi, M., Higashi, M.: Bijections for the numeric representation of labeled graphs. In: IEEE International Conference on Systems, Man and Cybernetics, pp. 447–452 (2014)
30. Parque, V., Miyashita, T.: On succinct representation of directed graphs. In: IEEE International Conference on Big Data and Smart Computing, pp. 199–205 (2017)
31. Parque, V., Miyashita, T.: On k-subset sum using enumerative encoding. In: 2016 IEEE International Symposium on Signal Processing and Information Technology (ISSPIT), pp. 81–86 (2016)

Complex Systems Modeling and Simulation

Using Gazebo to Generate Use Case Based Stimuli for SystemC

Thomas W. Pieber$^{(\boxtimes)}$, Thomas Ulz, and Christian Steger

Institute for Technical Informatics, Graz University of Technology, Graz, Austria
{thomas.pieber,thomas.ulz,steger}@tugraz.at

Abstract. Realistic simulations of new hardware are of utmost importance to achieve good results. The current approach to such simulations is that the Device under Test is exposed to stimuli that are either generated randomly, or that are generated by engineers reverse engineering the use cases and extending the inputs by some extreme cases. In this paper we describe an approach to generate useful stimuli for a SystemC simulation directly from a simulation of the use case. In this approach the use case is simulated using the Gazebo simulator. The stimuli for the Device under Test are then extracted and sent to the SystemC simulation.

1 Introduction

In the development of new systems simulations need to be performed to find errors early. With such simulations the developed system (or Device Under Test "DUT") can be tested extensively and optimizations and error corrections can be implemented quickly and inexpensively. These simulations are usually stimulated with events that occur in the expected use case as well as some extreme cases. These tests are designed by engineers reworking the scenarios and defining the inputs and the expected behaviour. In addition to these tests, random input sequences can be applied to test the DUT's reactions to faulty or unexpected inputs as random input is unlikely to be valid. All together that means that the current test procedure consists of valid inputs designed by the system engineers and (mostly) invalid inputs generated by random testing. We therefore propose an architecture for a generator that can produce valid inputs to the DUT design which can also be evaluated according to the expectations of the engineers. Such system can decrease the effort needed to design tests for the DUT, as only the valid scenarios need to be described. These will then automatically generate valid input data and the expected output.

For such simulation the environment in which the DUT should operate can be simulated. This environmental description only needs to describe the essential parameters that can affect the DUT. Such simulations can be performed in a simulator such as the Gazebo simulator [1,2]. This simulator is designed for robotic use cases and is designed to handle complex systems and generate accurate sensor information of any kind. This open source simulator also allows

© Springer Nature Switzerland AG 2019
M. S. Obaidat et al. (Eds.): SIMULTECH 2017, AISC 873, pp. 241–256, 2019.
https://doi.org/10.1007/978-3-030-01470-4_13

modifications to be as useful and accurate as we want it to be. These modifications are done by implementing plugins for the environment (world), the models, the sensors, the simulation core, the visuals, and the GUI. This simulator operates in discrete time steps of 1 ms. This degree of simulation accuracy is enough to simulate the movement of robots and sparse enough that the robot's operating system can handle most commands in this time step.

To simulate our DUT another tool such as SystemC [3] can be used. With this tool a complex microsystem can be designed and tested. Furthermore, the component parts of the system can be modelled in various degrees of detail. This allows for accurate simulations or even synthetization of the newly developed parts and efficient simulation of existing hardware. SystemC operates in discrete time intervals as small as 1 fs.

When combining these two simulations, this difference in simulation speed poses a major problem. The execution of a test scenario can last for many minutes. In combination with the fine grained simulation time steps of SystemC this can generate huge amounts of data which need to be handled. This problem needs to be considered when choosing the traced signals and information that should be transferred between the simulations. Additionally, the testbench of the SystemC simulation must be altered to include the communication between the simulations.

This leads to the issue of the communication itself. The simulations need to exchange data such as the generated input and output of the simulation step, as well as status information about the simulations itself. The difference in time steps also introduces the problem that SystemC requires data in more detail from the Gazebo simulation. This data is to be estimated and extrapolated from the existing inputs. It also produces output data that is filtered to allow Gazebo to work with the resulting data.

This paper is based on the work done by Pieber et al. [4]. It expands the ideas behind that publication, gives more detail on the design and implementation of the combination of the simulations. It furthermore expands the evaluation by constructing a detailed simulation run and interpreting the results.

The remainder of this paper is structured as follows: In Sect. 2 other works that combine SystemC or Gazebo with other simulators are described. Section 3 explains the motivation for our design, states the requirements that need to be implemented, and gives details on the solution for the requirements. An evaluation of the design is described in Sect. 4. This is done by constructing and analyzing a sample simulation run. Following that, Sect. 5 mentions ideas on how to further improve the proposed design. This paper concludes with Sect. 6.

2   Related Work

As Gazebo is an open source simulator for robotics it is primarily combined with a robot operating system such as ROS [5] or YARP [6] to control the simulated robots [7].

There are approaches to combine the Gazebo simulator to software for robotics and computational intelligence [8]. There are further works that connect other tools that can simulate hardware [9], but the main approach in these works is to use the interface from Gazebo to ROS and implement ROS nodes to connect to the rest.

A design process for SystemC is given by Panda [10]. With this language a model for complex systems can be described and executed. There are many publications that implement interfaces to SystemC as it provides a good basis for simulations [11–14]. In these approaches the functionality of SystemC is extended to provide the functions needed by the researchers.

An interface from SystemC to Matlab/Simulink was designed by Bouchhima et al. Here the SystemC simulation was stimulated by a continuous environment simulation written in Matlab/Simulink. SystemC was also connected to analog circuit simulators like SPICE [15] and VHDL [16] to improve on flexibility and simulation performance.

Mueller-Gritschneder et al. developed a robot simulation platform in SystemC [13]. They simulate the behavior of the robot on the transaction layer and forward the results to an environment simulation written in Java. They do this in order to simulate the movement of the robot as accurate as possible. In this paper the robot is simulated in SystemC, while in this proposal the robot's behavior is the input to the SystemC simulation to simulate parts of the environment.

In summary SystemC was connected to many other simulations. It is then used as core for other simulations or to generate more accurate results. In the context of robotics, SystemC has been used to simulate the movement of the robot. In contrast to that, this publication uses the robotic simulation to stimulate SystemC components with inputs from the environment to automatically generate valid stimuli.

In this paper, a use-case is evaluated where a sensor measures data from the environment, and is read out and charged via Near Field Communication (NFC) by a robot. Some publications describe the techniques used to transmit data alongside energy and storing the excess energy in small batteries or capacitors [17–19].

To connect simulations a common interface must be created over which data can be exchanged. In this approach the common interface used is a POSIX (Portable Operating System Interface) pipe where XML (Extensible Markup Language) formatted data is sent. Another possibility to format data efficiently is the JSON (JavaScript Object Notation) format. This would be more efficient than XML [20], but due to other reasons, explained below, the XML format is chosen. Gazebo uses Googles Protocol Buffer (ProtoBuf) as formatting method to transmit data internally. Sumaray and Makki compare the efficiency of this protocol to XML and JSON [21].

Based on [4], this publication expands on the detail of the design and implementation of the developed connection of the simulations. Main focus of the expansion is on details concerning the Gazebo plugin. Additionally the evaluation of the system is expanded. Here, detailed information on the traces produced by SystemC is given.

3   Design and Implementation

The goal of the presented design is to find a method to generate stimuli for a SystemC simulation automatically and being able to see how the simulated system behaves in the specific use-case. To enable this, a connection between SystemC and a high-level simulation is established. This high-level simulation (in this approach the Gazebo simulator) represents the surroundings of the newly developed system (a sensor in this use-case). Using the data from the Gazebo simulator, stimuli for the sensor can be created. That means that the environment of the sensor becomes the de-facto testbed. With this method the stimuli for the SystemC simulation are generated by the interaction of the sensor with the environment. This generates the stimuli not only faster than an engineer could, but also only small variations in the environment can generate a wide variety of different test scenarios such as more noise in the communication or energy fluctuations due to movements of the reader or changed mutual inductance due to small changes in the distance between the antennas.

To connect two simulations successful, both must support the interfaces necessary. To do so, an overall structure for the communication was developed. This structure is shown in Fig. 1. This plan visualizes how the simulations are connected, how they communicate and when operations are performed. This figure also shows the minimum requirements that this approach needs to work. In this example the SystemC process is forked from the Gazebo simulation. Here some initial configuration concerning the communication can be set up. One step of the Gazebo simulation then invokes one from SystemC. A return message from SystemC informs Gazebo that the simulation step of SystemC is properly executed. During the execution of the SystemC step additional messages for Gazebo may be sent that need to be captured from the Gazebo environment.

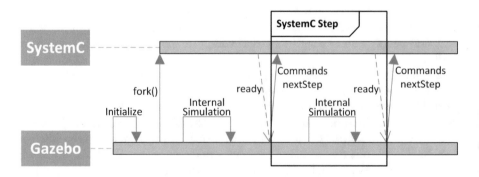

Fig. 1. States of the execution of the implemented plugin.

Gazebo can be extended by the use of plugins. To apply the input of the Gazebo environment to a SystemC simulation, one such plugin that handles the communication needs to be developed. The structure for that plugin can be seen in Fig. 2. This structure implements the following five operations.

O.1 The required data must be collected from the environment. That includes data that the sensor can measure as well as communication data.

O.2 The collected data needs to be packed into messages that can be sent to the SystemC simulation.

O.3 The plugin needs to halt the simulation of the environment until the SystemC simulation step finishes.

O.4 During the execution of the SystemC step, all messages needed for the remaining simulation(s) need to be received, stored, and ordered.

O.5 The collected information needs to be distributed to the remaining simulation(s). This can be communication data, visual data, or status information.

The operation defined in O.1 is needed to generate valid input to the sensor that can be evaluated. To generate this information a world plugin may be needed to gather the information. This operation can then be achieved by generating a Gazebo internal communication from this plugin to the plugin connecting the SystemC simulation. Additionally a communication path between the communicating entities must be established. This communication may also be altered in order to simulate the effect of the channel.

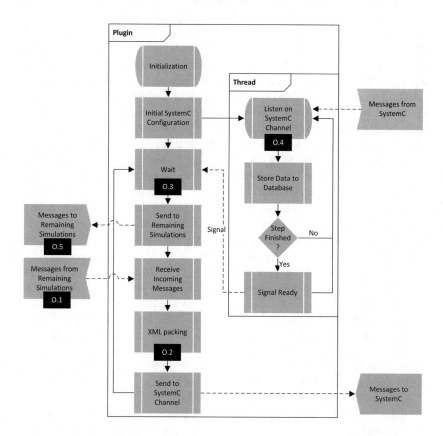

Fig. 2. States of the execution of the implemented plugin.

To send the gathered information to SystemC operation O.2 is required. This operation formats the data in a way suitable for transport to SystemC. To send arbitrary data the data items are converted into string format and packed using an XML structure.

To properly synchronize the two simulations Gazebo needs to be able to wait for SystemC to finish calculating the current time step. That implies that the plugin needs to be able to halt the Gazebo simulation until it receives the signal indicating the completion. This operation is referred to as O.3.

While Gazebo is waiting for SystemC to finish, SystemC may send different messages that are needed for the rest of the Gazebo simulation. This information needs to be processed and stored until Gazebo can simulate the next time step. When this happens, the plugin needs to forward the information received from SystemC to the rest of the simulation. This operation is described in O.4.

Operation O.5 describes the correct distribution of the gathered data. As SystemC can send different data (e.g.: visual updates such as LEDs, or data for communicating with other entities), the information needs to be split into the topics and sent to their destination using an internal communication mechanism.

The difference in simulation speeds is one of the biggest hurdles in connecting the two simulations. As a tool for simulating interactions and movements of robots, Gazebo works with time steps of 1 ms. On the other hand, SystemC can handle steps as small as 1 fs. This is needed for simulating hardware components as also a "slow" computer which only works with 50 MHz performs $5 \cdot 10^4$ operations in one time step of Gazebo. This difference of twelve orders of magnitude of the simulations can result in massive amounts of data generated by SystemC which is hard to evaluate in Gazebo. Therefore, some measures to limit the amounts of data that are transferred need to be implemented. This can be done best when defining the requirements for the connection between the simulators.

As the two simulations must be compatible in their interfaces to each other, the structure of a SystemC simulation needs to be adopted. Figure 3 shows the overall structure of such SystemC simulation. The messages coming from the Gazebo simulator are received and analyzed. If some parameters need changing, the adaptations are done. To reduce the simulation time, it is evaluated if the changes require instant action. Should that not be the case, the time that should be simulated is added as a debt in comparison to the Gazebo simulation. An action is required if the time debt is too large or if the sensor receives messages that require an answer. To simulate these actions in the correct order, the old parameters and commands are reset and the simulation is run to balance the time debt. In this run the conditions at the current time instance are estimated. After that the new parameters are set and the simulation is started for this time step. When a stable condition is reached the simulation is halted and the conditions for the end of the step is calculated. Should the calculations need more time than one time step, the checking if action is required evaluates to "yes" in the next time step.

To send arbitrary messages between the two simulation environments, an easy-to-implement approach is used. As the SystemC simulation gets forked from

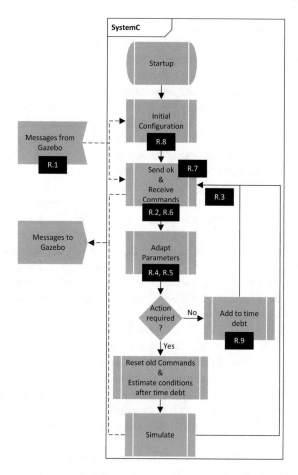

Fig. 3. Structure of the SystemC simulation.

the Gazebo plugin (Fig. 1) the standard input and output can be redirected to a POSIX pipe. The Gazebo plugin uses that pipe to exchange messages with the SystemC simulation. This interface allows the transmission of string-type messages. So the commands and parameters need to be packed in a format that can be sent as string and the string received needs to be parsed in order to get the information back. An easy to implement method is the use of XML (Extensible Markup Language) data structure to pack the information in strings. That means that the interesting values are encased in XML-style tags. With these tags the string can be split in parts containing different types of data, which are then evaluated. The use of XML also allows the SystemC components themselves to send their information as soon as it is available and Gazebo can receive and process the data in the order it was produced. Although JSON would be more efficient than XML, XML was chosen because this makes the composing of messages inside SystemC more easy. With XML the messages can be sent whenever the information is available, with JSON on the other hand the Information needs to be packed in smaller JSON objects which can be sent. This can introduce additional computational overhead, leading to JSON loose its better performance.

The data sent by the SystemC simulation is received by the Gazebo plugin and parsed. The incoming data stream is split into data chunks according to the XML-tags, preprocessed and stored in a fitting datastructure. When the SystemC simulation halts for the time step, a tag for synchronizing the simulations is sent to Gazebo. The reception of this tag signalizes the Gazebo plugin to transmit the collected data to the rest of the simulation and resume its work. Gazebo is capable of using many types of communication, but Google's Protocol Buffer (Protobuf) is the most efficient communication it supports [21]. To fully use this method, custom messages can be defined that can hold various types of data. The messages from the sensor that are intended for the robot (in this scenario) are for example transmitted to a simulation of the transmission channel. There environmental information is used to simulate signal degradation due to free space loss and multi-path interference. The channel is simulated separately from the rest to be able to test different communication channels such as WiFi, Bluetooth, Zigbee, or in this case NFC. The calculations of the transmission statistics are based on [17,19]. The channel then modifies the transmitted data and forwards it to the robot, where it can be accessed.

Communication from the robot to the sensor follows the same rules. The robot sends the data to the channel simulation. There, errors are introduced and the signal strength and received energy is calculated. This information is then sent to the sensor plugin which also collects the data for the sensor system itself. The collected data is packed in an XML message and sent to the SystemC simulation.

The SystemC simulation receives the commands, parses the XML data and performs the actions needed to simulate the sensor accurately. To accomplish this, the original testbed must be modified to accommodate the interface to the Gazebo simulation. In order for that to work properly, the following requirements need to be fulfilled:

R.1 Gazebo must be able to transfer information about the simulation, commands and parameters for the DUT to SystemC.

R.2 To be able to react to changes in the environment, the SystemC simulation needs to operate in steps. Between the steps the information exchange can occur.

R.3 The SystemC interface for the communication must be able to parse and distribute the received information.

R.4 To support different types of simulation, the simulation time step of SystemC should be variable at each step.

R.5 Parts of the SystemC simulation that need special information need to be able to get it directly.

R.6 Commands received from the Gazebo simulator need to be executed in the order they arrive.

R.7 The two simulations need to be synchronized during the execution.

R.8 To reduce the memory required to run the simulation for extended periods, traces from the simulation should be able to be deactivated.

(R.9 The simulation should be done as quickly as possible, while maintaining the accuracy where needed.)

In addition to the changes in the testbed - now the interface to the Gazebo simulation - some adaptations in the rest of the simulation have been made. The most notable adaptation is the insertion of messages that are sent to the Gazebo simulator. These changes need to be made as the bulk of information is evaluated during the simulation by the Gazebo simulator instead of after the simulation. These changes are made in a similar fashion as the changes needed for requirement R.7.

Figure 3 additionally shows the parts of the SystemC simulation where the implementations of the requirements are mostly located.

To be able to fulfill requirement R.1 a POSIX pipe can be used. This allows the transport of information in string format between SystemC and Gazebo. For that, the standard input and output of the SystemC simulation are rerouted.

An XML parser is needed to split the gathered information into the data chunks needed for the different settings and commands. With the use of a global datastructure the distribution to all parts of the simulation can be performed. These measures fulfill requirement R.3.

The call of the start procedure for the SystemC simulation ("sc_start(...)") is inside a loop. The loop is halted when the simulation waits for input from the Gazebo simulator and the condition to exit the loop is sent from Gazebo when exiting or resetting the Gazebo simulation.

To allow a multitude of simulations the time step size of SystemC may need to be changed during runtime. This is represented in requirement R.4. To meet this requirement two special commands are added. One to set the time unit and one to set the numerical value of the time step.

To support requirement R.5 we modified not only the testbed, but also parts of the simulation. For every module we want a direct communication path to, we need a special tag to extract the data. This data is then written into a global datastructure to be accessible by all modules. The interface from the module itself retrieves the data from this datastructure and can perform the needed operations without the need to communicate the information through the layers of the module. The finished data is then stored in FIFO (fist in, first out) structures to wait for the simulation to need it. This requirement is especially useful for this scenario as we modeled a sensor system and the data the sensor measures is represented in the Gazebo simulation.

Requirement R.6 can be met with the same strategy. This time the FIFO buffer storage is located in the control unit of the sensor. Now the commands that have arrived can then be loaded and executed in the order they should have arrived. To properly execute the commands an additional field that stores the time when the command should have arrived is needed to simulate a serialized channel.

To get a proper time synchronization (R.7) the SystemC simulation needs to send a marker message back to Gazebo indicating that the step has been processed. This is done by declaring an extra XML-tag. After receiving this signal the Gazebo simulation is allowed to execute another step - again triggering the SystemC simulation.

To use a simulated environment as generator for system stimuli it is necessary to simulate longer periods of time before the system is initially triggered. This is done in order to generate different edge conditions on the simulated device. As we do not know what exactly the sensor measures during this time, it is necessary to also simulate this time in SystemC. Furthermore, the environment simulations can be run for an extended period of time between stimuli. When storing all generated data large files are created. Requirement R.8 refers to that problem. This can be solved by activating only traces that are needed. The information which traces are needed can be received during the initialization process. This information does not only contain which traces are needed, but also if the traces should be active at all, and what the file name should be.

To further decrease the need for memory, a detector system is be implemented that determines weather a simulation step can be stopped prematurely, or even needs to be started. This detector has the potential to reduce the simulation time significantly and corresponds to the optional requirement R.9.

Such detector can only be implemented if detailed knowledge of the inner workings of the simulation and the system it simulates is available. This also requires some major appendices to the existing simulation.

When pausing the simulation prematurely, two challenges emerge. The first one is the desynchronization of the two simulations. As SystemC offers no methods to change the simulation time when it is stopped, we introduced flag signals that get triggered if the simulation time should get changed. With the help of these signals and some post processing of the generated traces the synchronization can be restored afterwards. Figure 4 shows a trace before (above) and after

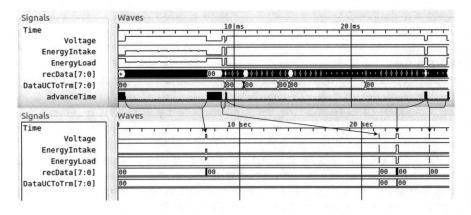

Fig. 4. Compression of idle-time.

the decompression of idle-times is performed. The markers between the traces indicate the compression.

The second challenge is the estimation of values that change during the skipping of time, such as the remaining energy in the battery. To estimate the values at the end of the time jump, detailed information about the process is required. As the time between two activations can be arbitrary, the error is unbounded. To mitigate that, a maximum time skip is defined. When linearizing the behavior of these transient values in the last instant, we can estimate the new value after the skip is completed. Depending on the maximum skip size, the final estimation can be very accurate.

4   Evaluation

The evaluation of the developed system is performed using the simulation of a smart sensor that charges the internal battery and communicates its information using NFC technology. The Gazebo simulator provides the context of the simulation. That is the environment in which the sensor is placed. During the startup of Gazebo, the developed sensor plugin is loaded and starts the SystemC simulation. A robot is placed outside the communication range of the sensor. The evaluation plan is to move the robot such that the sensor can be charged and communication can occur. When this is done, the robot requests data from the sensor.

For this evaluation we modeled the communication channel as a separate world plugin that can calculate the noises and signal attenuation due to the environment. This also allows us to change the channel parameters by swapping the plugin. This also allows the reuse of the system for other communication technologies.

The communication between the two simulation environments is performed with strings encased in XML-tags. The received information is then stored in a data structure that groups the data blocks by the XML tag and stores the blocks in the order they arrive.

Different stimuli for the SystemC simulation can be combined, processed and sent by the developed plugin. Figure 5 shows the concept for this evaluation. The robot wants to send commands and data to the smart sensor via some channel (in this case NFC). To do so, it approaches the sensor, thereby changing the environment. This change influences the channel parameters. When the antennas are in close proximity to each other, data and energy can be transferred. The channel can, based on the parameters, change the data that is sent (e.g.: introducing bit errors). This message is then transmitted to the developed plugin. Furthermore, the plugin receives status information from the simulation, and parameters from the environment that the sensor can measure. This information is relayed to the SystemC simulation. The modified testbench processes the data to be used in the simulation. The simulation returns the results to the plugin. This plugin can manipulate the appearance of the sensor in the world (e.g.: switching on an LED), and transmitting the return messages to the channel. The channel again modifies the message according to the parameters and forwards it to the robot.

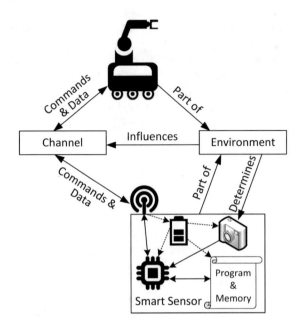

Fig. 5. Concept of the evaluation design.

As the global simulation is done with a robotic simulator, a new test case can be implemented by changing the start position of the robot or introducing some randomness in the movement of the arm with the antenna attached to it. The rest of the simulation does not need to be altered in order to get new results. This simulation approach furthermore allows the testing of the interaction between the newly developed system and an existing (robotic) system. The evaluation of the correctness of the new system can also be done directly in the simulation as the robot expects certain answers.

As mentioned before, Fig. 4 shows the compression of the idle time of the sensor showing the results of requirement R.9. Here the "advanceTime" trace is used to restore the time synchronicity of the two simulations after the simulation is finished. Whenever this trace peaks the simulation was stopped prematurely. The height of the peak indicates the time that is skipped. In this trace the first 7.1 s are condensed to about 0.6 ms. This means a reduction of memory and time usage of approximately 12000:1. Also the simulation can be stopped prematurely if the needed operations are finished before the time step is passed. This can be seen with the third drop of the "advanceTime" trace. This part is stored in 16.5 ms but refers to 0.15 s. This is a reduction of approximately 10:1.

Figure 6 shows the results of one simulation. Here the robot approaches the sensor until second 17. The antenna is activated from second 6.1 to 9. Here two messages are received (recData) but no return is generated, indicating that the channel has introduced errors in the sent message. This claim is substantiated by the fact that the received energy (EnergyIntake) is low and fluctuates. The

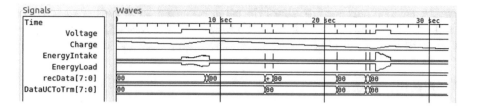

Fig. 6. Results of a completed simulation.

fluctuation comes from the movement of the arm during the approach to the sensor.

The message at second 14.2 is being answered (DataUCToTrm). This means that the antennas are in a range where the communication can be successful. But the message at second 15.05 is not answered.

At second 21 the received energy is larger than before, indicating that the robot is close enough to communicate efficiently. In second 24, two messages are exchanged.

After second 25 the robot moves away. To show this the antenna is activated. During the retreat the received energy decreases.

The energy usage (EnergyLoad) of the sensor shows the combined energy budged of the sensor. As the energy gathered by the NFC antenna is much larger than the energy required for the calculations, it mirrors the energy intake. This also shows that the charging of smart sensors using NFC can be effective. The usage of the sensor itself can be seen in the trace of the remaining charge inside the capacitor (Charge). During the phases where the antenna is active, the charge rises, indicating that the capacitor is being loaded. In the mean time, the charge is slowly depleted as the sensor performs its operations with the energy stored in the capacitor.

Furthermore, the voltage available to the sensor (Voltage) is shown in this figure. Because the module used to charge the capacitor supplies the system with the maximum voltage allowed for this capacitor, the voltage rises rapidly to this value. In times where no energy is harvested the voltage is calculated using the charge of the capacitor, the energy currently used, and the serial resistance of the capacitor.

5   Future Work

A possible expansion of this system is another world plugin comparing the measurements of the sensor to the ground truth evaluated by the environment. This can be done by combining the information of the environment, the measured data from SystemC, as well as some of the messages received by the robot.

One drawback from forking the SystemC simulation from the Gazebo simulator is that the two processes are running on the same computer. If a simulation is created that encompasses multiple entities that are simulated using SystemC the simulations may need to share the same processor core, further slowing the simulation. A solution to this would be to spread the SystemC simulation over a network and performing the communication using network sockets.

6   Conclusions

We presented an approach to connect SystemC simulations to the Gazebo simulator in order to automatically generate stimuli. This paper shows the difficulties that arise when connecting simulators that are designed to operate using different time steps. We showed a mechanism that can connect the simulations, proposed a mechanism that allows the interaction of the simulations, and formed requirements that need to be implemented on both sides to overcome the hurdles that we were presented by the simulations.

There are some core requirements that need to be changed we want to emphasize again. These include:

- *Synchronization* between the simulators is of utmost importance. SystemC operates usually more detailed and therefore needs longer to simulate one step. Gazebo must be halted while SystemC is running, otherwise the communication between the simulations can have unbounded delay.
- *Reduction of memory and time consumption* is important on all computers. Using our time reduction mechanisms, the realtime-factor of gazebo was optimized by a factor of 10^3 and the memory footprint reduced by up to 10^2.

This approach was first described by Pieber et al. [4]. In this publication, we extend the detail of the developed Gazebo plugin. Furthermore, we redefined some of the requirements needed for the SystemC adaptions to emphasize their purpose. Additionally, the evaluation describes how the systems interact and gives a detailed example of a complete simulation and what the created traces can look like.

Acknowledgements. This project has received funding from the Electronic Component Systems for European Leadership Joint Undertaking under grant agreement No 692480. This Joint Undertaking receives support from the European Union's Horizon 2020 research and innovation programme and Germany, Netherlands, Spain, Austria, Belgium, Slovakia.

IoSense is funded by the Austrian Federal Ministry of Transport, Innovation and Technology (BMVIT) under the program "ICT of the Future" between May 2016 and May 2019. More information https://iktderzukunft.at/en/.

References

1. Koenig, N., Howard, A.: Design and use paradigms for Gazebo, an open-source multi-robot simulator. In: 2004 IEEE/RSJ International Conference on Intelligent Robots and Systems (IROS) (IEEE Cat. No. 04CH37566). Institute of Electrical and Electronics Engineers (IEEE) (2004)
2. Open Source Robotics Foundation: Gazebo simulator (2004). http://www.gazebosim.org. Accessed 03 Jan 2017
3. Accelera: SystemC (2000). http://accellera.org/downloads/standards/systemc. Accessed 17 Jan 2017
4. Pieber, T.W., Ulz, T., Steger, C.: SystemC test case generation with the gazebo simulator. In: Proceedings of the 7th International Conference on Simulation and Modeling Methodologies, Technologies and Applications - Volume 1: SIMULTECH, pp. 65–72. INSTICC. SciTePress (2017)
5. Garage, W., Stanford Artificial Intelligence Laboratory: Robot Operating System (2007). http://www.ros.org/. Accessed 15 Feb 2018
6. Metta, G., Fitzpatrick, P., Natale, L.: YARP: yet another robot platform. Int. J. Adv. Robot. Syst. **3**, 8 (2006)
7. Meyer, J., Sendobry, A., Kohlbrecher, S., Klingauf, U., von Stryk, O.: Comprehensive simulation of quadrotor UAVs using ROS and Gazebo. In: Noda, I., Ando, N., Brugali, D., Kuffner, J.J. (eds.) Simulation, Modeling, and Programming for Autonomous Robots, pp. 400–411. Springer, Heidelberg (2012)
8. Zamora, I., Lopez, N.G., Vilches, V.M., Cordero, A.H.: Extending the OpenAI Gym for robotics: a toolkit for reinforcement learning using ROS and Gazebo. arXiv preprint arXiv:1608.05742 (2016)
9. Mathworks: Get Started with Gazebo and a Simulated TurtleBot (2016). https://de.mathworks.com/help/robotics/examples/get-started-with-gazebo-and-a-simulated-turtlebot.html. Accessed 03 Jan 2017
10. Panda, P.R.: SystemC - a modelling platform supporting multiple design abstractions. In: Proceedings of the 14th international symposium on Systems synthesis - ISSS. Association for Computing Machinery (ACM) (2001)
11. Bouchhima, F., Briere, M., Nicolescu, G., Abid, M., Aboulhamid, E.: A SystemC/Simulink co-simulation framework for continuous/discrete-events simulation. In: 2006 IEEE International Behavioral Modeling and Simulation Workshop. Institute of Electrical and Electronics Engineers (IEEE) (2006)
12. Huang, K., Bacivarov, I., Hugelshofer, F., Thiele, L.: Scalably distributed SystemC simulation for embedded applications. In: 2008 International Symposium on Industrial Embedded Systems. Institute of Electrical and Electronics Engineers (IEEE) (2008)
13. Mueller-Gritschneder, D., Lu, K., Wallander, E., Greim, M., Schlichtmann, U.: A virtual prototyping platform for real-time systems with a case study for a two-wheeled robot. In: Design, Automation and Test in Europe Conference and Exhibition (DATE). EDAA (2013)
14. Possadas, H., Adamez, J.A., Villar, E., Blasco, F., Escuder, F.: RTOS modeling in SystemC for real-time embedded SW simulation: a POSIX model. Des. Autom. Embed. Syst. **10**, 209–227 (2005)
15. Kirchner, T., Bannow, N., Grimm, C.: Analogue mixed signal simulation using Spice and SystemC. In: Proceedings of the Conference on Design, Automation and Test in Europe, DATE 2009, Leuven, Belgium, pp. 284–287. European Design and Automation Association (2009)

16. Bombana, M., Bruschi, F.: SystemC-VHDL co-simulation and synthesis in the HW domain. In: 2003 Design, Automation and Test in Europe Conference and Exhibition. IEEE Computer Society (2003)
17. Lee, W.S., Son, W.I., Oh, K.S., Yu, J.W.: Contactless energy transfer systems using antiparallel resonant loops. IEEE Trans. Ind. Electron. **60**, 350–359 (2013)
18. Strommer, E., Jurvansuu, M., Tuikka, T., Ylisaukko-oja, A., Rapakko, H., Vesterinen, J.: NFC-enabled wireless charging. In: 2012 4th International Workshop on Near Field Communication. Institute of Electrical and Electronics Engineers (IEEE) (2012)
19. Wireless Power Consortium, et al.: System description wireless power transfer. Volume I: Low Power, Part 1 (2010)
20. Nurseitov, N., Paulson, M., Reynolds, R., Izurieta, C.: Comparison of JSON and XML data interchange formats: a case study. Caine **2009**, 157–162 (2009)
21. Sumaray, A., Makki, S.K.: A comparison of data serialization formats for optimal efficiency on a mobile platform. In: Proceedings of the 6th International Conference on Ubiquitous Information Management and Communication, ICUIMC 2012, pp. 48:1–48:6. ACM, New York (2012)

Modeling Performance and Energy Efficiency of Virtualized Flexible Networks

Raffaele Bolla[1,2], Roberto Bruschi[2], Franco Davoli[1,2(✉)], and Jane Frances Pajo[1,2]

[1] Department of Electrical, Electronic and Telecommunications Engineering, and Naval Architecture (DITEN), University of Genoa, Genoa, Italy
{raffaele.bolla,franco.davoli}@unige.it,
jane.pajo@tnt-lab.unige.it
[2] CNIT (National Inter-University Consortium for Telecommunications) National Laboratory of Smart, Sustainable and Secure Internet Technologies and Infrastructures (S3ITI), Genoa, Italy
roberto.bruschi@unige.it

Abstract. We examine some aspects of modelling and control in modern telecommunication networks, in the light of their evolution toward a completely virtualised paradigm on top of a flexible physical infrastructure. The trade-off between performance indicators related to user satisfaction of services (e.g., in terms of perceived quality, delay and ease of the interaction) and the energy consumption induced on the physical infrastructure is considered with some attention. In this respect, we provide a discussion of potential problems and ways to face them, along with a short description of the approaches taken in some European project activities.

1 Introduction

In recent years, a significant shift in data networking paradigms and in resource allocation mechanisms in networking has been gaining increasing momentum. Whereas in the past bandwidth, among other resources, used to be considered a potential bottleneck to be administered carefully, especially in the user access area (and still is, to some extent, in wireless access), with the increase in available transmission and processing speed, paralleled by an unprecedented increase in user-generated traffic, other factors that were previously concealed have become evident: the legacy networking infrastructure makes use of a large variety of hardware appliances, dedicated to specific tasks, which typically are inflexible, energy-inefficient, unsuitable to sustain reduced Time to Market of new services.

In this context, the search for ways of making resource allocation in telecommunication networks more *dynamic*, *performance-optimized* and *cost-effective* has brought forth the characterizing features of *flexibility*, *programmability* and *energy-efficiency*. The first two aspects are addressed by Software Defined Networking (SDN) [1–4] and Network Functions Virtualisation [5, 6]. In particular, the latter leverages "…standard IT virtualisation technology to consolidate many network equipment types onto industry standard high volume servers, switches and storage, which could be located in

M. S. Obaidat et al. (Eds.): SIMULTECH 2017, AISC 873, pp. 257–273, 2019.
https://doi.org/10.1007/978-3-030-01470-4_14

Datacentres, Network Nodes and in the end user premises" [7]. The expected benefits are improved equipment consolidation, reduced time-to-market, single platform – multiple applications, users, and tenants, improved scalability, multiple open eco-systems and, last but not least, exploitation of the economy of scale of the Information Technology (IT) industry: according to [8], the 2016 market for data-centre servers has reached $32 billion worldwide, with a growth rate of 6%, against $27 billion world-wide for routers and switches, with growth rate of 1%; in any case, "the main disruption to the market is being provided by the growth of cloud and hosted solutions, which are redefining markets and enabling new competitors to emerge" [8].

SDN and NFV, along with Cloud and Fog Computing (or, more generally Multi-access Edge Computing – MEC [9]) paradigms create the basis for the "softwarisation" phenomenon that is going to find its full development in the 5G ecosystem [10, 11]. The creation of network slices in this context [12] provides the mechanisms to hierarchically abstract and orchestrate resources (both real and virtual) to eventually offer a complete, flexible, isolated and manageable networking environment to vertical industries for the deployment and dynamic instantiation of their applications.

However, it should be kept in mind that the certainly meritorious and notable effort behind the development of architectural concepts, abstractions, and standardised interfaces, as well as of the (most often open source) software constructs enabling their implementation, which has characterised the evolution of such innovative networking ecosystem, does not provide by itself the control and management mechanisms necessary for its proper operational functionality. The intelligence to perform dynamic resource allocation in such complex multi-actor environment must come from data- or model-based control strategies that operate at multiple levels, interact in non-mutually-obstructive fashion, and concur to the accomplishment of common as well as conflicting goals, within well-defined constraints. This consideration brings forward the other two aspects that we mentioned earlier, regarding *performance* and *energy-efficiency*. In this framework, the purpose of the paper is to highlight some of the issues concerning the trade-off between these two aspects in the virtualised networking framework.

The paper is organized as follows. We recall some of the energy-related issues in the next section. In Sect. 3, we summarise the results of some approaches to the problem of joint performance-energy optimisation in softwarised networks, and in Sect. 4 we briefly introduce two recent European projects that address some specific aspects in this general perspective. Section 5 contains the conclusions.

2   Energy-Efficiency Modelling and Control Aspects

Information and Communication Technology (ICT) has been historically and fairly considered as a key objective to reduce and monitor "third-party" energy wastes and achieve higher levels of efficiency. A classic example in this respect has been the use of video-conferencing services; more recent ones are Intelligent Transport Systems (ITS) and, directly affecting the energy sector, the Smart Grid. However, until relatively recently, ICT had not applied the same efficiency concepts to itself; consideration of energy consumption issues started first with wireless networks (see, e.g., [13]) and datacentres ([14, 15], among others), and later extended to fixed networks and the Internet in general ([16–18], among others).

There are two main motivations that drive the quest for "green" ICT: the environmental one, which is related to the reduction of wastes, in order to impact on the carbon footprint; the economic one, which stems from the reduction of operational costs (OpEx) of ICT services. Indeed, according to the Global e-Sustainability Initiative (GeSI) [19], global ICT emissions (including datacentres, voice and data networks, and end-user devices) of greenhouse gases (GHG) are bound to reach about 1.3 GtCO2e/y (Gtons of CO2 equivalent per year), amounting to 2.3% of overall GHG emissions. On the other hand, it is interesting to observe that ICT's abatement potential is estimated to be 7 times higher (16.1%).

Today's (and future) network infrastructures are characterized by a design capable to deal with strong requests and constraints in terms of resources and performance (large loads, very low delay, high availability, …), and by services that exhibit high variability of load and resource requests along time (burstiness, rush hours, …). The current feasible approach to cope with energy consumption is centred on smart power management (energy consumption should follow the dynamics of the service requests) and on flexibility in resource usage (virtualization to obtain an aggressive sharing of physical resources).

In [16] we have introduced a taxonomy of approaches to energy efficiency in fixed networks, where two broad families of techniques are identified to adapt the consumption to load variations, acting on different time scales: *dynamic adaptation* and *smart standby*. The first one can be further divided into Adaptive Rate (AR) and Low Power Idle (LPI), which aim at adjusting the processor's speed (by adjusting frequency, voltage, or both) according to the load, and at putting part of the hardware into lower-consumption states during idle periods, respectively. The second family of techniques is usually referred to in conjunction with longer "sleeping" periods, and can be used effectively in virtualised environments (e.g., to consolidate functionalities to execute onto a smaller group of servers and to shut down the unused physical machines). Such techniques have been long used in computing devices, where the Advanced Configuration and Power Interface (ACPI, maintained since 2013 by the Unified Extensible Firmware Interface Forum – UEFI) [20] provides a standardized interface between the hardware and the software layers; however, only relatively recently they have found application in networking devices.

The ACPI introduces two power saving mechanisms, which can be individually employed and tuned for each core: Power States (C-states, where C0 is the active power state, and C1 through Cn are processor sleeping or idle states, where the processor consumes less power and dissipates less heat), and Performance States (P-states; while in the C0 state, ACPI allows the performance of the core to be tuned through P-state transitions, by altering the working frequency and/or voltage, or throttling the clock, to perform AR). The adoption of similar concepts in the framework of networking devices (e.g., switches and routers), spawned by the ECONET project [21], among others, has led to the development of the Green Abstraction Layer (GAL) [22, 23], later adopted as ETSI standard 203 237 [24]. The GAL allows power-aware devices, or parts thereof, to communicate their power-performance adaptation capabilities to network control and management entities, and to receive parameter settings and commands from them, effectively enabling power-performance trade-off, on the basis of suitable optimisation techniques (see, e.g., [25–28]).

In this framework, the application of control and optimisation methods to manage the above-mentioned trade-off has been considered both at the device-level, with the application of Local Control Policies (LCPs), and the network-level, concerning Network Control Policies (NCPs). In many cases, the two can be applied in a hierarchical fashion, where NCPs perform a kind of periodic or event-driven parametric optimisation, in order to adaptively set LCP model parameters (e.g., in terms of the choice of C- and P-states). The various techniques adopted may differ according to the type of model used to represent the physical processes to be dealt with, which basically entail queue and flow dynamics. A very general scheme to highlight the main components and their interaction is represented in Fig. 1.

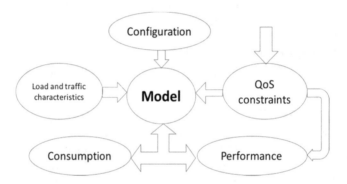

Fig. 1. Identification and interaction of models, inputs, goals and constraints.

In our opinion, two broad categories of modeling techniques are of particular interest here:

- Models based on classical queueing theory, possibly augmented with the explicit consideration of setup times (which stem from the different wakeup periods associated with different C-states of the processors), and taking into account the bursty nature of traffic at the packet level (e.g., the $M^X/G/1/SET$ adopted in [25, 26]) lend themselves to performance analysis or parametric optimization for adaptive control and management policies over longer (with respect to queueing dynamics) time scales;
- Fluid models suitable for real-time control can stem from the first category, as was shown in the pioneering work of J. Filipiak in [29], or even from simpler, measurement-based, stochastic continuous fluid approximations (see, e.g., [30]); very interesting models and techniques for the dynamic control of queues can be found in the Lyapunov optimization approach proposed by Neely [31].

Moreover, control techniques in this setting may be quasi-centralized (owing to the presence of SDN controllers, which can supervise a certain number of underlying switches) or hierarchical (considering LCP-NCP interactions [25–28]), or even based on more sophisticated completely distributed control techniques stemming from game [32] or team theory [33]. On this basis, and also taking into account the new capabilities offered by network softwarisation in terms of flexibility and programmability, which were mentioned in the Introduction, one would be led to conclude that the premises are there for a – technically and operationally – easier way to apply complex control and management strategies (with the latter operating on longer time scales, but often tightly integrated with the former and autonomic) for truly dynamic Traffic Engineering, accounting for both energy and Quality of Service/Quality of Experience (QoS/QoE) Key Performance Indicators (KPIs).

However, the introduction of NFV changes the perspective quite a bit with respect to "legacy" networking equipment:

- The hardware (HW) that consumes energy belongs to the Infrastructure Provider (InP), which in general may not coincide with the Network Service Providers (NSPs) in a multi-tenant environment;
- The HW is shared by multiple Virtual Machines (VMs) or by Network Slices, through a virtualization environment;
- Queueing models can be identified and used to assess the performance of VMs as function of the virtual resources assigned to them (as well as to control their assignment), but the relation between the performance of the VMs and their energy consumption is not straightforward, involving the virtualization layer, Infrastructure SDN Controllers (IC), Virtual Infrastructure Managers (VIMs), Wide-area Infrastructure Managers (WIM), Resource Orchestrators (ROs), Network Service Orchestrators (NSOs), and Tenant SDN Controllers (TCs) in the overall resource allocation process [12].

In the next Section, we will mention some initial approaches to the problem.

3   Managing the QoS-Energy Trade-off in Virtualised Networks

We have already noted that a significant reduction in Capital Expenditure (CapEx) should be realised by the economy of scale achievable with the adoption of general-purpose HW; which are then the main OpEx sources that can be reduced by technological advancement? They appear to be the ones related to energy consumption and network management, which roughly account for a figure equivalent to the entire infrastructure CapEx. Whereas it is true that local virtualisation of Base Stations can provide significant energy saving [34], it is not so straightforward to determine whether virtualisation in the backhaul network would reach the same result, unless specific energy-aware solutions are included in future 5G technologies. In [35] we have considered a use case corresponding to the virtualisation of a Serving Gateway (SGW) in the Evolved Packet Core (EPC), sketched in Fig. 2.

A number of aspects have been taken into account to cope with energy consumption: (i) appropriate downsizing of the parts of SGW that cannot be completely virtualized; (ii) dynamic activation of the minimum number of VMs to support the current traffic and their consolidation to the minimum number of servers;

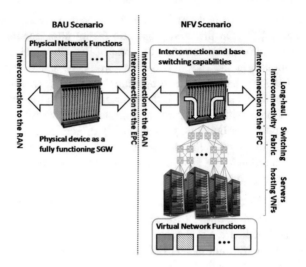

Fig. 2. SGW in the Business-as-Usual (BAU) and in the virtualization scenario [35].

(iii) optimisation of the interconnection switches' topology and enablement of energy-awareness capabilities; (iv) scaling of the throughput of a server through Amdahl's law [36], to account for parallelization. Even under these favourable assumptions, the power consumption of the virtualized Service Router (vSR), at the same target delay, results to be at least twice that of the "traditional" SR of the BAU solution.

Indeed, once fixed the silicon technology, energy consumption largely depends on the number of gates in the network device/chip hardware. The number of gates is generally directly proportional to the flexibility and programmability levels of HW engines. If we fix a target number of gates by using General Purpose CPUs, we obtain maximum flexibility, but reduced performance/power ratio; on the other hand, by using very specialized Application-Specific Integrated Circuits (ASICs), one would obtain minimum flexibility, but greatly enhanced performance/power ratio. Other technologies (e.g., network/packet processors) provide performance between these boundaries.

Essentially, there are three basic enablers at chip/system level: (i) dynamically programmable resources able to perform multi-purpose services; (ii) specialized HW for offloading to speed-up basic functionalities; (iii) standby capabilities to save energy if a resource is unused. The presence of general-purpose HW offers the possibility of moving services among the components of a node, or among nodes in a network. When the workload is low, many services can aggressively share single general-purpose HW resources. Thus, even if a general-purpose/programmable resource consumes more energy than ASIC-based solutions, a smaller number of HW elements can be left active, in order to effectively handle the current workload. Then, if programmability *for* energy efficiency is sought, two main issues need to be considered:

- which basic (sub-)functionalities need to be moved (and "frozen") to the offloaded specialized engines (best performance in terms of bps/W)?

- which ones have to remain in the programmability space (lower performance but stronger sharing and more evolution opportunities)?

The solutions need to be identified by considering and effectively supporting the newest trends in Internet technology evolutions. With these considerations in mind, we can turn back to the modelling aspects, and to the related control strategies that can be devised to jointly optimise performance and energy consumption. In this respect, as a possible example, we briefly summarise here the approach that has been taken in [37]. The scenario addressed is represented in Fig. 3. We consider a set of VMs dedicated to perform certain (virtualized) network functions (VNFs) on incoming traffic streams of various nature. For the sake of simplicity, a one-to-one correspondence is assumed between VNFs and VMs; the rationale behind this is that for a VNF consisting of multiple VMs the overall VNF performance can be derived from the individual VMs' performances, according to the chaining defined by the VNF provider. In any case, the VNF consolidation reduces to a VM consolidation problem. The VMs are initially placed among a given set of multicore servers through a First-Fit Decreasing (FFD) bin-packing algorithm [38] based on the workloads specified in the Service Level Agreement (SLA). Since such specifications are generally derived from peak workloads, the main goal here is to dynamically manage VM consolidation in each server according to actual workload variations, by jointly tuning the ACPI configuration and minimizing the number of active cores.

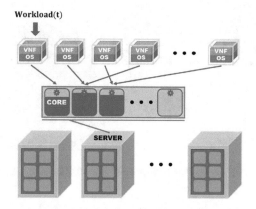

Fig. 3. Reference framework for VNF consolidation [37].

An $M^X/G/1/SET$ queueing model can be effectively used to represent the energy consumption, when considering the *aggregate workload* (see Fig. 4) produced by all VMs insisting on the core.

Then, two ways can be considered to enforce performance constraints: (i) coarsely, by imposing a limit on the maximum utilization for each core; (ii) more precisely, by computing the average system latency on the basis of the model. This requires the knowledge of the second moments of the batch size (obtainable from measurements of the second moment of the busy period) and of the packet service time (directly

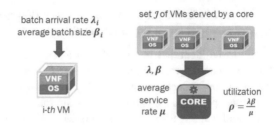

Fig. 4. Single VNF/VM pair and aggregate core workload.

measurable). Based on the model, decision rules can be defined to design an energy- and performance-aware consolidation policy in the space $(\lambda, \lambda\beta)$ of all pairs of aggregate batch arrival rate $\lambda = \sum_{i=1}^{|\mathcal{J}|} \lambda_i$ and aggregate workload $\lambda\beta$, where $\beta = (1/\lambda)\sum_{i=1}^{|\mathcal{J}|} \lambda_i\beta_i$ (see Fig. 4). The resulting policy is sketched in Fig. 5, where C_x and P_y are the C- and P-state values, respectively. The scheme includes energy- and performance-aware workload classification rules that define the most energy efficient configuration to be applied to the serving core.

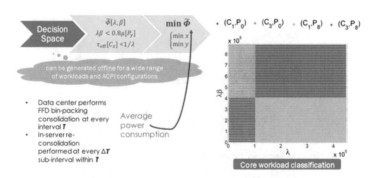

Fig. 5. Dynamic and long-term decision strategies.

Evaluations performed in [37] show that, despite VMs' workload variations, the total system workload is kept relatively stable, at approximately 18% below SLA specifications, and the policy applying both in-server consolidation and power scaling can gain about 10% with respect to baseline scenarios without the same capabilities. In a scaled-down datacenter example with 500 servers (with 2 octa-core processors each) and 10,000 VMs, at the average European Union electricity prices for industrial consumers of 0.12 €/kWh during the second half of 2014, this can turn to an annual saving of approximately 19,000 €.

We conclude the discussion in this Section with some further remarks about the evolution of the GAL. As we have already noted, a main challenge in the virtualised environment is that the correspondence between the HW that consumes energy (and belongs to the Infrastructure Provider) and the virtualized objects (VMs, containers, …) that execute Network Functions (and belong to the Network Service Provider) is not so straightforward as in the legacy networking infrastructure: the execution mediated by the hypervisor and its scheduling policies, the resource allocation performed through multiple functional modules, the presence of multiple tenants, among other factors, do not allow establishing a "direct" relation between virtual resources (e.g., vCPU) and Energy Aware States of the HW. As we noted in [39] in a more specific context, possible lines of action may comprise: (i) the use of queueing models for aggregated traffic only (per server/core); (ii) the adoption of simpler aggregated models for HW energy consumption (e.g., Generalized Amdhal's Law [40]), and of more detailed queueing models for execution machines; (iii) the introduction of "virtualized" Energy Aware States as backpressure from the Infrastructure Provider to create incentives toward tenants to become energy-aware (currently under investigation in ETSI for a second version of the GAL).

4   The Experience of Two H2020 European Projects

In conjunction with our previous discussion, two projects that we are currently coordinating, funded by the European Commission under the Horizon 2020 program, touch some of the issues we have raised and are attempting to provide some answers.

4.1   INPUT – In-Network Programmability for Next-Generation Personal Cloud Service Support

The INPUT project [41], started in January 2015, aims at designing a novel infrastructure and paradigm to support Future Internet personal cloud services in a more scalable and sustainable way and with innovative added-value capabilities. The INPUT technologies will enable next-generation cloud applications to go beyond classical service models, and even to replace physical Smart Devices SDs), usually placed in users' homes (e.g., network-attached storage servers, set-top-boxes, video recorders, home automation control units, etc.) or deployed around for monitoring purposes (e.g., sensors), with their virtual images, providing them to users "as a Service" (SD as a Service – SDaaS; see Fig. 6).

The INPUT Project defines a *virtual image* as a software instance that dematerializes a physical network-connected device, by providing its virtual presence in the network and all its functionalities. Virtual images are meant to realize smarter, always and everywhere accessible, performance-unlimited virtual devices into the cloud. They can be applied both to provide all the functionalities of fully dematerialized physical devices through the cloud, and to add potentially unlimited smartness and capacity to devices with performance- and functionality-constrained hardware platforms.

Virtual and physical SDs can be made available to users at any time and at any place by means of virtual cloud-powered *Personal Networks* (PNs), which constitute an underlying secure and trusted service model (Personal Network as a Service – PNaaS). PNs provide users with the perception of always being in their home Local Area Network with their own (virtual and physical) SDs, independently of their location.

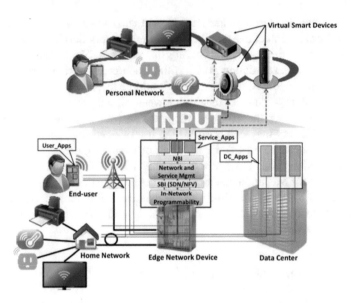

Fig. 6. The INPUT project general concept and structuring.

To achieve these ultimate objectives, the INPUT Project overcomes current limitations on the cloud service design due to the underlying obsolete network paradigms and technologies, by:

- introducing computing and storage capabilities to edge network devices ("in-network" programmability) in order to allow users/telecom operators to create/manage private clouds "in the network";
- moving cloud services closer to end-users and smart devices, in order both to avoid pointless network infrastructure and datacentre overloading, and to provide lower latency reactiveness to services;
- enabling personal and federated cloud services to natively and directly integrate themselves with the networking technologies close to end-user SDs, in order to provide new service models (e.g., the PN concept);
- assessing the validity of the proposed in-network cloud computing model through appropriately designed use cases and related proof-of-concept implementations.

As a side effect, the INPUT Project aims at fostering future-proof Internet infrastructures that will be "smarter," fully virtualized, power vs. performance optimised, and vertically integrated with cloud computing, with a clear impact on Telecom Operators', Service Providers' and end-users' CapEx and OpEx. In this respect, the INPUT Project is extending the programmability of network devices, to make them able to host cloud service applications, which cooperate with those in users' terminals and datacentres to realize the aforementioned cloud services. The INPUT approach and its infrastructural impact can contribute to the top line growth of European Telecom Operators and help increasing their revenue opportunities, enabling them to offer their infrastructure in support of novel value-added personal cloud services with reduced investments and operating expenses. To this purpose, "in-network" programmable network devices have been designed on top of state-of-the-art off-the-shelf hardware with advanced power management capabilities, and suitable consolidation and orchestration mechanisms have been developed to optimize energy consumption and user-perceived QoE.

Central to the INPUT architecture are the concepts, illustrated in Fig. 6, of cloud applications (*Service_Apps*) hosted in network edge devices, and of their capability of cooperating with and of offloading corresponding applications residing in the users' smart objects (*User_Apps*) and in datacentres (*DC_Apps*), to realize innovative personal cloud services.

The presence of such Service_Apps allows user requests to be manipulated before crossing the network and arriving at datacentres in ways that enhance performance. Such manipulations can include pre-processing, decomposition and proxying. Moreover, Service_Apps take advantage of a vertical integration in the network environment, where applications can benefit from network-cognitive capabilities to intercept traffic or to directly deal with network setup configurations and parameters. The integration of Service_Apps at the network edge level is a fundamental aspect, since this level is the one where the Telecom Operator terminates the user network access, and a direct trusting/control on user accounts and services is performed. Therefore, this level is the best candidate to host personal Service_Apps, and to provide novel network-integrated capabilities to the cloud environment in a secure and trusted fashion. To achieve this purpose, the INPUT Project has been also focused on the evolution of network devices acting at this level beyond the latest state-of-the-art SDN and NFV technologies, and on how to interface them with the "in-network" programmability. This approach enables the reduction of reaction times of cloud applications, by exploiting the ability to directly access network primitives, and by providing improved scalability in the interactions of the network with users and datacentres.

As shown in Fig. 7, the INPUT Project is centred on a multi-layered framework that allows, on one hand, multiple Personal Cloud Providers to request IT (e.g., in terms of computing, storage, caching, etc.) and network resources of the Telecom Infrastructure Provider via an extended Service Layer Agreement. On the other hand, in order to minimize the OpEx and increase the sustainability of its programmable network infrastructure, the Telecom Infrastructure Provider can make use of advanced Consolidation criteria that allow Service_Apps to be dynamically allocated and seamlessly migrated/split/joined on a subset of the available hardware resources. The unused hardware components can enter low-power standby states. The presence of these power management criteria and schemes is a key aspect for the maximisation of the Return on Investment (RoI) of the INPUT technology to Telecom Infrastructure Providers.

More detailed presentations of the INPUT outcomes can be found in the project-related publications [42]. Instrumental to the realisation of the project demonstrators and prototypes has been the development of OpenVolcano (Open Virtualization Operating Layer for Cloud/fog Advanced NetwOrks) [43, 44], a comprehensive open-source platform for Fog and Mobile Edge Computing.

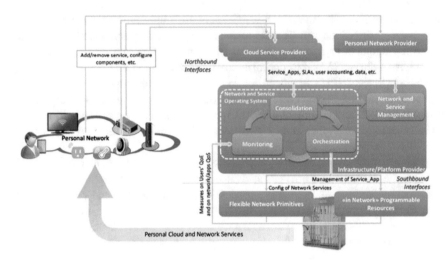

Fig. 7. INPUT architectural elements and their interaction.

4.2   MATILDA – A Holistic, Innovative Framework for Design, Development and Orchestration of 5G-Ready Applications and Network Services Over Sliced Programmable Infrastructure

MATILDA [45, 46] is an H2020 5G PPP (Public Private Partnership) project started in July 2017. Its vision is to design and implement a holistic 5G end-to-end services operational framework tackling the lifecycle of design, development and orchestration of 5G-ready applications and 5G network services over programmable infrastructure, following a unified programmability model and a set of control abstractions. MATILDA aims to devise and realize a radical shift in the development of software for 5G-ready applications, as well as virtual and physical network functions and network services, through the adoption of a unified programmability model, the definition of proper abstractions and the creation of an open development environment that may be used by application and network functions developers.

Intelligent and unified orchestration mechanisms will be applied for the automated placement of the 5G-ready applications and the creation and maintenance of the required network slices. Deployment and runtime policies enforcement is provided through a set of optimisation mechanisms realising deployment plans based on high-level objectives and a set of mechanisms supporting runtime adaptation of the application components and/or network functions based on policies defined on behalf of a services provider.

Multi-site management of the cloud/edge computing and IoT (Internet of Things) resources is supported by a multi-site virtualized infrastructure manager, while the lifecycle management of the supported VNF Forwarding Graphs (VNF-FGs), as well as a set of network management activities, are provided by a multi-site NFV Orchestrator (NFVO). Network and application-oriented analytics and profiling mechanisms are supported based on both real-time and a posteriori processing of the collected data from a set of monitoring streams. The developed 5G-ready application components, applications, virtual network functions and application-aware network services are made available for open-source or commercial purposes, re-use and extension through a 5G marketplace.

The MATILDA project envisions the design and development of a holistic framework that supports the tight interconnection among the creation of 5G-ready applications and of the on-demand required networking and computational infrastructure, in the form of an application-aware network slice, and the activation of the appropriate networking mechanisms for the support of industry-vertical applications.

The MATILDA layers along with the main artefacts and key technological concepts comprising the MATILDA framework per layer are depicted in Fig. 8.

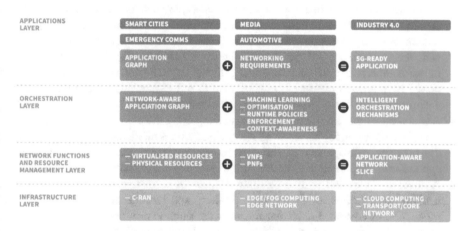

Fig. 8. The MATILDA general architectural framework.

The Applications layer corresponds to the Business Service and Business Function layer and regards the design and development of the 5G-ready applications per industry-vertical, along with the specification of the associated networking requirements. The Orchestration Layer regards the support of deployment and optimisation mechanisms of the 5G-ready applications over the available multi-site programmable infrastructure. Orchestration refers to both the application components and the attached virtual network functions and includes a set of intelligent mechanisms for optimal deployment, runtime policies enforcement, data mining and analysis, and context awareness support. The Network Functions and Resource Management Layer regards the implementation of the resource management functionalities over the available programmable infrastructure, as well as the lifecycle management of the activated virtual network functions. The Infrastructure Layer consists of the data communication network spanning a set of cloud computing and storage resources.

The key technological concepts and artefacts comprising the proposed MATILDA framework and constituting its unique selling points are:

- A conceptual architecture to support the provision of 5G end-to-end services tackling the overall lifecycle of design, development and orchestration of 5G-ready applications and 5G network services over a programmable infrastructure.
- A set of meta-models representing the vertical industry applications' components and graphs, the virtual – and physical – network functions and forwarding graphs.
- An innovative collaborative environment supporting the design and development of 5G-ready applications and VNF-FGs, including a web-based integrated development environment (IDE), verification and graphs' composition mechanisms.
- An orchestrator that is in charge of the optimal deployment and orchestration of the developed applications over the available programmable infrastructure – taking into account a set of objectives and constraints, as well as the defined policies, along with the instantiation of the required network functions for the support of the infrastructural-oriented functionalities. Policies enforcement is going to be supported by a context-awareness engine, able to infer knowledge based on a set of data monitoring, analytics and profiling production streams.
- A multi-site virtual infrastructure manager supporting the multi-site management of the allocated resources per network slice, along with a multi-site NFVO supporting the lifecycle management of the network functions embedded in the deployed application's graph and a set of network monitoring and management mechanisms.
- A novel analytics and unified profiling framework consisting of a set of machine learning mechanisms, of design time profiling and runtime profiling, toward the production of advanced analytics and software runtime profiling.
- A marketplace including an applications' and virtual network functions' repository and a set of mechanisms for the support of diverse 5G stakeholders.

5   Conclusions

The evolution of networks in the light of their "softwarisation" and virtualisation process and of the integration of diverse forms of access and transport paradigms has gained even greater impulse with the advent of 5G, the fifth-generation mobile network. In this framework, flexibility and programmability have become of paramount relevance, and new challenging KPIs have been set. Our attention has been focused on the aspect of network control and management of this complex heterogeneous environment, which appears to be sometimes hidden behind the architectural and operational constructs that form the basis for the virtualisation of resources and the orchestration of the physical and virtual elements. At the same time, we have tried to highlight the interaction of resource allocation policies performed for the purpose of attaining QoE/QoS-related KPIs with energy consumption of the physical network elements. The relation between certain network operations and their impact on energy consumption of the infrastructure is somehow blurred by the numerous mediating software and orchestration levels necessary to achieve the virtualised functionalities that ensure the two desired (and by now non-renounceable) aspects of network and services flexibility and programmability. In this respect, we have examined some potentially critical aspects, pointed out possible ways of coping with them, and briefly described some current project approaches.

References

1. Software-defined networking: the new norm for networks. In: Open Networking Foundation Whitepaper (2012). https://www.opennetworking.org/images/stories/downloads/sdn-resources/white-papers/wp-sdn-newnorm.pdf
2. Nunes, B.A.A., Mendonça, M., Nguyen, X.-N., Obraczka, K., Turletti, T.: A survey of software-defined networking: past, present, and future of programmable networks. IEEE Commun. Surv. Tut. 16(3), 1617–1634 (2014)
3. Kreutz, D., Ramos, F.M.V., Veríssimo, P.E., Rothenberg, C.E., Azodolmolky, S., Uhlig, S.: Software-defined networking: a comprehensive survey. Proc. IEEE 103(1), 14–76 (2015)
4. ONF TR-521, SDN Architecture (2016). https://www.opennetworking.org/images/stories/downloads/sdn-resources/technical-reports/TR-521_SDN_Architecture_issue_1.1.pdf
5. ETSI GS NFV 002 V1.1.1 (2013-10) Network Functions Virtualization (NFV); Architectural Framework. http://www.etsi.org/deliver/etsi_gs/nfv/001_099/002/01.01.01_60/gs_nfv002v010101p.pdf
6. Mijumbi, R., Serrat, J., Gorricho, J.-L., Bouten, N., De Turck, F., Boutaba, R.: Network function virtualization: state-of-the-art and research challenges. IEEE Commun. Surv. Tut. 18(1), 236–262 (2016)
7. Network functions virtualisation. In: Introductory White Paper, ETSI (2012). https://portal.etsi.org/nfv/nfv_white_paper.pdf
8. Synergy Research Group (2016). https://www.srgresearch.com/articles/enterprise-spending-nudged-downwards-2016-cisco-maintains-big-lead
9. http://www.etsi.org/technologies-clusters/technologies/multi-access-edge-computing
10. Manzalini, A., et al.: Towards 5G software-defined ecosystems – technical challenges, business sustainability and policy issues. In: IEEE SDN Initiative Whitepaper (2016). http://resourcecenter.fd.ieee.org/fd/product/white-papers/FDSDNWP0002
11. Expert Advisory Group of the European Technology platform Networld 2020, Strategic Research and Innovation Agenda (2016). Pervasive Mobile Virtual Services. https://www.networld2020.eu/wp-content/uploads/2014/02/SRIA_final.pdf
12. Ordonez-Lucena, J., Ameigeiras, P., Lopez, D., Ramos-Munoz, J.J., Lorca, J., Folgueira, J.J.: Network slicing for 5G with SDN/NFV: concepts architectures and challenges. IEEE Commun. Mag. 55(5), 80–87 (2017)
13. Suarez, L., Nuaymi, L., Bonnin, J.-M.: An overview and classification of research approaches in green wireless networks. EURASIP J. Wirel. Commun. Netw. 2012, 142 (2012)
14. Orgerie, A.C., Dias de Assunção, M., Lefevre, L.: A survey on techniques for improving the energy efficiency of large scale distributed systems. ACM Comp. Surv. 46(4), 1–35 (2014)
15. Moghaddam, F.A., Lago, P., Grosso, P.: Energy-efficient networking solutions in cloud-based environments: a systematic literature review. ACM Comp. Surv. 47(4), 64.1–64.32 (2015)
16. Bolla, R., Bruschi, R., Davoli, F., Cucchietti, F.: Energy efficiency in the future internet: a survey of existing approaches and trends in energy-aware fixed network infrastructures. IEEE Commun. Surv. Tut. 13(2), 223–244 (2011)

17. Bianzino, A.P., Chaudet, C., Rossi, D., Rougier, J.-L.: A survey of green networking research. IEEE Commun. Surv. Tut. **14**(1), 3–20 (2012)
18. Idzikowski, F., Chiaraviglio, F.L., Cianfrani, A., López Vizcaíno, A.J., Polverini, M., Ye, M. Y.: A survey on energy-aware design and operation of core networks. IEEE Commun. Surv. Tut. **18**(2), 1453–1499 (2016)
19. Global e-Sustainability Initiative (GeSI), SMARTer2020: The Role of ICT in Driving a Sustainable Future. http://gesi.org/SMARTer2020
20. http://uefi.org/specifications
21. Bolla, R., Bruschi, R., Davoli, F., Cucchietti, F.: Setting the course for a green internet. Science **342**(6164), 1316 (2013)
22. Bolla, R., et al.: The green abstraction layer: a standard power management interface for next-generation network devices. IEEE Intern. Comp. **17**(2), 82–86 (2013)
23. Bolla, R., et al.: A northbound interface for power management in next generation network devices. IEEE Commun. Mag. **52**(1), 149–157 (2014)
24. Green Abstraction Layer (GAL): Power management capabilities of the future energy telecommunication fixed network nodes (2013) ETSI Std. 203 237 version 1.1.1. http://www.etsi.org/deliver/etsi_es/203200_203299/203237/01.01.01_60/es_203237v010101p.pdf
25. Bolla, R., Bruschi, R., Carrega, A., Davoli, F.: Green networking with packet processing engines: modeling and optimization. IEEE/ACM Trans. Netw. **22**(1), 110–123 (2014)
26. Bolla, R., Bruschi, R., Carrega, A., Davoli, F., Pajo, J.F.: Corrections to: "Green Networking with Packet Processing Engines: Modeling and Optimization". IEEE/ACM Trans. Netw. (2017). https://doi.org/10.1109/TNET.2017.2761892
27. Niewiadomska-Szynkiewicz, E., Sikora, A., Arabas, P., Kołodziej, J.: Control system for reducing energy consumption in backbone computer network. Concurr. Computat. Pract. Exper. **25**, 1738–1754 (2013)
28. Kamola, M., Niewiadomska-Szynkiewicz, E., Arabas, P., Sikora, A.: Energy-saving algorithms for the control of backbone networks: a survey. J. Telecommun. Inform. Technol. **2**, 13–20 (2016)
29. Filipiak, J.: Modelling and Control of Dynamic Flows in Communication Networks. Springer-Verlag, Berlin (1988)
30. Bruschi, R., Davoli, F., Mongelli, M.: Adaptive frequency control of packet processing engines in telecommunication networks. IEEE Commun. Lett. **18**(7), 1135–1138 (2014)
31. Neely, M.J.: Stochastic Network Optimization with Application to Communication and Queueing Systems. Morgan & Claypool, San Rafael (2010)
32. Bruschi, R., Carrega, A., Davoli, F.: A game for energy-aware allocation of virtualized network functions. J. Elec. Comp. Eng. **2016**, 4067186 (2016)
33. Aicardi, M., Bruschi, R., Davoli, F., Lago, P.: A decentralized team routing strategy among telecom operators in an energy-aware network. In: Proceedings of SIAM Conference on Control and its Application, Paris, France, July 2015, pp. 340–347 (2015)
34. Rahman, M.M., Despins, C., Affes, S.: Analysis of CAPEX and OPEX benefits of wireless access virtualization. In: IEEE Internat Conference Communications (ICC), Workshop on Energy Efficiency in Wireless Networks & Wireless Networks for Energy Efficiency (E2Nets), Budapest, Hungary (2013)
35. Bolla, R., Bruschi, R., Davoli, F., Lombardo, C., Pajo, J.F., Sanchez, O.R.: The dark side of network functions virtualization: a perspective on the technological sustainability. In: IEEE Internat Conference on Communications (ICC), Paris, France, 2017, pp. 1–7 (2017)
36. Woo, D., Lee, H.-S.: Extending Amdahl's Law for energy-efficient computing in the many-core era. IEEE Comput. **41**(12), 24–31 (2008)

37. Bruschi, R., Davoli, F., Lago, P., Pajo, J.F.: Joint power scaling of processing resources and consolidation of virtual network functions. In: 5th IEEE International Conference on Cloud Networking (CloudNet), Pisa, Italy, October 2016, pp. 70–75 (2016)
38. Coffman Jr., E.G., Garey, M.R., Johnson, D.S.: Approximation algorithms for bin packing: a survey. In: Hochbaum, D.S. (ed.) Approximation Algorithms for NP-Hard Problems. PWS Publishing, Boston (1996)
39. Bolla, R., Bruschi, R., Davoli, F., Depasquale, E.V.: Energy-efficient management and control in video distribution networks: "Legacy" hardware based solutions and perspectives of virtualized networking environments. In: Popescu, A. (ed.) Guide to Greening Video Distribution Networks - Energy-Efficient Internet Video Delivery, pp. 25–57. Springer (2018)
40. Cassidy, A.S., Andreou, A.G.: Beyond Amdahl's Law: an objective function that links multiprocessor performance gains to delay and energy. IEEE Trans. Comp. **61**(8), 1110–1126 (2012)
41. http://www.input-project.eu/
42. http://www.input-project.eu/index.php/outcomes/publications
43. Bruschi, R., Lago, P., Lombardo, C., Mangialardi, S.: OpenVolcano: an open-source software platform for fog computing. In: 28th International Teletraffic Congress (ITC 28) 1st International Workshop on Programmability for Cloud Networks and Applications (PROCON), Wuerzburg, Germany, September 2016
44. http://openvolcano.org/
45. Bolla, R., et al.: Design, development and orchestration of 5G-ready applications over sliced programmable infrastructure. In: 29th International Teletraffic Congress (ITC 28) 1st International Workshop on Softwarized Infrastructures for 5G and Fog Computing (Soft5 2017), Genoa, Italy, September 2017, pp. 13–18 (2017)
46. http://www.matilda-5g.eu/index.php

Fast Algorithms for Computing Continuous Piecewise Affine Lyapunov Functions

Sigurdur Hafstein[(⊠)]

University of Iceland, Dunhagi 5, 107 Reykjavik, Iceland
shafstein@hi.is
https://notendur.hi.is/shafstein/

Abstract. Algorithms that parameterize continuous and piecewise affine Lyapunov functions for nonlinear systems, both in continuous and discrete time, have been proposed in numerous publications. These algorithms generate constraints that are linear in the values of a function at all vertices of a simplicial complex. If these constraints are fulfilled for certain values at the vertices, then they can be interpolated on the simplices to deliver a function that is a Lyapunov function for the system used for their generation. There are two different approaches to find values that fulfill the constraints. First, one can use optimization to compute appropriate values that fulfill the constraints. These algorithms were originally designed for continuous-time systems and their adaptation to discrete-time systems and control systems poses some challenges in designing and implementing efficient algorithms and data structures for simplicial complexes. Second, one can use results from converse theorems in the Lyapunov stability theory to generate good candidates for suitable values and then verify the constraints for these values. In this paper we study several efficient data structures and algorithms for these computations and discuss their implementations in C++.

Keywords: Simplicial complex · Algorithm · Lyapunov function
Nonlinear system

1 Introduction

A Lyapunov function V for a dynamical system is a continuous function from the state-space to the real numbers that has a minimum at an equilibrium and is decreasing along the system's trajectories. For a continuous-time system given by a differential equation $\mathbf{x}' = \mathbf{f}(\mathbf{x})$ the decrease along solution trajectories can be ensured by the condition

$$D_{\mathbf{f}}V(\mathbf{x}) := \nabla V(\mathbf{x}) \cdot \mathbf{f}(\mathbf{x}) < 0. \tag{1}$$

For a discrete-time system $\mathbf{x}_{k+1} = \mathbf{g}(\mathbf{x}_k)$ the corresponding condition is

$$\Delta_{\mathbf{g}}V(\mathbf{x}) := V(\mathbf{g}(\mathbf{x})) - V(\mathbf{x}) < 0. \tag{2}$$

© Springer Nature Switzerland AG 2019
M. S. Obaidat et al. (Eds.): SIMULTECH 2017, AISC 873, pp. 274–299, 2019.
https://doi.org/10.1007/978-3-030-01470-4_15

In [13,19,32] novel algorithms for the computation of Lyapunov functions for nonlinear discrete-time systems were presented. In these algorithms the relevant part of the state-space is first triangulated, i.e. subdivided into simplices, and then a continuous and piecewise affine (CPA) Lyapunov function is parameterized by fixing its values at the vertices of the simplices. These algorithms resemble earlier algorithms for the computation of Lyapunov functions for nonlinear continuous-time systems, cf. e.g. [4,15,16,24,25,33,34], referred to as the CPA algorithm. The essential idea is to formulate the conditions for a Lyapunov function as linear constraints in the values of the Lyapunov function to be computed at the vertices of the simplices of the simplicial complex.

The implementation of these algorithms for discrete-time systems can be done similarly to the continuous-time case. First a simplicial complex is constructed that triangulates the relevant part of the state-space. Then an appropriate linear programming problem for the system at hand is generated, of which any feasible solution parameterizes a Lyapunov function for the system. Then one either uses a linear programming solver, e.g. GLPK or Gurobi, to search for a feasible solution, or one uses results from converse theorems in the Lyapunov stability theory to compute values that can be expected to fulfill the constraints and then verifies if these computed values constitute a feasible solution to the linear programming problem.

The non-locality of the dynamics in the discrete-time case, however, poses an additional challenge in implementing the construction of a suitable simplicial complex and the generation of the linear constraints in an efficient way. Namely, whereas the condition (1) for a continuous-time system is a local condition that can be formulated as linear constraints for each simplex, the condition (2) for a discrete-time system is not local. For a vertex $\mathbf{x}$ of a simplex $\mathfrak{S}_\nu$ in the triangulation $\mathcal{T}$ we must be able find a simplex $\mathfrak{S}_\mu \in \mathcal{T}$ such that $\mathbf{g}(\mathbf{x}) \in \mathfrak{S}_\mu$ to formulate this condition as linear constraints and we must know the barycentric coordinates of $\mathbf{g}(\mathbf{x})$ in $\mathfrak{S}_\mu$. For triangulations consisting of many simplices a linear search is very inefficient and therefore more advanced methods are called for.

The first contribution of this paper is an algorithm that efficiently searches for a simplex $\mathfrak{S}_\mu \in \mathcal{T}$ such that $\mathbf{x} \in \mathfrak{S}_\mu$ and computes its barycentric coordinates for fairly general simplicial complexes, that were specially designed for our problem of computing Lyapunov functions.

The CPA algorithm has additionally been adapted to compute Lyapunov functions for differential inclusions [2] and control Lyapunov functions [3] in the sense of Clarke's subdifferential [9]. The next logical step is to compute control Lyapunov functions in the sense of the Dini subdifferential, a work in progress with promising first results. For these computations one needs information on the common faces of neighbouring simplices in the simplicial complex and detailed information on normals of the hyperplanes separating neighbouring simplices. Efficient algorithms and data structures for these computations are presented. This is the second contribution of this paper.

The third contribution is an algorithms to compute circumscribing hyperspheres of the simplices of the simplicial complex. These can be used to implement more advanced algorithms for the computation of Lyapunov functions for discrete-time systems, also a work in progress.

The fourth contribution is an efficient algorithm combining the four-step Adam-Bashforth method for initial-value problems and Simpson's Rule for numerical integration to approximate values of a Lyapunov function from a converse theorem in the Lyapunov stability theory [35] at the vertices of the simplicial complex. We also undertake a detailed error analysis of our approach.

The first three contributions are the same as in [18] but improved and advanced in numerous ways. We discuss more general functions **F** in Sect. 3 than in [18] and in all three the algorithms and data structures have been tweaked for performance. For example, Stroustrup's statements, e.g. [42], motivated us to replace linked lists with vectors in several places, and because the old code relied on member functions of stl::list the code had to be adapted to use stl::vector and/or suitable functions from stl::algorithm where appropriate. Further, care must be taken to avoid methods that are inheritably inefficient for vectors. We try to keep the discussion largely self-contained, but to keep it at a reasonable length we avoid repetitions of material presented in [18] that is not necessary to understand the approach here. We thus refer to [18] for numerous issues and more detailed results and keep the same notation. The fourth contribution has not been published in any form before.

Before we describe our algorithms in Sects. 2.1 and 3, we first discuss suitable triangulations for the computation of CPA Lyapunov functions in Sect. 2. In Sect. 4 we describe, analyze and give the implementation of our method to approximate a Lyapunov function from a converse theorem at the vertices of a simplicial complex and in Sect. 5 we sum up the contributions give a few concluding remarks.

1.1   Notation

We denote by $\mathbb{Z}$, $\mathbb{N}_0$, $\mathbb{R}$, and $\mathbb{R}_+$ the sets of the integers, the nonnegative integers, the real numbers, and the nonnegative real numbers respectively. For integers $r, s \in \mathbb{Z}$, $r < s$, we write $r : s$ for $r, r+1, \ldots, s$. We write vectors in boldface, e.g. $\mathbf{x} \in \mathbb{R}^n$ and $\mathbf{y} \in \mathbb{Z}^n$, and their components as $x_1, x_2, \ldots, x_n$ and $y_1, y_2, \ldots, y_n$. All vectors are assumed to be column vectors unless specified otherwise. An inequality for vectors is understood to be component-wise, e.g. $\mathbf{x} < \mathbf{y}$ means that all the inequalities $x_1 < y_2$, $x_2 < y_2, \ldots, x_n < y_n$ are fulfilled. The null vector in $\mathbb{R}^n$ is written as $\mathbf{0}$ and the standard orthonormal basis as $\mathbf{e}_1, \mathbf{e}_2, \ldots, \mathbf{e}_n$, i.e. the i-th component of $\mathbf{e}_j$ is equal to $\delta_{i,j}$, where $\delta_{i,j}$ is the Kronecher delta, equal to 1 if $i = j$ and 0 otherwise. The scalar product of vectors $\mathbf{x}, \mathbf{y} \in \mathbb{R}^n$ is denoted by $\mathbf{x} \cdot \mathbf{y}$, the Euclidean norm of $\mathbf{x}$ is denoted by $\|\mathbf{x}\|_2 := \sqrt{\mathbf{x} \cdot \mathbf{x}}$, and the maximum norm of $\mathbf{x}$ is denoted by $\|\mathbf{x}\|_\infty := \max_{i=1:n} |x_i|$. The transpose of a vector $\mathbf{x}$ is denoted by $\mathbf{x}^T$ and similarly the transpose of a matrix $A \in \mathbb{R}^{n \times m}$ is denoted by A^T. For a nonsingular matrix $A \in \mathbb{R}^{n \times n}$ we denote its inverse by A^{-1} and the inverse of its transpose by A^{-T}. This should not lead to misunderstandings since $(A^{-1})^T = (A^T)^{-1}$. In the rest of the paper n and in the code the global variable const int n is the dimension of the Euclidean space we are working in.

We write sets $\mathcal{K} \subset \mathbb{R}^n$ in calligraphic and we denote the closure, interior, and the boundary of $\mathcal{K}$ by $\overline{\mathcal{K}}$, $\mathcal{K}^\circ$, and $\partial\mathcal{K}$ respectively.

The *convex hull* of an $(m + 1)$-tuple $(\mathbf{v}_0, \mathbf{v}_1, \ldots, \mathbf{v}_m)$ of vectors $\mathbf{v}_0, \mathbf{v}_1, \ldots, \mathbf{v}_m \in \mathbb{R}^n$ is defined by

$$\mathrm{co}(\mathbf{v}_0, \mathbf{v}_1, \ldots, \mathbf{v}_m) := \left\{ \sum_{i=0}^m \lambda_i \mathbf{v}_i : 0 \leq \lambda_i, \sum_{i=0}^m \lambda_i = 1 \right\}.$$

If $\mathbf{v}_0, \mathbf{v}_1, \ldots, \mathbf{v}_m \in \mathbb{R}^n$ are affinely independent, i.e. the vectors $\mathbf{v}_1 - \mathbf{v}_0, \mathbf{v}_2 - \mathbf{v}_0, \ldots, \mathbf{v}_m - \mathbf{v}_0$ are linearly independent, the set $\mathrm{co}(\mathbf{v}_0, \mathbf{v}_1, \ldots, \mathbf{v}_m)$ is called an *m-simplex*. For a subset $\{\mathbf{v}_{i_0}, \mathbf{v}_{i_1}, \ldots, \mathbf{v}_{i_k}\}$, $0 \leq k < m$, of affinely independent vectors $\{\mathbf{v}_0, \mathbf{v}_1, \ldots, \mathbf{v}_m\}$, the k-simplex $\mathrm{co}(\mathbf{v}_{i_0}, \mathbf{v}_{i_1}, \ldots, \mathbf{v}_{i_k})$ is called a *k-face* of the simplex $\mathrm{co}(\mathbf{v}_0, \mathbf{v}_1, \ldots, \mathbf{v}_m)$. Note that simplices are usually defined as convex combinations of vectors in a set and not of ordered tuples, i.e. $\mathrm{co}\{\mathbf{v}_0, \mathbf{v}_1, \ldots, \mathbf{v}_m\}$ rather than $\mathrm{co}(\mathbf{v}_0, \mathbf{v}_1, \ldots, \mathbf{v}_m)$. For the implementation of the simplicial complexes below it is however very useful to stick to ordered tuples. A function $\rho : \mathbb{R}_+ \to \mathbb{R}_+$ is said to be of class $\mathcal{K}_\infty$ if it is continuous, strictly increasing, and fulfills $\rho(0) = 0$ and $\lim_{x \to \infty} \rho(x) = \infty$.

In the implementations of the algorithms we make heavy use of the Standard C++ Library and the Armadillo linear algebra library [40]. We thus always assume in the code that using namespace std and using namespace arma have been declared. Further, we assume that all the necessary libraries are accessible. Very good documentation on Armadillo is available at http://arma. sourceforge.net and some comments on its use for the implementation of the basic simplicial complex in Sect. 2 are also given in [17]. The vector and matrix types of Armadillo we use in this paper are ivec, vec, and mat, which represent a column vector of int, a column vector of double, and a matrix of double respectively.

2   The Simplicial Complex $\mathcal{T}_{N,K}^{\mathrm{std}}$

In [17] the simplicial complex $\mathcal{T}_{N,K}^{\mathrm{std}}$ and its efficient implementation is elaborately described. For completeness we recall its definition but refer to [17] for the details. To define the simplicial complex $\mathcal{T}_{N,K}^{\mathrm{std}}$ we first need several preliminary definitions.

An *admissible triangulation* of a set $\mathcal{C} \subset \mathbb{R}^n$ is the subdivision of $\mathcal{C}$ into n-simplices, such that the intersection of any two different simplices in the subdivision is either empty or a common k-face, $0 \leq k < n$. Such a structure is often referred to as a *simplicial n-complex*.

For the definition of $\mathcal{T}_{N,K}^{\mathrm{std}}$ we use the set S_n of all permutations of the set $\{1 : n\}$, the characteristic functions $\chi_{\mathcal{J}}(i)$ equal to one if $i \in \mathcal{J}$ and equal to zero if $i \notin \mathcal{J}$. Further, we use the functions $\mathbf{R}^{\mathcal{J}} : \mathbb{R}^n \rightarrow \mathbb{R}^n$, defined for every $\mathcal{J} \subset \{1 : n\}$ by

$$\mathbf{R}^{\mathcal{J}}(\mathbf{x}) := \sum_{i=1}^{n} (-1)^{\chi_{\mathcal{J}}(i)} x_i \mathbf{e}_i.$$

Thus $\mathbf{R}^{\mathcal{J}}(\mathbf{x})$ puts a minus in front of the i-th coordinate of $\mathbf{x}$ whenever $i \in \mathcal{J}$.

To construct the triangulation $\mathcal{T}_{N,K}^{\mathrm{std}}$, we first define the triangulations $\mathcal{T}_{N}^{\mathrm{std}}$ and $\mathcal{T}_{K,\mathrm{fan}}^{\mathrm{std}}$ as intermediate steps.

Definition of $\mathcal{T}_{N,K}^{\mathrm{std}}$

1. For every $\mathbf{z} \in \mathbb{N}_0^n$, every $\mathcal{J} \subset \{1 : n\}$, and every $\sigma \in S_n$ define the simplex

$$\mathfrak{S}_{\mathbf{z}\mathcal{J}\sigma} := \mathrm{co}(\mathbf{x}_0^{\mathbf{z}\mathcal{J}\sigma}, \mathbf{x}_1^{\mathbf{z}\mathcal{J}\sigma}, \ldots, \mathbf{x}_n^{\mathbf{z}\mathcal{J}\sigma}) \tag{3}$$

where

$$\mathbf{x}_i^{\mathbf{z}\mathcal{J}\sigma} := \mathbf{R}^{\mathcal{J}}\left(\mathbf{z} + \sum_{j=1}^{i} \mathbf{e}_{\sigma(j)}\right) \quad \text{for } i = 0 : n. \tag{4}$$

2. Let $\mathbf{N}^m, \mathbf{N}^p \in \mathbb{Z}^n$, $\mathbf{N}^m < \mathbf{0} < \mathbf{N}^p$, and define the hypercube $\mathcal{N} := \{\mathbf{x} \in \mathbb{R}^n : \mathbf{N}^m \leq \mathbf{x} \leq \mathbf{N}^p\}$. The simplicial complex $\mathcal{T}_{N}^{\mathrm{std}}$ is defined by

$$\mathcal{T}_{N}^{\mathrm{std}} := \{\mathfrak{S}_{\mathbf{z}\mathcal{J}\sigma} : \mathfrak{S}_{\mathbf{z}\mathcal{J}\sigma} \subset \mathcal{N}\}. \tag{5}$$

3. Let $\mathbf{K}^m, \mathbf{K}^p \in \mathbb{Z}^n$, $\mathbf{N}^m \leq \mathbf{K}^m < \mathbf{0} < \mathbf{K}^p \leq \mathbf{N}^p$, and consider the intersections of the n-simplices $\mathfrak{S}_{\mathbf{z}\mathcal{J}\sigma}$ in $\mathcal{T}_{N}^{\mathrm{std}}$ and the boundary of the hypercube $\mathcal{K} := \{\mathbf{x} \in \mathbb{R}^n : \mathbf{K}^m \leq \mathbf{x} \leq \mathbf{K}^p\}$. We are only interested in those intersections that are $(n-1)$-simplices, i.e. intersections that can be written as $\mathrm{co}(\mathbf{v}_1, \mathbf{v}_2, \ldots, \mathbf{v}_n)$ with exactly n-vertices. For every such intersection add the origin as a vertex to it, i.e. consider $\mathrm{co}(\mathbf{0}, \mathbf{v}_1, \mathbf{v}_2, \ldots, \mathbf{v}_n)$. The set of such constructed n-simplices is denoted $\mathcal{T}_{K,\mathrm{fan}}^{\mathrm{std}}$. It is a triangulation of the hypercube $\mathcal{K}$.

4. Finally, we define our main simplicial complex $\mathcal{T}_{N,K}^{\mathrm{std}}$ by letting it contain all simplices $\mathfrak{S}_{\mathbf{z}\mathcal{J}\sigma}$ in $\mathcal{T}_{N}^{\mathrm{std}}$, that have an empty intersection with the interior $\mathcal{K}^\circ$ of $\mathcal{K}$, and all simplices in the simplicial fan $\mathcal{T}_{K,\mathrm{fan}}^{\mathrm{std}}$. It is thus a triangulation of $\mathcal{N}$ having a simplicial fan in $\mathcal{K}$.

The triangulation $\mathcal{T}_{K,\mathrm{fan}}^{\mathrm{std}}$ of the hypercube $\mathcal{K} := \{\mathbf{x} \in \mathbb{R}^n : \mathbf{K}^m \leq \mathbf{x} \leq \mathbf{K}^p\}$ is a straightforward extension of the 3D graphics primitive *triangular fan* to arbitrary dimensions. Therefore the term *simplicial fan*. For a graphical presentation of the complex $\mathcal{T}_{N,K}^{\mathrm{std}}$ with $n = 2$ see Fig. 1 taken form [18]. For figures of the complex with $n = 3$ see Figs. 2 and 3 in [17].

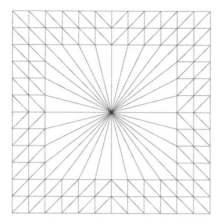

Fig. 1. The simplicial complex $\mathcal{T}_{N,K}^{\mathrm{std}}$ in two dimensions with $\mathbf{K}^m = (-4, -4)^T, \mathbf{K}^p = (4, 4)^T, \mathbf{N}^m = (-6, -6)^T$, and $\mathbf{N}^p = (6, 6)^T$.

The class T_std_NK implements the basic simplicial complex $\mathcal{T}_{N,K}^{\mathrm{std}}$ is. It is defined as follows:

```
 1  class T_std_NK {
 2  public:
 3      ivec Nm,Np,Km,Kp;
 4      Grid G;
 5      int Nr0;
 6      vector<ivec> Ver;
 7      vector<vector<int>> Sim;
 8      vector<zJs> NrInSim;
 9      vector<int> Fan;
10      int InSimpNr(vec x);    // returns -1 if not found
11      bool InSimp(vec x,int s);
12      T_std_NK(ivec Nm,ivec Np,ivec Km,ivec Kp);
13      // added since (Hafstein, 2013) below
14      vector<vector<int>> SimN;
15      vector<vector<int>> SCV;
16      vector<vector<vector<int>>> Faces;
17      vector<int> BSim;
18      int SVerNr(int s,int i);
19      ivec SVer(int s,int i);
20  };
```

Nm = $\mathbf{N}^m$ and Np = $\mathbf{N}^p$ define the hypercube $\mathcal{N}$ and Km = $\mathbf{K}^m$ and Kp = $\mathbf{K}^p$ define the hypercube $\mathcal{K}$. vector<ivec> Ver is a vector containing all the vertices of all the simplices in the complex, the value of int Nr0 is such that Ver[Nr0] is the zero vector, and vector<vector<int>> Sim is a vector containing all the simplices of the complex. Each simplex is stored as a vector of $(n + 1)$-integers, the integers referring to the positions of the vertices of the simplex in vector<ivec> Ver. G is a Grid initialized by ivec Nm,Np that is used to enumerate the vertices of T_std_NK and vector<int> Fan and vector<zJs> NrInSim are auxiliary structures that allow for a given $\mathbf{x} \in \mathbb{R}^n$ to efficiently compute an int s such that for $\mathfrak{S}_\nu$ =Sim[s] we have $\mathbf{x} \in \mathfrak{S}_\nu$. Their properties and implementation is described in detail in [17]. vector<vector<int>> SimN, vector<vector<int>> SCV, vector<vecotr<vector<int>>> Faces, and vector<int> BSimp are variables in the class T_std_NK that have been added since [17]. Their purpose and initialization is described in the next section.

2.1   Added Functionality in T_std_NK

In this section we describe the functionality that has been added to the class T_std_NK since [17]. The added functionality is as described in [18] but the implementation has been made more efficient. To simplify the notation we write $\mathcal{T}$ for the basic simplicial complex $\mathcal{T}_{N,K}^{\text{std}}$ from now on. Further we denote the set of all its vertices by $\mathcal{V}_\mathcal{T}$ and its domain by $\mathcal{D}_\mathcal{T} := \bigcup_{\mathfrak{S}_\nu \in \mathcal{T}} \mathfrak{S}_\nu$.

In [17] a fast algorithm is given to compute a simplex $\mathfrak{S}_\nu \in \mathcal{T}$ such that $\mathbf{x} \in \mathfrak{S}_\nu$, but fell back on linear search if $\mathbf{x}$ was in the simplicial fan $\mathcal{T}_{K,\text{fan}}^{\text{std}}$ of $\mathcal{T}$. Under the premise that the simplicial fan is much smaller than the rest of the simplicial complex this is a reasonable strategy. Therefore we did not add a fast search structure zJs to vector<zJs> NrInSim for simplices in the fan. However, for a simplicial complex for which this premise is not fulfilled, an improved strategy shortens the search time considerably. This might also be of importance to different computational methods for Lyapunov functions that use conic partitions of the state-space, a topic that has obtained considerable attention cf. e.g. [1, 8, 22, 23, 28–31, 36–39, 44].

We achieve this by first adding the simplices in $\mathcal{T}_{K,\text{fan}}^{\text{std}}$ with appropriate values for $\mathbf{z}$, $\mathcal{J}$, and σ to vector<zJs> NrInSim. This is very simple to make: In CODE BLOCK 1 in [17] directly after Fan.push_back(SLE); simply add

```
NrInSim.push_back(zJs(z,J,sigma,SLE));.
```

To actually find such simplices fast through their $\mathbf{z}$, $\mathcal{J}$, σ values a little more effort is needed. If $\mathbf{x} \in \mathbb{R}^n$ is in the simplicial fan of $\mathcal{T}$, i.e. if $\mathbf{K}^m < \mathbf{x} < \mathbf{K}^p$, we project $\mathbf{x}$ radially just below the boundary of the hypercube $\mathcal{K} := \{\mathbf{y} \in \mathbb{R}^n : \mathbf{K}^m \le \mathbf{y} \le \mathbf{K}^p\}$. Thus if originally $\mathbf{x} \in \text{co}(\mathbf{0}, \mathbf{v}_1, \mathbf{v}_2, \dots, \mathbf{v}_n)$ its radial projection $\mathbf{x}_r$, $\mathbf{x}_r := r\mathbf{x}$ with an appropriate $r > 0$, will be in $\text{co}(\mathbf{v}_0, \mathbf{v}_1, \mathbf{v}_2, \dots, \mathbf{v}_n)$, where $\mathbf{v}_0$ is the vertex that was replaced by $\mathbf{0}$ as in step 3 in the definition of $\mathcal{T} = \mathcal{T}_{N,K}^{\text{std}}$. When we compute the appropriate $\mathbf{z}$, $\mathcal{J}$, and σ for $\mathbf{x}_r$ we will actually get the position of the simplex $\text{co}(\mathbf{0}, \mathbf{v}_1, \mathbf{v}_2, \dots, \mathbf{v}_n)$, because of the changes described above in CODE BLOCK 1.

```
1   int T_std_NK::InSimpNr(vec x){
2     vec origx=x;
3     if(!(min(Np-x)>=0.0 && min(x-Nm)>=0.0)){
4       // not in the simplicial complex
5       return -1;
6     }
7     if(min(Kp-x)>0.0 && min(x-Km)>0.0){
8       // in the fan
9       double eps = 1e-15;
10      if(norm(x,"inf")>eps) {
11        double r=numeric_limits<double>::max();
12        for(int i=0;i<n;i++){
13          if(abs(x(i))>eps){
14            r=min(r,(x(i)>0 ? Kp(i):Km(i))/x(i));
15          }
16        }
17        x *= r*(1-eps);
18      }
19      else{
20        // be careful, use linear search
21        for(int i=0;i<Fan.size();i++){
22          if(InSimp(x,Fan[i])){
23            return Fan[i];
24          }
25        }
26      }
27    }
28    // compute the zJs of the simplex,
29    // for details cf. (Hafstein, 2013)
30    int J=0;
31    ivec z(n),sigma;
32    for(int i=0;i<n;i++){
33      if(x(i)<0){
34        x(i)=-x(i);
35        J|=1<<i;
36      }
37      z(i)=static_cast<int>(x(i));
38    }
39    sigma=conv_to<ivec>::from(sort_index(x-z,1));
40    // find and return the appropriate simplex
41    auto found=
42    equal_range(NrInSim.begin(),NrInSim.end(),zJs(z,J,sigma));
43    // If one wants to be sure everything is OK
44    assert(found.first!=found.second);
45    assert(InSimp(origx,found.first->Pos));
46    return found.first->Pos;
47  }
```

The simplices are stored as a `vector` of `vector<int>`, the integers being indices of vertices in `vector<ivec>` Ver. Thus `vector<vector<int>>` Sim contains the simplices in $\mathcal{T}$ and Sim[s][i] is the index of the i-th vertex of simplex number s in Ver. To make this access more transparent the member functions int SVerNr(int s,int i) and ivec SVer(int s,int i) were added:

```
1   int T_std_NK::SVerNr(int s,int i){
2   // returns a j such that Ver[j] is the
3   // i-th vertex of simplex Sim[s]
4      return Sim[s][i];
5   }
6
7   ivec T_std_NK::SVer(int s,int i){
8   // returns the i-th vertex of simplex Sim[s]
9      return Ver[SVerNr(s,i)];
10  }
```

To be able to use the Standard C++ Library functions set_intersection and set_difference we sort each `vector<int>` Sim[s]. This is implemented in the constructor of T_std_NK in the trivial way:

```
1   for(int s=0;s<Sim.size();s++){
2      sort(Sim[s].begin(),Sim[s].end());
3   }
```

The `vector<vector<int>>` SimN contains the neighbouring simplices for each simplex and `vector<vector<int>>` SCV contains all simplices, of which a particular vertex in $\mathcal{V}_\mathcal{T}$ is a vertex of. More exactly SimN[s] is a sorted vector of the indices in Sim of the simplices neighbouring simplex Sim[s] (not including s itself) and SCV[i] is a sorted vector of the indices in Sim of the simplices, of which Ver[i] is a vertex. They are constructed as follows in the constructor of T_std_NK:

```
1   SCV.resize(Vertices.size());
2   for(int s=0;s<Sim.size();s++){
3      for(int i=0;i<=n;i++){
4         SCV[SVerNr(s,i)].push_back(s);
5      }
6   }
7
8   vector<vector<int>>::iterator pSCV;
9   for(pSCV=SCV.begin();pSCV!=SCV.end();pSCV++){
10     sort((*pSCV).begin(),(*pSCV).end());
11  }
12
13  SimN.resize(Sim.size())
14  for(int s=0;s<SimN.size();s++){
```

```
15    list<int> lSimN;
16    for(int i=0;i<=n;i++){
17      lSimN.insert(lSimN.end(),
18        SCV[SVerNr(s,i)].begin(),
19          SCV[SVerNr(s,i)].end());
20    }
21    lSimN.sort();
22    lSimN.unique();
23    lSimN.remove(s);
24    SimN[s].assign(lSimN.begin(),lSimN.end());
25  }
```

For every neighbouring simplex Sim[k] of the simplex Sim[s] we keep track of the common face. The member vector<vector<vector<int>>> Faces of T_std_NK was added for this purpose. A face is stored as a vector<int> of the indices of its vertices in vector<ivec> Ver. Each Faces[s] is a vector containing the faces the simplex Sim[s] shares with other simplices in $\mathcal{T}$ and they are listed in the same order as in SimN[s], i.e.

```
1  vector<int>::iterator pSN=SimN[s].begin();
2  vector<vector<int>>::iterator p=Faces[s].begin();
3  for(;pSN!=SimN[s].end();pSN++,p++){
4    // here (*p) is a vector<int> containing the
5    // indices in Ver of the vertices of the
6    // common face of Sim[s] and (*pSN).
7  }
```

The vector Faces is built as follows in the constructor of T_std_NK:

```
1  Faces.resize(Simp.size());
2  for(int s=0;s<Faces.size();s++){
3    for(auto p=SimN[s].begin();p!=SimN[s].end();p++){
4      vector<int> F(n);    // Face
5      auto Fend=set_intersection(Sim[*p].begin(),Sim[*p].end(),
6        Sim[s].begin(),Sim[s].end(),
7          F.begin());
8      F.resize(Fend-F.begin());
9      Faces[s].push_back(F);
10   }
11 }
```

One application of storing the common faces is when one uses the CPA method to compute control Lyapunov functions as in [3]. The faces can also be used to identify the simplices of $\mathcal{T}$ that build the boundary $\partial \mathcal{D}_{\mathcal{T}}$ of $\mathcal{D}_{\mathcal{T}}$. A face is said to be maximal if it is an $(n-1)$-simplex and thus spanned by exactly n of its vertices. We define a simplex $\mathfrak{S}_\nu$ to be an *interior simplex* in $\mathcal{T}$ if all of its maximal faces are common with other simplices in $\mathcal{T}$. Otherwise, we define $\mathfrak{S}_\nu$ to be a *boundary simplex* in $\mathcal{T}$. Note that an n-simplex has $\binom{n+1}{n} = n+1$ number of maximal faces. We can thus identify a boundary simplex Sim[s] by simply counting the number of its maximal faces in Faces[s]. The boundary simplices of $\mathcal{T}$ are stored sorted in vector<int> BSim, which is build as follows in the constructor of T_std_NK:

```
1  for(int s=0;s<Sim.size();s++){
2    int NrMax=0;
3    for(auto pF=Faces[s].begin();pF!=Faces[s].end();pF++){
4      if((*pF).size()==n){
5        NrMax++;
6      }
7    }
8    if(NrMax<n+1){
9      BSim.push_back(s);
10   }
11 }
12 sort(BSim.begin(),BSim.end());
```

The linear program from the CPA method always posses a feasible solution if the system $\mathbf{x}' = \mathbf{f}(\mathbf{x})$ has an exponentially stable equilibrium at the origin and if the simplices used have a small enough diameter and are not too degenerated, cf. e.g. [15]. For discrete time systems $\mathbf{x}_{k+1} = \mathbf{g}(\mathbf{x}_k)$ an analogous proposition holds true [13]. When generating such linear programming problems it is convenient to map the basic simplicial complex $\mathcal{T}$ to a simplicial complex $\mathcal{T}^{\mathbf{F}}$ with smaller simplices using a map $\mathbf{F} : \mathbb{R}^n \to \mathbb{R}^n$. A simplex $\mathfrak{S}_\nu := \mathrm{co}(\mathbf{v}_0, \mathbf{v}_1, \dots, \mathbf{v}_n)$ in $\mathcal{T}$ is mapped to the simplex $\mathfrak{S}_\nu^{\mathbf{F}} = \mathrm{co}(\mathbf{F}(\mathbf{v}_0), \mathbf{F}(\mathbf{v}_1), \dots, \mathbf{F}(\mathbf{v}_n))$ in $\mathcal{T}^{\mathbf{F}}$. This is implemented by the class FT, which is the subject of the next section.

3   The Simplicial Complex $\mathcal{T}^{\mathbf{F}}$

The simplicial complex $\mathcal{T} = \mathcal{T}_{N,K}^{\mathrm{std}}$ is not adequate for the generation of linear programming problems for the computation of Lyapunov functions because its simplices are too large. Our solution to this issue is the simplicial complex $\mathcal{T}^{\mathbf{F}}$, which is implemented in class FT. An instance SC (Simplicial Complex) of class FT holds a pointer T_std_NK *pBC to an underlying basic simplicial complex $\mathcal{T}$ and a mapping $\mathbf{F} : \mathbb{R}^n \to \mathbb{R}^n$ that maps the vertices of $\mathcal{T}$ to the vertices of $\mathcal{T}^{\mathbf{F}}$.

The relationship between $\mathcal{T}$ and $\mathcal{T}^{\mathbf{F}}$ is that

$$\mathrm{co}(\mathbf{F}(\mathbf{v}_0), \mathbf{F}(\mathbf{v}_1), \dots, \mathbf{F}(\mathbf{v}_n)) \in \mathcal{T}^{\mathbf{F}}, \quad \text{if and only if} \quad \mathrm{co}(\mathbf{v}_0, \mathbf{v}_1, \dots, \mathbf{v}_n) \in \mathcal{T}.$$

For $\mathfrak{S}_\nu = \mathrm{co}(\mathbf{v}_0, \mathbf{v}_1, \dots, \mathbf{v}_n)$ in $\mathcal{T}$ we denote the corresponding simplex in $\mathcal{T}^{\mathbf{F}}$ by $\mathfrak{S}_\nu^{\mathbf{F}} := \mathrm{co}(\mathbf{F}(\mathbf{v}_0), \mathbf{F}(\mathbf{v}_1), \dots, \mathbf{F}(\mathbf{v}_n))$. Clearly the collection of the vertices of the simplices in $\mathcal{T}^{\mathbf{F}}$ are the set $\mathcal{V}_{\mathcal{T}^{\mathbf{F}}} := \mathbf{F}(\mathcal{V}_{\mathcal{T}})$. Note that the mapping $\mathbf{F} : \mathbb{R}^n \to \mathbb{R}^n$ must be chosen such that $\mathcal{T}^{\mathbf{F}}$ is an admissible triangulation and in general this is not true, even for a homeomorphism $\mathbf{F} : \mathbb{R}^n \to \mathbb{R}^n$, and in general neither $\mathbf{F}(\mathfrak{S}_\nu) \neq \mathfrak{S}_\nu^{\mathbf{F}}$ nor $\mathbf{F}^{-1}(\mathfrak{S}_\nu^{\mathbf{F}}) = \mathfrak{S}_\nu$.

Suitable candidates for the mapping $\mathbf{F}$ are discussed in [18]. A mapping that generalizes the mapping suggested there is given by the generic form

$$\mathbf{F}(\mathbf{x}) = \frac{\rho(\|\mathbf{x}\|_\infty)}{\|\mathbf{x}\|_*} A\mathbf{x}, \tag{6}$$

where $\| \cdot \|_*$ is any norm on $\mathbb{R}^n$, $\rho : \mathbb{R}_+ \to \mathbb{R}_+$ is a function of class $\mathcal{K}_\infty$, and $A \in \mathbb{R}^{n \times n}$ is a nonsingular matrix.

Note that for $\mathbf{F}$ as in (6) with $A = I$ we have $\|\mathbf{F}(\mathbf{x})\|_* = \rho(\|\mathbf{x}\|_\infty)$, i.e. $\mathbf{F}$ maps the hypercube $[-a, a]^n$ injectively onto the set $\{\mathbf{x} \in \mathbb{R}^n : \|\mathbf{x}\|_* \le \rho(a)\}$. By fixing the norm $\| \cdot \|_*$ as the energetic norm $\|\mathbf{x}\|_* = \|\mathbf{x}\|_Q := \sqrt{\mathbf{x}^T Q \mathbf{x}}$ for a symmetric, positive definite matrix $Q \in \mathbb{R}^{n \times n}$, we obtain by setting $A := Q^{-\frac{1}{2}}$, i.e. $A = O^T \operatorname{diag}(\mu_1^{-\frac{1}{2}}, \mu_2^{-\frac{1}{2}}, \ldots, \mu_n^{-\frac{1}{2}})O$ where $Q = O^T \operatorname{diag}(\mu_1, \mu_2, \ldots, \mu_n)O$ is the eigendecomposition of Q with an orthogonal $O \in \mathbb{R}^{n \times n}$, that $\|\mathbf{F}(\mathbf{x})\|_Q = \rho(\|\mathbf{x}\|_\infty)$. That is, $\mathbf{F}$ maps the hypercube $[-a, a]^n$ injectively onto the closed hyperellipsoid centered at the origin and of which the lengths of the principal axes are $\rho(a)/\mu_i^{\frac{1}{2}}$, $\mu_i > 0$ an eigenvalue of Q. See Fig. 2 for a picture of $\mathcal{T}^{\mathbf{F}}$ with $\mathbf{F}$ as in (6) with

$$Q = \begin{pmatrix} 3 & 1 \\ 1 & 3 \end{pmatrix} \quad \text{and} \quad A = Q^{-\frac{1}{2}} = \frac{1}{4} \begin{pmatrix} 1 + \sqrt{2} & 1 - \sqrt{2} \\ 1 - \sqrt{2} & 1 + \sqrt{2} \end{pmatrix}. \tag{7}$$

Note especially that the axes of the simplicial complex are rotated to better fit the level-sets of the energetic norm $\| \cdot \|_Q$. Such $\mathbf{F}$ have been used, for example, in [2–4,6]. As shown later in this section some algorithms can be made much

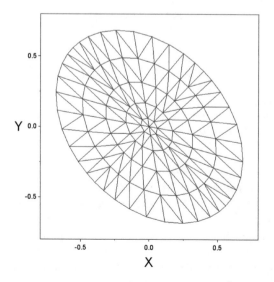

Fig. 2. The complex $\mathcal{T}^{\mathbf{F}}$ with $\mathbf{F}$ as in (6) with $\rho(x) = 0.1x^{\frac{3}{2}}$, Q and A as in (7) and $\| \cdot \|_* = \| \cdot \|_Q$.

more efficient if a formula for the inverse $\mathbf{F}^{-1}$ of $\mathbf{F}$ is available. If $\mathbf{F}$ is as in (6), then its inverse is easily verified to be given by

$$\mathbf{F}^{-1}(\mathbf{x}) = \frac{\rho^{-1}(\|A^{-1}\mathbf{x}\|_*)}{\|A^{-1}\mathbf{x}\|_\infty} A^{-1}\mathbf{x}. \tag{8}$$

Note that $\rho \in \mathcal{K}_\infty$ implies $\rho^{-1} \in \mathcal{K}_\infty$. In Sect. 3.2 we describe a fast algorithm that given an $\mathbf{x} \in \mathbb{R}^n$ and the inverse $\mathbf{F}^{-1}$ of $\mathbf{F}$ searches for a simplex $\mathfrak{S}_\nu^{\mathbf{F}} \in \mathcal{T}^{\mathbf{F}}$ such that $\mathbf{x} \in \mathfrak{S}_\nu^{\mathbf{F}}$.

3.1   Implementation of `class FT`

The data structure `class FT` implements the simplicial complex $\mathcal{T}^{\mathbf{F}}$. Its definition is:

```
1  class FT {
2  public:
3      T_std_K *pBC;
4      function<vec(vec)> pF, ipF;
5      vec F(vec x);
6      vec F(ivec x);
7      vec iF(vec x);
8      vector<vec> xVer;
9      vector<mat> XmT;
10     vector<vec> SCC;
11     vector<double> SCR;
12     vector<vector<mat>> Fnor;
13     FT(T_std_K *_pBC, function<vec(vec)> _pF,
14         function<vec(vec)> _ipF=nullptr);
15     ~FT();
16     int InSimpNrSlow(vec x,vec &L);
17     int InSimpNrFast(vec x,vec &L,int guess);
18     int InSimpNrAppr(vec x);
19     bool InSimp(vec x,int s,vec &L);
20     int SVerNr(int s,int i);
21     vec SVer(int s,int i);
22     bool CFSS;   // default value "true"
23  };
```

We first discuss the constructor of `FT` declared in lines 13 and 14 in the code above. Its first argument is a pointer `T_std_NK *_pBC` to the underlying basic simplicial complex, which is assigned to the member variable `T_std_NK *pBC`. The second argument of the constructor is the mapping $\mathbf{F} : \mathbb{R}^n \to \mathbb{R}^n$ that is used to map the vertices of the basic simplicial complex. It is assigned to the member variable `function<vec(vec)> pF` and can be called with the member functions `vec FT::F(vec x)` and `vec FT::F(ivec x)`. Their implementation is trivial:

```
1  vec F(vec x){
2    return pF(x);
3  }
4
5  vec F(ivec x){
6    return pF(conv_to<vec>::from(x));
7  }
```

The third argument of the constructor is the inverse $\mathbf{F}^{-1}$ of $\mathbf{F}$. If it is available it can be used to decrease the computational complexity of several algorithms considerably. It is stored in the member variable function<vec(vec)> ipF. If the inverse is not available we initialize ipF=nullptr. The implementation of the member function vec iF(vec x) is analogous to the implementation of vec F(vec x), just with pF replaced by ipF. To avoid that one delivers to function<vec(vec)> ipF a function that is not the inverse of $\mathbf{F}$ we verify $\mathbf{F}^{-1}(\mathbf{F}(\mathbf{x})) = \mathbf{x}$ for a random sample in the domain $\mathcal{D}_{\mathcal{T}}$ and we verify $\mathbf{F}^{-1}(\mathbf{0}) = \mathbf{0}$. If these tests are not passed we set ipF=nullptr and warn the user. This is implemented as follows in the constructor of FT:

```
1  if(ipF!=nullptr){
2    arma_rng::set_seed_random();
3    int NrRandVec=1000;
4    double tol=1e-10;
5    if(norm(iF(F(zeros<vec>(n))),"inf")>tol){
6      cerr<<"iF(F(0)) != 0"<<endl;
7      ipF=nullptr;
8    }
9    vec m=F(pBC->Nm);
10   vec M=F(pBC->Np);
11   for(int i=0;i<NrRandVec;i++){
12     vec r=randu<vec>(n);
13     vec x=r%(M-m)+m;
14     if(norm(iF(F(x))-x,2)>tol*norm(x,2)){
15       cerr<<"iF(F(x)) != x for x="<<x.t();
16       ipF=nullptr;
17       break;
18     }
19   }
20 }
```

Note that for vec x and vec y in Armadillo x%y denotes element-by-element multiplication, similar to x.*y in Matlab, Scilab, and Octave.

We store the vertices vector<vec> xVer of $\mathcal{T}^{\mathbf{F}}$ in the same order as the corresponding integer coordinate vertices vector<ivec> Ver of $\mathcal{T}$ in the basic simplicial complex pointed to by T_std_NK *pBC. This is implemented in the constructor of FT in the obvious way:

```
1  for(auto p=pBC->Ver.begin();p!=pBC->Ver.end();p++){
2    xVer.push_back(F(*p));
3  }
```

We can thus refer to the simplex pBC->Sim[s] or just the simplex Sim[s] in $\mathcal{T}^{\mathbf{F}}$. It is the simplex with vertices xVer[pBC->Sim[s][i]] for i= 0 : n. The implementation of the member function int FT::SVerNr(int s,int i) simply calls the homonymous member function in the underlying basic simplicial complex and the implementation of vec FT::SVer(int s,int i) is trivial.

```
1  int FT::SVerNr(int s,int i){
2    return pBC->SVerNr(s,i);
3  }
4
5  vec FT::SVer(int s,int i){
6    return xVer[SVerNr(s,i)];
7  }
```

The class FT stores for each simplex $\mathfrak{S}_\nu^{\mathbf{F}}$ the transpose of the inverse X_ν^{-T} of the so-called shape-matrix X_ν of $\mathfrak{S}_\nu^{\mathbf{F}}$, cf. [21]. The shape-matrix X_ν of $\mathfrak{S}_\nu^{\mathbf{F}} = \mathrm{co}(\mathbf{v}_0, \mathbf{v}_1, \ldots, \mathbf{v}_n)$ is defined by writing the vectors $\mathbf{v}_1 - \mathbf{v}_0$, $\mathbf{v}_2 - \mathbf{v}_0$, $\ldots$, $\mathbf{v}_n - \mathbf{v}_0$ consecutively in its rows. Note that because the vectors $\mathbf{v}_0, \mathbf{v}_1, \ldots, \mathbf{v}_n$ are affinely independent the matrix X_ν is invertible. The matrices X_ν^{-T} are stored as vector<mat> XmT in the same order as the simplices vector<int> pBC->Sim in the basic simplicial complex. Thus XmT[s] is the transpose the inverse of the shape-matrix of the simplex pBC->Sim[s] in $\mathcal{T}^{\mathbf{F}}$. The reason why we store X_ν^{-T} rather than X_ν^{-1} or simply X_ν is that X_ν^{-T} is the most useful form for bool FT::InSimp(vec x,int s,vec &L).

Further information we want to have available for the simplices $\mathfrak{S}_\nu^{\mathbf{F}}$ in $\mathcal{T}^{\mathbf{F}}$ are the centers of their circumscribing hyperspheres and their radii, stored in vector<vec> SCC and vector<double> SCR respectively. Again, they are stored in the same order as the simplices in the basic simplicial complex, thus SCC[s] is the center and SCR[s] is the radius of the circumscribing hypersphere of the simplex pBC->Sim[s] in $\mathcal{T}^{\mathbf{F}}$. The formulas

$$\mathbf{c} = \mathbf{v}_0 + \frac{1}{2}X_\nu^{-1}\mathbf{b}, \text{ with } b_i = \|\mathbf{v}_i - \mathbf{v}_0\|_2^2 \text{ for } i = 1:n \text{ and } r = \|\mathbf{c} - \mathbf{v}_0\|_2$$

for the center $\mathbf{c} \in \mathbb{R}^n$ of the circumscribing hypersphere of $\mathfrak{S}_\nu^{\mathbf{F}} = \mathrm{co}(\mathbf{v}_0, \mathbf{v}_1, \ldots, \mathbf{v}_n)$ and its radius r were derived in [18]. The construction of vector<mat> XmT, vector<vec> SCC, and vector<double> SCR is implemented in the constructor of FT as follows.

```
1  for(int s=0;s<pBC->Sim.size();s++){
2    mat XT(n,n);
3    vec v0=SVer(s,0);
```

```
4    vec b(n);
5    for(int i=1;i<=n;i++){
6       XT.col(i-1)=SVer(s,i)-v0;
7       b(i-1)=pow(norm(XT.col(i-1),2),2);
8    }
9    XmT.push_back(XT.i());
10   vec c=x0+0.5*XmT[s].t()*b;
11   SCC.push_back(c);
12   SCR.push_back(norm(c-v0,2));
13 }
```

Using the matrices `vector<mat>` XmT one can easily check whether a vector $\mathbf{x} \in \mathbb{R}^n$ is in a particular simplex $\mathfrak{S}_\nu^{\mathbf{F}}$ =Sim[s] or not. In [18] it was shown that $\mathbf{x} \in \mathbb{R}^n$ is in $\mathfrak{S}_\nu^{\mathbf{F}} = \mathrm{co}(\mathbf{v}_0, \mathbf{v}_1, \mathbf{v}_2, \ldots, \mathbf{v}_n)$, if and only if

$$(\lambda_1, \lambda_2, \ldots, \lambda_n)^T = X_\nu^{-T}(\mathbf{x} - \mathbf{v}_0) \in \mathbb{R}_+^n \quad \text{and} \quad \lambda_0 := 1 - \sum_{i=1}^{n} \lambda_i \geq 0.$$

Then $\lambda_0, \lambda_1, \ldots, \lambda_n$ are the barycentric coordinates of $\mathbf{x}$ in $\mathfrak{S}_\nu^{\mathbf{F}}$. The implementation is:

```
1   bool FT::InSimp(vec x,int s,vec &L){
2      vec mu=XmT[s]*(x-SVer(s,0));
3      if(min(mu)>= 0 && sum(mu)<=1){
4         L(0)=1.0-sum(mu);
5         L(span(1,n))=mu;
6         return true;
7      }
8      else{
9         return false;
10     }
11  }
```

If $\mathbf{x}$ is in the simplex Sim[s] in $\mathcal{T}^{\mathbf{F}}$ the function returns true and assigns the barycentric coordinates of $\mathbf{x}$ to vec &L, i.e. L = $(\lambda_0, \lambda_1, \ldots, \lambda_n)^T$.

Finally, normals to the hyperplanes defining and separating neighbouring simplices are stored. Although we have changed some data structures used from list to vector the implementation given in [18] works and has not been changed. Since the discussion is quite involved and the normals are not needed in the rest of this paper, we refer the interested reader to [18].

3.2   Fast Search for $\mathfrak{S}_\nu^{\mathbf{F}}$ Such that $\mathbf{x} \in \mathfrak{S}_\nu^{\mathbf{F}}$

The following problem is often of interest: Given an $\mathbf{x} \in \mathbb{R}^n$ find an $\mathfrak{S}_\nu^{\mathbf{F}} \in \mathcal{T}^{\mathbf{F}}$ such that $\mathbf{x} \in \mathfrak{S}_\nu^{\mathbf{F}}$. Additionally, one often then needs the barycentric coordinates of $\mathbf{x}$ in $\mathfrak{S}_\nu$, i.e. the λ_i such that $\mathbf{x} = \sum_{i=0}^{n} \lambda_i \mathbf{v}_i$ is the convex combination of the vertices $\mathbf{v}_i$ of $\mathfrak{S}_\nu^{\mathbf{F}}$. Without any additional information one must rely on linear search, implemented as:

```
1  int FT::InSimpNrSlow(vec x,vec &L){
2    for(int s=0;s<pBC->Sim.size();s++){
3      if(InSimp(x,s,L)==true){
4        return s;
5      }
6    }
7    return -1;
8  }
```

Note that the vec &L corresponds to the vector $(\lambda_0, \lambda_1, \ldots, \lambda_n)^T$ and if a simplex containing $\mathbf{x}$ is not found the impossible value -1 is returned from the function.

If a formula for the inverse mapping $\mathbf{F}^{-1}$ of $\mathbf{F}$ is available a much faster search is possible. For some applications, e.g. when plotting a computed Lyapunov function, it might be sufficient to know a simplex $\mathfrak{S}_\nu^\mathbf{F}$ such that $\mathbf{x}$ is close to $\mathfrak{S}_\nu^\mathbf{F}$. If the mappings $\mathbf{F}$ and $\mathbf{F}^{-1}$ are not too exotic a simplex $\mathfrak{S}_\nu^\mathbf{F} \in \mathcal{T}^\mathbf{F}$, such that $\mathbf{y} := \mathbf{F}^{-1}(\mathbf{x}) \in \mathfrak{S}_\nu$, is often a good candidate. Such an $\mathfrak{S}_\nu$ can be computed very efficiently in the member function T_std_NK::InSimpNr(vec x) as described in [17]. This is implemented as

```
1  int FT::InSimpNrAppr(vec x){
2    assert(ipF!=nullptr);
3    return pBC->InSimpNr(iF(x));
4  }
```

For many applications, however, one needs a simplex $\mathfrak{S}_\nu^\mathbf{F}$ such that truly $\mathbf{x} \in \mathfrak{S}_\nu^\mathbf{F}$ and one needs the barycentric coordinates of $\mathbf{x}$ in $\mathfrak{S}_\nu^\mathbf{F}$. An important example is when a linear programming problem for discrete-time dynamical systems is constructed, cf. [13, 19, 32].

The idea behind the fast search for a simplex $\mathfrak{S}_\nu^\mathbf{F} \in \mathcal{T}^\mathbf{F}$ such that $\mathbf{x} \in \mathfrak{S}_\nu^\mathbf{F}$ is as follows: Given an $\mathbf{x} \in \mathbb{R}^n$, for which we need a simplex $\mathfrak{S}_\nu^\mathbf{F} \in \mathcal{T}^\mathbf{F}$ such that $\mathbf{x} \in \mathfrak{S}_\nu^\mathbf{F}$, start with a simplex $\mathfrak{S}_\xi^\mathbf{F} \in \mathcal{T}^\mathbf{F}$ that is a good guess. This guess can either be delivered by the caller of the search function or one can compute an $\mathfrak{S}_\xi \in \mathcal{T}$ such that $\mathbf{F}^{-1}(\mathbf{x}) \in \mathfrak{S}_\xi$. If $\mathbf{x} \in \mathfrak{S}_\xi^\mathbf{F}$ we are finished. If not check if $\mathbf{x} \in \mathfrak{S}_\mu^\mathbf{F}$ for neighbouring simplices $\mathfrak{S}_\mu^\mathbf{F}$ of $\mathfrak{S}_\nu^\mathbf{F}$. If an $\mathfrak{S}_\mu^\mathbf{F}$ is found such that $\mathbf{x} \in \mathfrak{S}_\mu^\mathbf{F}$ we are finished. If not check if $\mathbf{x} \in \mathfrak{S}_\eta^\mathbf{F}$ for neighbouring simplices $\mathfrak{S}_\eta^\mathbf{F}$ of the simplices $\mathfrak{S}_\mu^\mathbf{F}$ that have not already been checked. Repeat this procedure until a simplex is found.

One comment about another speedup of the search before we give the implementation. If $\mathbf{F}^{-1}(\mathbf{x}) \notin \mathcal{D}_\mathcal{T} := \bigcup_{\mathfrak{S}_\nu \in \mathcal{T}} \mathfrak{S}_\nu$ implies $\mathbf{x} \notin \mathcal{D}_\mathcal{T}^\mathbf{F} := \bigcup_{\mathfrak{S}_\nu \in \mathcal{T}} \mathfrak{S}_\nu^\mathbf{F}$ the member variable bool CFSS (careful simplex search), whose default value is true, should be assigned the value false. It was shown in [18] that this implication holds true for many important triangulations and suitable $\mathbf{F}$ and $\mathbf{F}^{-1}$ as in (6) and (8). This simple trick, and variants of it, can save a lot of computations because it is very laborious to search exhaustively for a simplex that is not in the complex $\mathcal{T}^\mathbf{F}$.

```
1  int FT::InSimpNrFast(vec x,vec &L,int guess=-1){
2    int s;
3    if(guess!=-1){
4      s=guess;
5    }
6    else{
7      s=InSimpNrAppr(x);
8    }
9    if(s==-1){
10     if(CFSS==true){ // might be in complex
11       s=InSimpNrSlow(x,L);
12     }
13     return s;
14   }
15   if(InSimp(x,s,L)==true){
16     return s;
17   }
18   vector<int> TC, CH, CN, CNtemp;
19   // TC = To Check
20   // CH = already CHecked
21   // CN = Check Next
22   // CNtemp = temporary values for CN
23   for(int is=0;is<pBC->SimN[s].size();is++){
24     if(pBC->Faces[s][is].size()==n){
25       TC.push_back(pBC->SimN[s][is]);
26     }
27   }
28   for(auto p=TC.begin();p!=TC.end();p++){
29     if(InSimp(x,*p,L)==true){
30       return s=*p;
31     }
32   }
33   // initialize the main loop
34   CH.push_back(s);
35   TC.push_back(s);
36   int MaxSweeps=3;
37   while(MaxSweeps--){
38     CN.clear();
39     for(p=TC.begin();p!=TC.end();p++){
40       vector<int> A2CN;    // add to CN
41       for(int is=0;is<pBC->SimN[*p].size();is++){
42         if(pBC->Faces[*p][is].size() == n){
43           A2CN.push_back(pBC->SimN[*p][is]);
44         }
45       CN.insert(CN.end(),A2CN.begin(),A2CN.end());
46     }
47     sort(CN.begin(),CN.end());
48     CNtemp.reserve(CN.size());
49     auto it=unique_copy(CN.begin(),CN.end(),CNtemp.begin());
```

```
50      CN.assign(CNtemp.begin(),it);
51      TC.clear();
52      set_difference(CN.begin(),CN.end(),CH.begin(),CH.end(),
53                                        back_inserter(TC));
54      if(TC.empty()==true){
55        return -1;
56      }
57      for(p=TC.begin();p!=TC.end();p++){
58        if(InSimp(x,*p,L)==true){
59          return *p;
60        }
61      }
62      CH.insert(CH.end(),TC.begin(),TC.end());
63      sort(CH.begin(),CH.end());
64    }
65    // give up on being clever
66    return InSimpNrSlow(x,L);
67  }
```

We have four remarks on this algorithm, which is similar to the one given in [18], but has been tweaked in several ways. First, using std::vector instead of std::list for TC, CH, CN, CNtemp as in [18] turned out to be about 5% faster in most cases and we therefore adapted the implementation accordingly. Second, if one wants α_1 in Eq. (9) in Sect. 4 to be a so-called CPA function defined on the simplicial complex $\mathcal{T}^{\mathbf{F}}$, cf. e.g. [15], then one sequentially searches for simplices for points $\mathbf{x}_i$, $i = 0 : N$, where $\mathbf{x}_i$ is close to $\mathbf{x}_{i+1}$. Thus it might significantly speed up computations storing the last simplex found by this algorithm in a static variable and check directly if the new point delivered to the algorithm is in the same simplex as the point before. This however has the major drawback that it interferes with searching for simplices in parallel using multithreading. We achieve the same speedup of the search by adding the variable int guess, which has the (impossible) default value -1. The caller of the function can deliver the number of the simplex in guess to the algorithm where the search should start. If nothing is delivered the search starts in the simplex whose number is delivered by FT::InSimpNrAppr(x) as in [18]. Third, if the main loop is used to search through the whole simplicial complex, then it is considerably slower than a linear search through the whole simplicial complex by FT::InSimpNrSlow(x,L). Thus, the implementation uses the parameter int MaxSweeps to decide when to give up on searching for neighbours and just do a linear search. Its appropriate value will depend on the kind of the problem at hand to be solved. We used MaxSweeps=3 which worked quite well for our examples. Fourth, instead of using all neighbours as in [18] it turned out to be about 10% faster in our examples to use only neighbours with common faces that are maximal, i.e. $(n-1)$-simplices (n common vertices). Therefore the implementation was changed accordingly.

4   Computing Values of $V(\xi)$ for $\xi \in \mathcal{V}_{\mathcal{T}^{\mathrm{F}}}$ Directly

Instead of solving the linear optimization problem generated to obtain a Lyapunov function, one can use a different method to make *educated guesses* of their values and then verify if the linear constraints are fulfilled for these values. This is the approach followed in, e.g. [4,5,7,10–12,19,20,32], and it generates the values much faster than solving the linear programming problem. An additional advantage is that one can localize the area where the constraints are not fulfilled, whereas a solver will simply state that the linear programming problem does not possess a feasible solution when that is the case.

For a continuous-time system $\mathbf{x}' = \mathbf{f}(\mathbf{x})$ with an exponentially stable equilibrium at the origin a Lyapunov function is given by

$$V(\xi) = \int_0^T \alpha_1(\phi(\tau, \xi)) d\tau, \tag{9}$$

where $\tau \mapsto \phi(\tau, \xi)$ is the solution to $\mathbf{x}' = \mathbf{f}(\mathbf{x})$ with initial-value ξ at time $\tau = 0$, $T > 0$ is a large enough constant, and $\alpha_1 : \mathbb{R}^n \to \mathbb{R}_+$ is continuous and positive definite function, i.e. $\alpha_1(\mathbf{0}) = \mathbf{0}$ and $\alpha_1(\mathbf{x}) > 0$ for $\mathbf{x} \neq \mathbf{0}$. The idea for constructing a Lyapunov function like this goes back to Massera [35], see also [26], and is discussed in many textbooks on nonlinear systems, e.g. [27,43]. Typically for our applications $\alpha_1(\mathbf{x}) = \alpha(\|\mathbf{x}\|_2)$, where α is of class $\mathcal{K}_\infty$, but there are other suitable choices, for example $\alpha_1(\mathbf{x}) = \|\mathbf{f}(\mathbf{x})\|_2$ is used in [7].

Note that although (9) gives an explicit formula for a Lyapunov function, this formula includes the solution $\phi(\tau, \xi)$ to the differential equation and the solution is usually not known. It can, however, be approximated at the vertices of the simplicial complex with a subsequent verification of the constraints of the linear programming problem.

To approximate (9) numerically we used the Adam-Bashforth four-step method for obtaining numerically a solution $t \mapsto \phi(t, \xi)$ to the initial-value problem

$$\mathbf{x}' = \mathbf{f}(\mathbf{x}), \quad \mathbf{x}(0) = \xi,$$

on $[0, T]$ and the composite Simpson's Rule to integrate $\alpha_1(\phi(t, \xi))$ over the same interval. Both are standard methods that are described in most textbooks on numerical analysis, cf. e.g. [41].

For completeness we give a short-description: First choose the number of steps N to be used, an even number ≥ 4, and define $h := T/N$, $t_i := ih$, and $\phi_i := \phi(t_i, \xi)$ for $i = 0 : N$. The composite Simpson's Rule approximates the integral in (9) with the sum

$$\frac{h}{3} \left[\alpha_1(\phi_0) + \alpha_1(\phi_N) + 4 \sum_{i=1}^{N/2} \alpha_1(\phi_{2i-1}) + 2 \sum_{i=1}^{N/2-1} \alpha_1(\phi_{2i}) \right] \tag{10}$$

and the error is bounded by $C_1 h^4$ for some constant $C_1 > 0$. The Adam-Bashforth four-step method approximates the ϕ_i with numbers w_i using the formula

$$w_{i+1} = w_i + \frac{h}{24}\left[55 f_i - 59 f_{i-1} + 37 f_{i-2} - 9 f_{i-3}\right], \quad \text{where } f_i := \mathbf{f}(w_i), \quad (11)$$

and an error bound is given by $|w_i - \phi_i| \leq C_2 (e^{Lhi} - 1)h^4$ for some constant $C_2 > 0$ and where $L > 0$ is a Lipschitz constant for $\mathbf{f}$ on the relevant subset of $\mathbb{R}^n$. We compute w_1, w_2, w_3 with the Runge-Kutta Method of fourth order (RK4), which has compatible error bounds. Thus, we compute

$$\frac{h}{3}\left[\alpha_1(w_0) + \alpha_1(w_N) + 4\sum_{i=1}^{N/2}\alpha_1(w_{2i-1}) + 2\sum_{i=1}^{N/2-1}\alpha_1(w_{2i})\right] \quad (12)$$

as an approximation to the integral in (9). With $K > 0$ as a Lipschitz constant for α_1 on the relevant subset of $\mathbb{R}^n$ the difference between (10) and (12) is bounded by

$$\frac{h}{3}\cdot 2K\left[\sum_{i=1}^{N}|w_i - \phi_i| + \sum_{i=1}^{N/2-1}|w_{2i-1} - \phi_{2i-1}|\right]$$

$$\leq \frac{2Kh}{3}C_2 h^4\left[\sum_{i=1}^{N}(e^{Lhi} - 1) + \sum_{i=1}^{N/2-1}(e^{Lh(2i-1)} - 1)\right]. \quad (13)$$

Now

$$\sum_{i=1}^{N}e^{Lhi} = e^{Lh}\frac{e^{LNh} - 1}{e^{Lh} - 1} \leq e^{Lh}\frac{e^{LT} - 1}{Lh}$$

and similarly

$$\sum_{i=1}^{N/2-1}e^{Lh(2i-1)} \leq e^{Lh}\frac{e^{LT} - 1}{2Lh}$$

and since $Nh = T$ we can bound (13) by

$$\frac{2K}{3}C_2 h^4\left[\frac{3e^{Lh}}{2L}(e^{LT} - 1) - \frac{3}{2}T + h\right]. \quad (14)$$

Thus we have shown that the error we make when we compute (12) as an approximation to the integral in (9) the error is bound by $Ce^{LT}h^4$ for some constant $C > 0$, or in big-O notation, the error is $O(e^{LT}h^4)$.

The computation of $V(\boldsymbol{\xi})$ was implemented through the class ODEsolInt with ODEsolInt::operator()(vec x) doing the actual evaluation. Note that one can evaluate $V(\boldsymbol{\xi})$ for many different $\boldsymbol{\xi}$ simultaneously using multithreading. The class ODEsolInt has the members function<vec(vec)> f, which is the right-hand-side of the differential equation $\mathbf{x}' = \mathbf{f}(\mathbf{x})$, double T,h and int N, where

T,h and N correspond to T, h, and N above. A further member is the function function<double(vec)> alpha1, which is the function α_1 in the integral (9). By default it is fixed to $\alpha_1(\mathbf{x}) = \|\mathbf{x}\|_2$. The definition of the class ODEsolInt and the implementation of its constructor are given below. Note that we need N to be an even integer ≥ 4 for our implementation for the numerical approximation of the integral to be correct, so an inadequate value for N is modified accordingly. Recall that for int i,j the command i%j gives the reminder when i is divided by j. Thus i%2 equals zero if i is an even number and i%2 equals one if i is odd.

```
1   double def_alpha1(vec x){ return norm(x, 2); }
2
3   class ODEsolInt{
4   public:
5       function<vec(vec)> f;
6       function<double(vec)> alpha1;
7       double T,h;
8       int N;
9       ODEsolInt(function<vec(vec)> _f,double _T,int _N,
10                      function<double(vec)> _alpha1=def_alpha1);
11      ODEsolInt(){};
12      double operator()(vec x);
13  };
14
15  ODEsolInt::ODEsolInt(function<vec(vec)> _f,double _T,
16      int _N,function<double(vec)> _alpha1) :
17          f(_f),T(_T),N(_N),alpha1(_alpha1) {
18      // N should be an even number >= 4
19      if(N<4){
20          N=4;
21      }
22      if(N%2==1){
23          N++;
24      }
25      h=T/N;
26  }
```

For an efficient implementation of formulas (11) and (12) we store as little information as possible and plug the values of w_i from (11) immediately into (12). We use the Runga-Kutta Method of fourth-order to compute the first three initial w_1, w_2, w_3 and then we use four-step Adam-Bashforth, which is multistep method that only needs the evaluation of $\mathbf{f}(\mathbf{x})$ at one new point $\mathbf{x}$ for each new value w_i, instead of at four points as in the Runge-Kutta Method. The four-step Adam-Bashforth Method achieves this by using already computed values as can be seen in (11). We thus store $f_i, f_{i-1}, f_{i-2}, f_{i-3}$ in the columns of the matrix mat AB(n,4), and after having used these values to compute w_{i+1} we compute f_{i+1} using w_{i+1}. Then we overwrite f_{i-3}, which is not needed anymore, with

f_{i+1}. This can be implemented by accessing the columns of mat AB(n,4) with i modulo 4, that is, f_i =mat AB.col(i%4). Simultaneously to computing the w_i we sum up

$$\alpha_1(w_0) + 4\alpha_1(w_1) + 2\alpha_1(w_2) + 4\alpha_1(w_3) + 2\alpha_1(w_4) + \ldots + 4\alpha_1(w_{N-1}) + 2\alpha_1(w_N)$$

and then subtract $\alpha_1(w_N)$ in the end and multiply the sum by $h/3$ to obtain the value from formula (12).

```
1   vec RK4(function<vec(vec)> f,vec w_i,double h){
2       vec s1(n),s2(n),s3(n),s4(n);
3       s1=f(w_i);
4       s2=f(w_i+h/2*s1);
5       s3=f(w_i+h/2*s2);
6       s4=f(w_i+h*s3);
7       return w_i+h/6*(s1+2*s2+2*s3+s4);
8   }
9
10  double ODEsolInt::operator()(vec x){
11      double ret=0.0;
12      mat AB(n,4);
13      vec x1,x0;
14      x0=x;
15      AB.col(0)=f(x0);
16      ret+=alpha1(x0);
17      for(int i=1;i<=3;i++){
18          x1=RK4(f,x0,h);
19          ret+=2.0*(1+i%2)*alpha1(x1);
20          AB.col(i)=f(x1);
21          x0=x1;
22      }
23      for(int i=3;i<N;i++) {
24          x1=x0+h/24.0*(55.0*AB.col(i%4)-59.0*AB.col((i-1)%4)
25              +37.0*AB.col((i-2)%4)-9.0*AB.col((i-3)%4));
26          AB.col((i+1)%4)=f(x1);
27          ret+=2.0*(2-i%2)*alpha1(x1);
28          x0=x1;
29      }
30      ret-=alpha1(x1);
31      ret*=h/3.0;
32      return ret;
33  }
```

5   Summary

We discussed the implementation in C++ of simplicial complexes using efficient algorithms and data structures for the computation of Lyapunov functions for nonlinear systems. This paper builds upon [6,14,17] and advances and improves the methods presented in [18] in various ways. The algorithms are designed for both continuous-time and discrete-time systems and the implementation for control Lyapunov functions using the Dini subdifferential is accounted for. Additionally, we gave a detailed description and error-analysis for the fast computation of an approximation to a Lyapunov function at the vertices of a simplicial complex using a Lyapunov function candidate from a converse theorem by Massera. This approach has been used in several publications, cf. e.g. [4,5,7,10–12,19,20,32], but its efficient implementation has not been discussed before.

Acknowledgement. The author's research is supported by the Icelandic Research Fund (Rannís) 'Complete Lyapunov functions: Efficient numerical computation' (163074-052) and 'Lyapunov Methods and Stochastic Stability' (152429-051), which is gratefully acknowledged.

References

1. Ambrosino, R., Garone, E.: Robust stability of linear uncertain systems through piecewise quadratic Lyapunov functions defined over conical partitions. In: Proceedings of the 51st IEEE Conference on Decision and Control, Maui (HI), USA, pp. 2872–2877, December 2012
2. Baier, R., Grüne, L., Hafstein, S.: Linear programming based Lyapunov function computation for differential inclusions. Discrete Contin. Dyn. Syst. Ser. B **17**(1), 33–56 (2012)
3. Baier, R., Hafstein, S.: Numerical computation of Control Lyapunov Functions in the sense of generalized gradients. In: Proceedings of the 21st International Symposium on Mathematical Theory of Networks and Systems (MTNS), Groningen, The Netherlands, pp. 1173–1180, no. 0232 (2014)
4. Björnsson, J., Giesl, P., Hafstein, S., Kellett, C., Li, H.: Computation of continuous and piecewise affine Lyapunov functions by numerical approximations of the Massera construction. In: Proceedings of the 53rd IEEE Conference on Decision and Control, Los Angeles (CA), USA, pp. 55056–5511 (2014)
5. Björnsson, J., Giesl, P., Hafstein, S., Kellett, C., Li, H.: Computation of Lyapunov functions for systems with multiple attractors. Discrete Contin. Dyn. Syst. Ser. A **35**(9), 4019–4039 (2015)
6. Björnsson, J., Gudmundsson, S., Hafstein, S.: Class library in C++ to compute Lyapunov functions for nonlinear systems. In: Proceedings of MICNON, 1st Conference on Modelling, Identification and Control of Nonlinear Systems, no. 0155, pp. 788–793 (2015)
7. Björnsson, J., Hafstein, S.: Efficient Lyapunov function computation for systems with multiple exponentially stable equilibria. Procedia Comput. Sci. **108**, 655–664 (2017)
8. Branicky, M.: Multiple Lyapunov functions and other analysis tools for switched and hybrid systems. IEEE Trans. **43**(4), 475–482 (1998)
9. Clarke, F.: Optimization and Nonsmooth Analysis. Classics in Applied Mathematics. SIAM, Philadelphia (1990)
10. Doban, A.: Stability domains computation and stabilization of nonlinear systems: implications for biological systems. Ph.D. thesis, Eindhoven University of Technology (2016)

11. Doban, A., Lazar, M.: Computation of Lyapunov functions for nonlinear differential equations via a Yoshizawa-type construction. IFAC-PapersOnLine **49**(18), 29–34 (2016). 10th IFAC Symposium on Nonlinear Control Systems NOLCOS 2016

12. Giesl, P., Hafstein, S.: Computation and verification of Lyapunov functions. SIAM J. Appl. Dyn. Syst. **14**(4), 1663–1698 (2015)

13. Giesl, P., Hafstein, S.: Computation of Lyapunov functions for nonlinear discrete systems by linear programming. J. Differ. Equ. Appl. **20**, 610–640 (2014)

14. Giesl, P., Hafstein, S.: Implementation of a fan-like triangulation for the CPA method to compute Lyapunov functions. In: Proceedings of the 2014 American Control Conference, Portland (OR), USA, pp. 2989–2994, no. 0202 (2014)

15. Giesl, P., Hafstein, S.: Revised CPA method to compute Lyapunov functions for nonlinear systems. J. Math. Anal. Appl. **410**, 292–306 (2014)

16. Hafstein, S.: An algorithm for constructing Lyapunov functions, volume 8 of Monograph. Electron. J. Diff. Eqns. (2007)

17. Hafstein, S.: Implementation of simplicial complexes for CPA functions in C++11 using the armadillo linear algebra library. In: Proceedings of the 2nd International Conference on Simulation and Modeling Methodologies, Technologies and Applications (SIMULTECH), Reykjavik, Iceland, pp. 49–57 (2013)

18. Hafstein, S.: Efficient algorithms for simplicial complexes used in the computation of Lyapunov functions for nonlinear systems. In: Proceedings of the 7th International Conference on Simulation and Modeling Methodologies, Technologies and Applications (SIMULTECH), Madrid, Spain, pp. 398–409 (2017)

19. Hafstein, S., Kellett, C., Li, H.: Computation of Lyapunov functions for discrete-time systems using the Yoshizawa construction. In: Proceedings of the 53rd IEEE Conference on Decision and Control (2014)

20. Hafstein, S., Kellett, C., Li, H.: Continuous and piecewise affine Lyapunov functions using the Yoshizawa construction. In: Proceedings of the 2014 American Control Conference, Portland (OR), USA, pp. 548–553, no. 0170 (2014)

21. Hafstein, S., Kellett, C., Li, H.: Computing continuous and piecewise affine Lyapunov functions for nonlinear systems. J. Comput. Dyn. **2**(2), 227–246 (2015)

22. Johansson, M.: Piecewise linear control systems. Ph.D. thesis, Lund University, Sweden (1999)

23. Johansson, M., Rantzer, A.: Computation of piecewise quadratic Lyapunov functions for hybrid systems. IEEE Trans. Automat. Control **43**(4), 555–559 (1998)

24. Julian, P.: A high level canonical piecewise linear representation: theory and applications. Ph.D. thesis, Universidad Nacional del Sur, Bahia Blanca, Argentina (1999)

25. Julian, P., Guivant, J., Desages, A.: A parametrization of piecewise linear Lyapunov functions via linear programming. Int. J. Control **72**(7–8), 702–715 (1999)

26. Kellett, C.: Converse theorems in Lyapunov's second method. Discrete Contin. Dyn. Syst. Ser. B **20**(8), 2333–2360 (2015)

27. Khalil, H.: Nonlinear Systems, 3rd edn. Pearson, London (2002)

28. Lazar, M.: On infinity norms as Lyapunov functions: alternative necessary and sufficient conditions. In: Proceedings of the 49th IEEE Conference on Decision and Control, Atlanta, USA, pp. 5936–5942, December 2010

29. Lazar, M., Doban, A.: On infinity norms as Lyapunov functions for continuous-time dynamical systems. In: Proceedings of the 50th IEEE Conference on Decision and Control, Orlando (Florida), USA, pp. 7567–7572 (2011)

30. Lazar, M., Doban, A., Athanasopoulos, N.: On stability analysis of discrete-time homogeneous dynamics. In: Proceedings of the 17th International Conference on Systems Theory, Control and Computing, Sinaia, Romania, pp. 297–305, October 2013

31. Lazar, M., Jokić, A.: On infinity norms as Lyapunov functions for piecewise affine systems. In: Proceedings of the Hybrid Systems: Computation and Control Conference, Stockholm, Sweden, pp. 131–141, April 2010

32. Li, H., Hafstein, S., Kellett, C.: Computation of continuous and piecewise affine Lyapunov functions for discrete-time systems. J. Differ. Equ. Appl. **21**(6), 486–511 (2015)

33. Marinósson, S.: Lyapunov function construction for ordinary differential equations with linear programming. Dyn. Syst. Int. J. **17**, 137–150 (2002)

34. Marinósson, S.: Stability analysis of nonlinear systems with linear programming: a Lyapunov Functions based approach. Ph.D. thesis, Gerhard-Mercator-University Duisburg, Duisburg, Germany (2002)

35. Massera, J.: Contributions to stability theory. Ann. Math. **64**, 182–206 (1956). Erratum. Ann. Math. **68**, 202 (1958)

36. Ohta, Y.: On the construction of piecewise linear Lyapunov functions. In: Proceedings of the 40th IEEE Conference on Decision and Control, vol. 3, pp. 2173–2178, December 2001

37. Ohta, Y., Tsuji, M.: A generalization of piecewise linear Lyapunov functions. In: Proceedings of the 42nd IEEE Conference on Decision and Control, vol. 5, pp. 5091–5096, December 2003

38. Polanski, A.: Lyapunov functions construction by linear programming. IEEE Trans. Automat. Control **42**, 1113–1116 (1997)

39. Polanski, A.: On absolute stability analysis by polyhedral Lyapunov functions. Automatica **36**, 573–578 (2000)

40. Sanderson, C.: Armadillo: an open source C++ linear algebra library for fast prototyping and computationally intensive experiments. Technical report, NICTA (2010)

41. Sauer, T.: Numerical Analysis, 2nd edn. Pearson, London (2012)

42. Stroustrup, B.: Software development for infrastructure. Computer **45**(1), 47–58 (2012)

43. Vidyasagar, M.: Nonlinear System Analysis. Classics in Applied Mathematics, 2nd edn. SIAM, Philadelphia (2002)

44. Yfoulis, C., Shorten, R.: A numerical technique for the stability analysis of linear switched systems. Int. J. Control **77**(11), 1019–1039 (2004)

A Resource Flow Based Multistage Dynamic Scheduling Method for the State-Dependent Work

Nobuaki Ishii[1]([⊠]), Yuichi Takano[2], and Masaaki Muraki[3]

[1] Faculty of Engineering, Kanagawa University,
3-27-1 Rokkakubashi, Kanagawa-ku, Yokohama 221-8686, Japan
n-ishii@kanagawa-u.ac.jp
[2] Faculty of Engineering, Information and Systems, University of Tsukuba,
1-1-1 Tennodai, Tsukuba, Ibaraki 305-8577, Japan
ytakano@sk.tsukuba.ac.jp
[3] Graduate School of Decision Science and Technology,
Tokyo Institute of Technology,
2-12-1 Ookayama, Meguro-ku, Tokyo 152-8550, Japan
m.muraki85ll@gmail.com

Abstract. We study a dynamic scheduling method for the state-dependent work based on the resource flow within a process of the work. Namely, we develop a multistage dynamic scheduling model consisting of N classes of activities and a three-layer control structure. Then, we devise a resource flow based order selection method and resource allocation method to provide successful results from the works under the limited resources. To demonstrate the effectiveness of our method via numerical examples, we apply the developed method to the project cost estimation process of the EPC (Engineering-Procurement-Construction) contractor for the purpose of determining acceptance of profitable projects in competitive bidding situations.

Keywords: Business process modeling · Order selection · Project management
Resource allocation

1 Introduction

Scheduling is a decision problem that develops a plan with reference to the sequence of required tasks and time allocated for each activity necessary to complete them under the constraints of a due date, the amount of resources available, and so on. The goal of scheduling is to optimize one or more objectives, such as cost, makespan, the number of tardy jobs, and so on.

Although there are many scheduling problems in practice, production scheduling problems [1, 2] have been studied extensively. Production scheduling is divided into several categories, based on the production system, characteristics of data used, assumption of order arrivals, and so on. However, in any case, the deliverables from the scheduled work are predetermined, and orders are usually scheduled so as to minimize

© Springer Nature Switzerland AG 2019
M. S. Obaidat et al. (Eds.): SIMULTECH 2017, AISC 873, pp. 300–316, 2019.
https://doi.org/10.1007/978-3-030-01470-4_16

the makespan or cost in the production scheduling. No valuable deliverables can be obtained if the activities are eventually terminated in the middle of the schedule.

In the state-dependent work, on the other hand, the work process, resources and time, and therefore the value of deliverables from the work, can be changed according to the work environments, such as work volume, resource availability, and so on. Namely, a certain volume of value can be gained from the deliverables even if the work is terminated in the middle of the work process. Sales, software testing, project cost estimation, research & development, training & education, and so on are typical examples of state-dependent work.

The above characteristics make the scheduling problem of the state-dependent work different from those of the ordinary production scheduling. Thus, different approaches should be taken in the scheduling of state-dependent work.

Based on the above observations, this paper examines the state-dependent work in dynamic order arrival situations. Then, we develop a dynamic scheduling method that dynamically selects orders and allocates resources to each selected order to maximize the expected profits. For this purpose, we begin by building a generic work process model of the state-dependent work with reference to previous studies by Ishii et al. [3–5]. Then we devise a resource flow based order selection method and resource allocation method to provide successful results from the work under the limited resources. To demonstrate the effectiveness of our method via numerical examples, we apply the developed method to the cost estimation work process of the EPC (Engineering-Procurement-Construction) [6, 7] contractor for the purpose of determining acceptance of profitable projects in competitive bidding situations.

2   Related Work

Regarding the dynamic scheduling problem for the state-dependent work, only a few studies have been conducted for the project estimation process. For example, Ishii et al. [3–5] studied a dynamic scheduling problem for estimating project cost. They then developed a simulation-based method that uses the threshold function to select orders to estimate project cost in dynamic order arrival situations. In addition, Ishii et al. [8] developed a resource flow based order selection method used in project cost estimation process.

Regarding order selection and resource allocation, several papers have analyzed the problem of allocating scarce resources in competitive bidding (see Rothkopf and Harstad [9] for detailed references). For example, Shafahi and Haghani [10] proposed an optimization model that combines project selection decisions and markup selection decisions in consideration of eminence and previous works as the non-monetary evaluation criterion used by owners for evaluating bids. Kortanek et al. [11] considered sequential bidding models where the obtained contracts require the use of restricted resources, such as production capacity, at the time of actual production. In addition, several studies that focus on the volume of MH (Man-Hour) for cost estimation and cost estimation accuracy have been conducted. For example, Ishii et al. [12, 13] developed an algorithm that determines the bidding prices under limited MH for cost estimation. Their algorithm allocates MH to the orders so as to maximize expected profits based on the cost estimation accuracy determined by allocated MH. In addition, Takano et al. [14] developed a stochastic dynamic programming model for establishing an optimal sequential bidding strategy in a competitive bidding situation. Their model determines the optimal markup in consideration of the effect of inaccurate cost estimates. Furthermore, Takano et al. [15] developed a multi-period resource allocation method for estimating project costs in a sequential competitive bidding situation. Their method allocates resources for cost estimation by solving a mixed integer programming problem that is formulated by making a piecewise liner approximation of the expected profit functions.

Based on the above literature review, in this paper, we develop a generalized resource flow based dynamic scheduling method in the state-dependent work by extending the method proposed by Ishii et al. [8]. It selects orders and determines resource allocation dynamically in consideration of the resource availability and the profitability of the orders based on change of the flow of required volume of resources and the expected profits through the process of the state-dependent work.

3   A Multistage Model for the State-Dependent Work

The state-dependent work has the following characteristics, (1) Work activities can be terminated in the middle of process, (2) Deliverables gained through the work have some value according to the amount of resources and time invested, no matter which work activities are terminated in the middle of process, (3) The value of the deliverables is evaluated based on the state when the work was terminated or completed. Because of these characteristics, the scheduling problems in the state-dependent work are different from those of the production process as shown in Table 1, thus an ordinary approach for the dynamic scheduling cannot be applied in the state-dependent work.

Table 1. Comparison of scheduling problems between state-dependent work and production process

	State-dependent work	Production process
Evaluation criteria	Expected profits, Amount of accepted orders, Customer satisfaction, etc.	Cost, Makespan, The number of tardy jobs
Process	Can change or skip by condition	Cannot change or skip. Entire process must be completed
Value of deliverables	Varies according to resource used	Does not vary

Based on the above background, Ishii et al. [3] developed a generic multistage model of dynamic scheduling for the state-dependent work as shown in Fig. 1

In the model, they suppose that the work can be divided into N classes of activities, i.e., Class 1, Class 2, …, and Class N activity, and are carried out from Class 1 to Class N, sequentially. Orders arrive randomly and are first filed in the queue for the Class 1 activity and wait for allocated resources to carry out the Class 1 activity by the

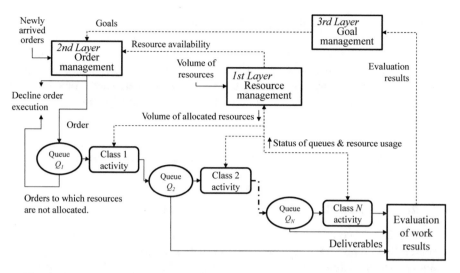

Fig. 1. Generic multistage model of dynamic scheduling for state-dependent work [3].

mechanism of resource management. If any resource is not allocated to the order by the due date, then no activity is carried out due to lack of resources. If the resources are allocated to the order, the activity is performed by creating the value via the Class1 activity. The order is then filed in the queue of the Class 2 activity and waits for resource allocation to the Class 2 activity. If the resources are not further allocated to the order by the due date, the deliverables are evaluated based on the results by the activity of Class 1. By contrast, if the resources are allocated to the order waiting in the queue of the Class 2 activity, the activity is complete, and then filed in the queue of the Class 3 activity. In the model, the same decision is made for the orders in the queue of each class until they complete the Class N activity or terminate the work on the way to Class N activity.

In the state-dependent work, the value of the deliverables gradually improves through the activities in each class because the scope of the activity is wider and more detailed according to the progress of activities. However, the work can be terminated on the way to Class N activity. In such a case, the value of deliverables created through the work is determined by the final class carried out. Then, the goals of the work process are modified according to the work results up to this point.

In addition, the generic multistage model of dynamic scheduling in the state-dependent work shown in Fig. 1 assumes to control works with a three-layer control structure as follows:

- 1st Layer (Resource management): allocates resources required to carry out the activities of the order waiting in a queue file within the available resource,
- 2nd Layer (Order management): decides whether to decline the activities on the newly arrived order based on the goals set in goal management,
- 3rd Layer (Goal management): sets and changes goals in an appropriate time based on the results of state-dependent work.

4 Resource Flow Based Dynamic Scheduling

This section shows the two methods, i.e., order selection and resource allocation. They are used for the order management and the resource management, respectively, in the multistage model of dynamic scheduling for the state-dependent work shown in Fig. 1. These two methods are developed based on the following assumptions:

Assumptions:

(1) Orders arrive randomly;
(2) Expected profit, required resource and periods required in each activity are predetermined.

4.1 Resource Flow Based Order Selection Method

For the order selection in the order management shown in Fig. 1, we develop a Resource Flow based order Selection Method (RFSM) that determines accepting or declining arrived orders according to the changes of MHR and EPR by the arrived orders. MHR_n and EPR_n are the flow rate of the required resource for the work and the total expected profits from orders, respectively, when the work is completed from Class 1 to class n activity within the state-dependent work process. Those are shown below as Eqs. (1) and (2).

$$MHR_n = \sum_{i \in UE} MH_i^n / D_i \qquad (1)$$

$$ERP_n = \sum_{i \in UE} EP_i^n / D_i \qquad (2)$$

where i is the order for carrying out the work in the process and n indicates the class of activity from 1 to N. MH_i^n, EP_i^n, and D_i are the volume of resources, the expected profit, and period for the activity of order i from the class 1 activity to the class n activity, respectively. In addition, UE is a set of orders within the work process.

Now, assume that $PE_n(MHR_n, EPR_n)$ indicates the coordinate point where activities of all the orders are carried out from class 1 to n activity, $PE_{n-1}(MHR_{n-1}, EPR_{n-1})$ indicates the coordinate point where activities of all the orders are carried out to the class $n - 1$, and MHR_{CP} is the maximum flow rate of resources available in the state-dependent work process.

Then, the rate of maximum expected profits EPR_{max} is calculated based on the magnitude relationship between MHR_n and MHR_{CP} as Eqs. (3) or (4). For example, Fig. 2 shows the relation among PE_n, PE_{n-1}, and EPR_{max} when the relation among flow rate of resource is as $MHR_{n-1} < MHR_{CP} < MHR_n$.

(1) If $MHR_n \leq MHR_{CP}$:

$$EPR_{max} = EPR_n \qquad (3)$$

(2) If $MHR_{n-1} < MHR_{CP} < MHR_n$:

$$EPR_{\max} = \frac{EPR_n - EPR_{n-1}}{MHR_n - MHR_{n-1}} MHR_{cp} + \frac{MHR_n EPR_{n-1} - MHR_{n-1} EPR_n}{MHR_n - MHR_{n-1}} \quad (4)$$

Equation (4) assumes that there is linearity between PE_n and PE_{n-1}, then EPR_{max} exists where MHR_{CP} intersects the line connecting the points of PE_n and PE_{n-1} as shown in Fig. 2.

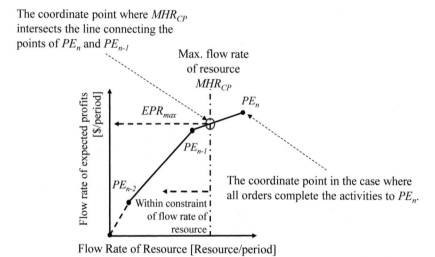

Fig. 2. Graphical explanation of the rate of maximum expected profits (EPR_{max}).

Next, if the new order nwd has arrived in the work process, $PE'_n \left(MHR'_n, EPR'_n \right)$ and $PE'_{n-1} \left(MHR'_{n-1}, EPR'_{n-1} \right)$, which indicate the coordinate points including nwd, are calculated in Eqs. (5) to (8), where MHR^{nwd} and EPR^{nwd} indicate MHR and EPR of a newly selected order for the work process, respectively in each class of activity. In addition, r is a coefficient to discount the flow rate by the next order arrival if the newly arrived order is not selected. It is calculated by Eq. (9) by the average work period of orders within the cost estimation process ED and the number of orders within the process NE, where $r = 0$ if $NE = 0$.

$$MHR'_n = r\,MHR_n + MHR_n^{nwd} \quad (5)$$

$$EPR'_n = r\,EPR_n + EPR_n^{nwd} \quad (6)$$

$$MHR'_{n-1} = r\,MHR_{n-1} + MHR_{n-1}^{nwd} \quad (7)$$

$$EPR'_{n-1} = r\, EPR_{n-1} + EPR^{nwd}_{n-1} \tag{8}$$

$$r = (ED - ED/NE)/ED = 1 - 1/NE \tag{9}$$

Then RFSM evaluates EPR'_{max} indicating the flow rate of maximum expected profits if *nwd* is selected by Eqs. (10) or (11). Equation (11) calculates the value where MHR_{cp} intersects the line connecting the points of PE'_n and PE'_{n-1} based on the assumption that there is linearity between PE'_n and PE'_{n-1} which is the same as shown the case of Eq. (4).

(1) If $MHR'_n < MHR_{CP}$:

$$EPR'_{max} = ERP'_n \tag{10}$$

(2) If $MHR'_{n-1} < MHR_{CP} < MHR'_{n-1}$:

$$EPR'_{max} = \frac{EPR'_n - EPR'_{n-1}}{MHR'_n - MHR'_{n-1}} MHR_{cp} + \frac{MHR'_n\, EPR'_{n-1} - MHR'_{n-1}\, EPR'_n}{MHR'_n - MHR'_{n-1}} \tag{11}$$

Finally, the order *nwd* is selected for cost estimation in the following case:

$$rEPR_{max} < EPR'_{max}$$
or
$$MHR'_n < MHR'_{cp}.$$

The former condition indicates the cases where the flow rate of expected profit EPR'_{max} gained by selecting *nwd* for cost estimation, i.e., point A in Fig. 3, is higher than the flow rate of expected profit $r\, EPR_{max}$ gained by cutting *nwd*, i.e., point B in Fig. 3. The latter condition indicates the case where the flow rate of the required resource including the resource for *nwd*, i.e., MHR'_n , is less than the maximum flow rate available in the process.

4.2 Man-Hour Allocation Method

For the allocation of resources in each activity under dynamic order arrival situations, we use a dispatching method, which is used in the dynamic scheduling problem in production systems [2]. Namely, when a resource is available, the resource allocation rule selects an order for allocating the resource from the orders waiting for the allocation. Specifically, when the resource is released from the activity of an order, this method selects an order based on the resource allocation rules, which prioritize orders in the queue of each activity. The selected order subsequently allocates the required resource for its activity. If the required volume of the resource is more than the resource available, the selected order waits in the queue until the required volume of the resource is released.

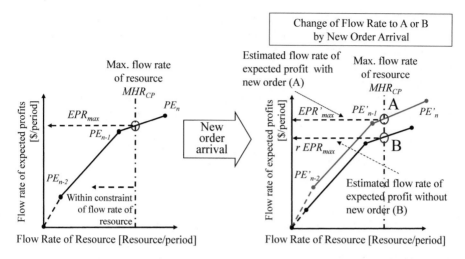

Fig. 3. Idea of resource flow based order selection method (RFSM).

Table 2. Potential resource allocation rules [8].

Rule	Description
FIFO	First-In First-Out: Order is selected on a first-in first-out basis
SDUF	Shortest DUe date First: Order remaining with the shortest estimation period is selected
SET	Shortest Estimation Time: Order having the shortest estimation period is selected
HEPF	Highest Expected Profit per resource First: Order having the highest expected profit per resource for activity is selected

Table 2 shows potential rules that could be applicable for dynamic resource allocation in the dynamic scheduling problem of the state-dependent work.

5 Multistage Model of Project Cost Estimation Process

We apply, in this paper, the developed methods to the project cost estimation process in section six to show the effectiveness of the methods.

The project cost estimation process can be recognized as a series of activities that starts with the arrival of bid invitations from the client and closes by the date of bidding [5]. A variety of orders arrive, and the contractor selects orders to estimate the project costs through the cost estimation process. Then, the contractor allocates the MH (Man Hour) as the required resource to the cost estimation activities of selected orders in consideration of the MH availability, expected profits, competitive bidding situations, and so on. When the available MH is not sufficient to estimate cost accurately, the contactor must allocate fewer MH, thereby reducing expected profit due to inaccurate cost estimation, or no-bid on the order. Namely, in the project cost estimation process, it is possible to see that the deliverable product is the estimated cost, and the value of deliverable product is the accuracy of the estimated cost which is related to the expected profits from the order. Because of the above characteristics, we can say that the project cost estimation process is related to the dynamic scheduling problem in the state-dependent work.

Based on the above observations, we propose a multistage model of the project cost estimation process which consists of pre-evaluation, order selection, MH allocation for cost estimation, and a series of cost estimation steps, as shown in Fig. 4, by referencing the model for the dynamic scheduling in the state-dependent work shown in Fig. 1. In the model, we assume that the cost is estimated through the cost estimation steps: E1, E2, and E3. Each step needs MH and a period of time for cost estimation, and the accuracy of the estimated cost increases through the cost estimation activities in each step. MH corresponds to the resource to carry out each activity in the work process. Table 3 shows the relationship between the generic multistage model of the dynamic scheduling for the state-dependent work (Fig. 1) and the model of the project cost estimation process (Fig. 4).

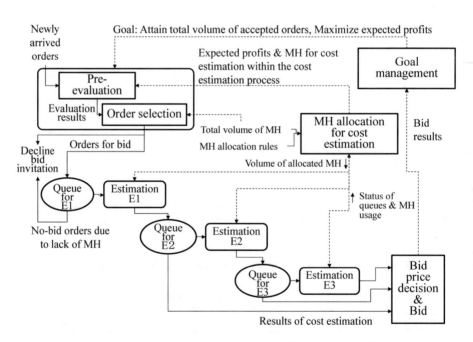

Fig. 4. Multistage model of project cost estimation process [8].

The cost estimate manuals, such as AACE classification matrix [16, 17], the classes of estimates by Kerzner [18], and so on, can be used for reference of the cost estimation accuracy and for the required MH in each step. For example, AACE classifies cost estimation into five classes and indicates the estimation methods, required data, and the accuracy of cost estimation in each class. Two of the five classes in AACE are in the

Table 3. Generic multistage model of the dynamic scheduling for the state-dependent work and the model of the project cost estimation process

		Generic model	Project cost estimation process model
Control structure		Resource management	MH allocation for cost estimation
		Order management	Pre-evaluation & Order selection
		Goal management	Goal management
Resource		Resource	MH (Man-Hour)
Work process		N activities (Class N to Class 1)	Three steps (E1 to E3)

order of magnitude type estimation for a project feasibility study. Thus, the developed model divides the cost estimate activities into three steps in reference to the AACE system. Namely, we assume that the project cost is estimated through a series of three cost estimation steps, i.e., E1, E2, and E3, and the accuracy of the estimated cost is improved in accordance with the steps.

In the model, the pre-evaluation and the order selection determine whether to select and bid the newly arrived order or not. Specifically, the pre-evaluation evaluates the resource flow of the process if the newly arrived orders are selected as explained in Sect. 4.1. The order selection determines whether to select orders for estimating costs or not from the viewpoint of changes of the resource flow, the volume of orders to be accepted, the expected profits, MH availability for cost estimation, and so on.

The selected order is first filed in the queue for the E1 estimate and waits to be allocated the MH for cost estimation by the mechanism of MH allocation for cost estimation. If no MH is allocated to the order until the bidding date, the contractor does not bid for it due to the lack of MH. If the MH is allocated to the order, its project cost is estimated with the accuracy of the E1 estimate. This order is then filed in the queue of the E2 estimate and waits for MH allocation for the E2 estimate. If the MH is not further allocated to the order until the bidding date, the contractor determines the bidding price based on the accuracy of the E1 estimate. By contrast, if the MH is allocated to the order waiting in the queue of the E2 estimate, its project cost is estimated with the accuracy of the E2 estimate, and the order is filed in the queue of the E3 estimate. The same decision is made for the orders in the queue of the E3 estimate.

6   Numerical Examples in Project Cost Estimation Process

This section evaluates the performance of the developed methods for controlling the state-dependent work effectively by applying it to the project cost estimation process with simulation experiments. We use the general-purpose simulation system AweSim! [19] for building a simulation model of the project cost estimation process.

In our experiments the following assumptions are made.

Assumptions:

1. Orders for cost estimation arrive randomly;
2. Expected profit, required MH and periods for cost estimation of each estimate class are predetermined;
3. Probability of a successful bid of each order is predetermined.

6.1   Design of Simulation Experiments

In our simulation experiments, the performance of the two order selection methods, i.e., the developed method in this paper RFSM, and the existing method, i.e., a Threshold Function Method (TFM) [3–5], is compared as two basic methods shown in Table 4. Namely, 100 simulation runs of a 120 period simulation length are performed by each method, and the average expected profits per 12 periods are compared.

Table 4. Basic simulation case [8].

	Order selection method	MH allocation rule
Case A	RFSM	HEPF
Case B	TFM	

The total volume of MH for cost estimation is set as 16,000 [MH/Period] in reference to a mid-size process plant EPC contractor. Furthermore, as the MH allocation rule, the HEPF rule, shown in Table 2, is used throughout all the simulation experiments, because it is reported that the higher expected profit is gained by the HEPF rule [3–5].

Three order arrival scenarios—scenario S1, scenario S2, and scenario S3—based on the order arrival intervals defined by the triangular distribution, as shown in Table 5, are determined. In each scenario, orders of the three sizes, i.e., Small, Medium, Large, arrive dynamically. The total periods for cost estimation, periods for cost estimation in each step, and the volume of MH for cost estimation are set as shown in Table 6. In addition, two scenarios of expected profit of accepted orders, i.e., scenarios I and II, are set as shown in Table 7. Furthermore, as the probability of order acceptance, the arrived orders are sorted into grade H: 70%, M: 40%, and L: 10%. Regarding the rate of the grade of arrival order, grade M is set as 40%, and grade H changes from 0% to 60%, and thus it changes from 60% to 0% in grade L accordingly in each simulation experiment. The expected profit of each order is computed by multiplying the value of the triangle distribution shown in Table 7 by the probability of order acceptance. For

Table 5. Order arrival interval [Orders/Period] [8].

Scenario of order arrival	Parameters of triangular distribution	Order size		
		Small	Medium	Large
S1	Min.	1.05	2.70	3.15
	Mode	1.50	3.00	4.50
	Max.	1.95	3.90	5.85
S2	Min.	0.84	1.68	2.52
	Mode	1.20	2.40	3.60
	Max	1.56	3.12	4.68
S3	Min.	0.70	1.40	2.10
	Mode	1.00	2.00	3.00
	Max	1.30	2.60	3.90

Table 6. Cost estimation conditions [8].

Total periods available for cost estimation		Order size		
		Small	Medium	Large
		Triangular distribution		
		(Min.: 4.0, Mode: 7.5, Max.:9.0)		
Periods for cost estimation	E1	1.0	1.0	1.0
	E2	1.5	1.5	1.5
	E3	2.0	2.0	2.0
MH for cost estimation [M MH]	E1	1.0	2.0	3.0
	E2	2.0	3.0	4.0
	E3	3.0	4.0	5.0

Table 7. Expected profit of accepted orders [MM$] (Mode of triangle distribution. Min. & Max. are −/+ 10% of the mode value.) [8].

Scenario of expected profit		Order size		
		Small	Medium	Large
I	E1	1	2	3
	E2	10	20	30
	E3	20	40	60
II	E1	1	2	3
	E2	15	30	45
	E3	20	40	60

example, if the grade of arrived order is M (40%) and the expected profit from the accepted order calculated by the triangle distribution is 20 [MM$], the expected profit is 8 [MM$].

Regarding the threshold function used for selecting orders in TFM, the order with the expected profit per MH 350 [$/MH] and the volume of MH under estimating cost 6,000 [MH] are set, i.e., the threshold function P(350, 6000). It is searched by using the algorithm developed by Ishii et al. [4, 5], under cost estimation conditions as follows:

(1) Order arrival interval: Scenario S2 in Table 5,
(2) The expected profit of orders: Scenario I in Table 7,
(3) The rate of probability of order acceptance in each grade: H:30-M:40-L:30 [%].

Namely, the newly arrived order is selected for estimating cost by the threshold function in TFM when its expected profit per MH is higher than 35.0 [$/MH] and the MH under estimating cost is less than 6,000 [MH].

6.2   Results of Simulation Experiments

As shown in Figs. 5 and 6, RFSM gains almost the same or higher expected profit than that by the existing TFM in scenarios S2.I and S3.I. RFSM performs especially well when the rate of probability of order acceptance on grade L is large. For example, in the case of grade H:0%-M:40%-L:60% in the rate of order acceptance probability of the arrival order, the expected profit by RFSM increases 17.1% compared to that by TFM as shown in Fig. 5. On the other hand, in the case of the expected profit by TFM being better than RFSM, its difference is less than 4.3% and 5.0% as shown in Figs. 5 and 6, respectively. TFM uses the fixed threshold function determined under the fixed cost estimation conditions throughout the simulation experiments. In our experiments, it is determined as P(350, 6000) as stated in Sect. 6.1. We can say that TFM could not maintain the performance when the cost estimation conditions changed dynamically.

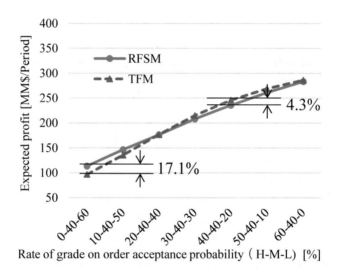

Fig. 5. Expected profits in scenario S2.I [8].

In addition, RFSM has almost the same or higher expected profit than that of TFM at all the rates of probability of the order acceptance in scenario S1.I where orders arrive fewer than in scenario S2.I as shown in Fig. 7. For example, in the case of H:0%-M:40%-L:60% in the rate of probability of order acceptance, the expected profit by RFSM increases 23.8% compared to that by TFM.

It is also obvious that the expected profit gained by RFSM is expanded compared to that of TFM when the conditions of the expected profit in each step are changed to scenario II as shown in Figs. 8 and 9. For example, in the case of H:0%-M:40%-L:60% in the rate of probability of order acceptance, the expected profit by RFSM in scenario S2.I increases 17.1% compared to that by TFM as shown in Fig. 5, however it increases 24.0% in scenario S2.II as shown in Fig. 8. In these cases, TFM cuts too many orders compared to that in RFSM. The threshold function of TFM is determined based on the lower ratio of low profit orders, i.e., H:30%-M:40%-L:30% through the

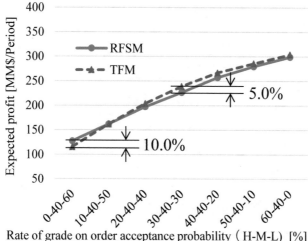

Fig. 6. Expected profits in scenario S3.I [8].

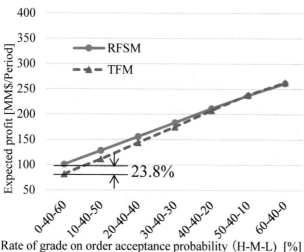

Fig. 7. Expected profits in scenario S1.I [8].

numerical experiments, thus it tends to cut low profit orders when the low profit orders arrive at a higher number than the rate assumed.

Furthermore, since RFSM determines the order selection based on the changes of the resource flow rate, which reflects the conditions of the cost estimation process, we can say that the resource flow based method is effective for selecting the order, especially when the conditions of cost estimation, such as order arrival intervals, the expected profit of accepted orders, and so on, change dynamically.

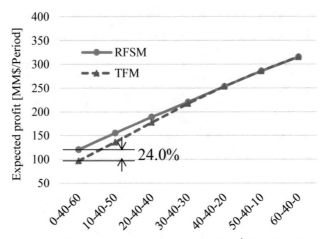

Fig. 8. Expected profits in scenario S2.II [8].

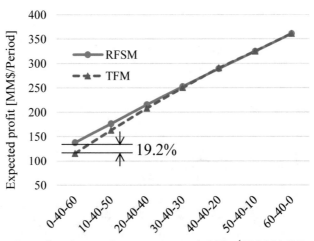

Fig. 9. Expected profits in scenario S3.II [8].

In addition, RFSM needs no complicated mechanism to determine the order selection rules as the threshold function TFM requires. Thus, RFSM can work using lower computational loads compared to that of TFM. We can say that the RFSM is simple and sufficient to be implemented as an order selection mechanism in the project cost estimation process in practical situations.

7 Conclusion

This paper explores the scheduling methods in dynamic order arrival situations for the state-dependent work. Namely, we develop a generic multistage dynamic scheduling model for the state-dependent work. Based on the model, we develop a resource flow based order selection method by extending the method proposed by Ishii et al. [8]. It selects orders for the work process at each order arrival according to the changes of the flow rate of the required resources for carrying out activities and that of the expected profits from the orders to maximize the total expected profits from orders. We analyze the effectiveness of the developed method in terms of the expected profit through numerical examples by applying the methods to the project cost estimation process.

The following conclusions can be drawn from the analysis of the numerical examples:

- For increasing the total expected profits from orders, the resource flow based order selection method is effective as an order selection mechanism in the state-dependent work.
- The performance of the resource flow based order selection method is obvious, especially in cases where the work process conditions change dynamically.

Several issues require further research. For example, a mechanism that changes the resource allocation rules dynamically in corresponding to the work process conditions, such as, resource availability, order arrival situations, and so on, should be developed. Regarding the expected profits from orders, the interrelationship of the order selection method and the resource allocation rule should be explored. Management mechanisms for an advanced work process model that changes the total volume of resource associated with the backlog of orders should also be explored.

In any case, the dynamic scheduling problem in the production process has been a leading edge research theme [20]. However, it has not been well studied in the state-dependent work so far. Because of the increasing demand to improve productivity in the state-dependent work, we hope that research in this field will be expanded.

Acknowledgements. This work was supported by JSPS KAKENHI Grant Number 16K01252

References

1. Pinedo, M.L.: Scheduling: Theory, Algorithms, and Systems, 4th edn. Springer, New York (2014)
2. Jacobs, F.R., Berry, W.L., Whybark, C.D., Vollmann, T.E.: Manufacturing Planning and Control for Supply Chain Management. McGraw-Hill, New York (2011)
3. Ishii, N., Takano, Y., Muraki, M.: A simulation-based dynamic scheduling method in project cost estimation process. In: Mohammad, S., Obaidat, M.S., Ören, T., Merkuryev, Y. (eds.) Simulation and Modeling Methodologies, Technologies and Applications, Advances in Intelligent Systems and Computing book series, vol. 676, pp. 261–279. Springer, Heidelberg (2017)

4. Ishii, N., Takano, Y., Muraki, M.: A dynamic scheduling problem for estimating project cost. In: Proceedings of Scheduling Symposium 2015, Tokyo, pp. 119–124 (2015)
5. Ishii, N., Takano, Y., Muraki, M.: A dynamic scheduling problem in cost estimation process of EPC projects. In: Proceedings of the 6th International Conference on Simulation and Modeling Methodologies, Technologies and Applications, pp. 187–194 (2016)
6. Pritchard, N., Scriven, J.: EPC Contracts and Major Projects, 2nd edn. Sweet & Maxwell, London (2011)
7. Ishii, N., Takano, Y., Muraki, M.: An order acceptance strategy under limited engineering man-hours for cost estimation in engineering-procurement-construction projects. Int. J. Project Manage. 32(3), 519–528 (2014)
8. Ishii, N., Takano, Y., Muraki, M.: Resource flow based order selection method in project cost estimation process. In: Proceedings of the 7th International Conference on Simulation and Modeling Methodologies, Technologies and Applications, pp. 155–162 (2017)
9. Rothkopf, M.H., Harstad, R.M.: Modeling competitive bidding: A critical essay. Manage. Sci. 40, 364–384 (1994)
10. Shafahi, A., Haghani, A.: Modeling contractors' project selection and markup decisions influenced by eminence. Int. J. Project Manage. 32(8), 1481–1493 (2014)
11. Kortanek, K.O., Sodeni, J.V., Sodaro, D.: Profit analyses and sequential bid pricing models. Manage. Sci. 20, 396–417 (1973)
12. Ishii, N., Takano, Y., Muraki, M.: A heuristic bidding price decision algorithm based on cost estimation accuracy under limited engineering man-hours in EPC projects. In: Obaidat, M., Koziel, S., Kacprzyk, J., Leifsson, L., Ören, T. (eds.) Simulation and Modeling Methodologies, Technologies and Applications, Advances in Intelligent Systems and Computing, vol. 319, pp. 101–118. Springer, Heidelberg (2015)
13. Ishii, N., Takano, Y., Muraki, M.: A revised algorithm for competitive bidding price decision under limited engineering man-hours in EPC projects. Oukan (J. Transdiscipl. Fed. Sci. Technol.) 10(1), 47–56 (2016)
14. Takano, Y., Ishii, N., Muraki, M.: A sequential competitive bidding strategy considering inaccurate cost estimates. OMEGA 42(1), 132–140 (2014)
15. Takano, Y., Ishii, N., Muraki, M.: Multi-period resource allocation for estimating project costs in competitive bidding. CEJOR 25(2), 303–323 (2017)
16. AACE International: Cost Estimate Classification System – As Applied in Engineering, Procurement, and Construction for the Process Industries. AACE International Recommended Practice No. 18R-97 (2011)
17. Turton, R., Bailie, R.C., Whiting, W.B., Shaeiwitz, J.A., Bhattacharyya, D.B.: Analysis, Synthesis and Design of Chemical Processes, 4th edn. Pearson, New Jersey (2012)
18. Kerzner, H.: Project Management: A Systems Approach to Planning, Scheduling, and Controlling. Wiley, New Jersey (2013)
19. Pritsker, A.A.B., O'Reilly, J.J.: AWESIM: the integrated simulation system. In: Proceedings of the 1998 Winter Simulation Conference, pp. 249–255 (1998)
20. Zhou, L., Zhang, L., Sarker, B.R., Laili, Y., Ren, L.: An event-triggered dynamic scheduling method for randomly arriving tasks in cloud manufacturing. Int. J. Comput. Integr. Manuf. 31(3), 318–333 (2018)

Application Domains

Simulation Based Behavior
for a Welfare Robot

Stefan-Daniel Suvei$^{(\boxtimes)}$, Leon Bodenhagen, Thomas N. Thulesen,
and Norbert Krüger

SDU Robotics, Maersk Mc-Kinney Moller Institute,
University of Southern Denmark, 5230 Odense M, Denmark
{stdasu,lebo,tnt,norbert}@mmmi.sdu.dk

Abstract. Learning the appropriate parameters which lead to a successful action, even under the effect of noise and inaccuracies stemming from faulty perception or poor sensor calibration is of great importance in service robotics, where the flexibility of the robotic systems is a main requirement. This paper proposes a method inspired by available industrial robotics solutions which makes use of a dynamic simulator and Kernel Density Estimation to find the parameter sets that lead to a successful Peg-in-Hole action. The obtained solution is successfully tested on a real service robot, where the action is used in the context of picking-up and inserting bottles in a crate for depositing.

Keywords: Service robot · Peg-in-hole · Pick-up action · Simulation

1 Introduction

In the industrial domain, robotic solutions are well established for decades. Here robots benefit of controlled environments, separated of direct human interference, where uncertainties and hence error sources are minimized. Current approaches focus on addressing uncertainties within the robot application e.g. by introducing additional sensors and probabilistic modeling of robot actions. Solutions for handling uncertainties in e.g. assembly processes include online pose estimation of relevant items [1], the use of compliant manipulators [2] or re-planning [3].

Service robots are often operating in rather unstructured environments shared with humans, such that the handling of uncertainties is considered to be essential. While some solutions developed for industrial applications can be transferred, additional effort can be required to handle changes in the environment or poor sensor readings to achieve a robust performance.

The process of inserting an item, e.g. a peg, into a tight fitting hole (peg-in-hole, PiH) is an example of a tasks that is found in industry [4,5] and can be generalized to many applications also outside it. The handling of deposit bottles, such as inserting them into a crate, e.g. in house hold scenarios or at institutions

© Springer Nature Switzerland AG 2019
M. S. Obaidat et al. (Eds.): SIMULTECH 2017, AISC 873, pp. 319–334, 2019.
https://doi.org/10.1007/978-3-030-01470-4_17

such as care centers, or the insertion of cutlery into a dishwasher compartments are just two such examples.

When performed on a real robot, PiH actions can be viewed as a stochastic process since inaccuracies and noise on the robot perception and the non-static nature of the setup can be expected to influence the outcome of the robot's action substantially. Therefore, manually finding and tuning the parameters determining the action is infeasible.

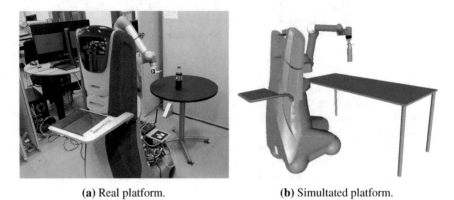

(a) Real platform.   (b) Simultated platform.

Fig. 1. The Care-O-Bot service robot.

Sørensen et al. [6] proposes an online action parameter optimization technique based on Kernel Density Estimation (KDE) for reducing the number of samples required to identify robust PiH actions. The particular action was of industrial nature, implying a well calibrated setup which is not available in a domestic context. However, parts of the proposed solution, such as the action parametrization, can be adapted.

In this paper we propose the use of dynamic simulations with artificial noise and the use of KDE to identify parameter sets that lead to PiH actions robust to noise. We make use of the detailed information available in the simulated environment, e.g. contact forces, to assess the quality of a PiH action, assuming that the optimal action is not only successful but also implies low contact forces. The result is evaluated on a real platform where detailed force information is not available. Similarly to the PiH action, the bottle pick-up action is also addressed.

2   Related Work

The increase in popularity for the service robotics application has brought new focus on classic robotic topics such as safe planning, visual servoing and manipulation, with the goal of adapting some for the well-known industrial methods to the challenges presented by home-like environments. Action-parameter learning, in particular, has been of great interest, specifically because it is applicable to a large array of tasks and it can lead to an increase of the system's flexibility.

The learning process can be done using the execution feedback data received from the sensors and adapting the action process according to the found environment constraints [7,8], by human demonstration [9] or by using dynamic simulation, thus facilitating an extended exploration of the space wherein learning is done [6,10,11].

Due to the nature of their work environment (which is unstructured and highly dynamic), service robots have to be very adaptable in order to properly function. As opposed to the industrial environment, creating a rigid calibration of the tested setup is not a viable solution in service robotics. For systems that are required to manipulate the environment and are equipped with a camera, a way of overcoming this issue is through position based visual servoing (PBVS) [12]. This method uses the difference between the pose of the desired target object and that of the manipulator to determine the control error. For the target object, the pose estimation can be determined by using a 3D model of the object and the homography, like in [13]. [14] proposes an alternative solution, by using a real-time model-based tracking method to estimate the pose of the gripper.

Compared to other approaches, we tackle the task of picking up bottles and inserting them into a deposit crate using a welfare robot, for which no special calibration, specific environment manipulation (i.e. light adjustment, position control, etc.) or additional sensors (such as force-torque sensors or compliant grippers) are used. PBVS and dynamic simulation are utilized in order to achieve the PiH task with minimum contact force. The parameter set that describes the tested Pick-Up and PiH actions, alongside the actual action paths are determined in simulation. The focus is on using the contact forces and the process uncertainty information together with Kernel Density Estimation to find promising parameter sets. We demonstrate that the solution found in simulation can be directly used on a real platform, where the PiH action is not only successful, but also generates low contact forces.

3   Experimental Setup and Methodology

The Care-O-Bot [15] is a research platform targeting the development of mobile service robots for the assistance of humans in domestic environments. The Care-O-Bot platform (see Fig. 1) is equipped with multiple sensors of different modalities and has a modular hardware setup, which makes it applicable for a large variety of tasks, such as: pick-and-place, object detection, human tracking, etc. The main components of the robot are: the mobile base platform with 4 omnidirectional wheels, an actuated torso with 3 DoF, the UR5 robotic arm, the UR connector extending the workspace of the arm, the tray with 1 DoF, the head (containing a Carmine 3D Sensor and a high resolution stereo camera) and three laser scanners. Using this platform, we investigate the possibility of handling off-the-shelf deposit bottles in terms of picking them up and stowing them in a crate.

(a) Original RGB data.   (b) Modified RGB data

Fig. 2. Visualization of the ARUCO traker output - the two markers are detected and the transformation between them is computed (Source: [16]).

3.1 Tracking and Visual Servoing

Using the PBVS method, the error between the pose estimation of the robot's end effector and the desired object (the bottle or the deposit crate) is determined and used as a control error, in order to move the arm towards the desired position. The control loop is an endpoint closed loop, where both the end effector and the object of interest have to be in the field of view.

To ease the process of pose estimation and tracking, AR (Augumneted Reality) markers are attached to the end effector, the bottle and the crate. The markers are detected and the transformation between them is determined. For the PiH action, the ARUCO [17] ROS tracker is used, while for the Pick-Up action, the ALVAR [18] software library is used for the tracking of the AR markers. The obvious downside of using this technique is that the markers must constantly be in the field of view of the camera. Additionally, due to the fact that the robot was placed in a laboratory with a lot of natural light, the level of illumination in the environment can become problematic (especially in the case of the ARUCO tracker), because the markers become undetectable due to reflections. To overcome this issue, we apply the Contrast Limited Adaptive Histogram Equalization algorithm [19] on the input RGB image data of the ARUCO tracker, thus improving the overall contrast of the image, as can be seen in Fig. 2b, and aiding the corner detection. The ALVAR tracker is less sensitive to the illumination level, so no additional image processing is required.

The computed transformation between the markers is used in a classic linear interpolation algorithm which will generate incremental displacements that will move the robotic arm with the gripper towards the desired position and will decrease the pose estimation error between the AR markers. The algorithm is repeated until the Euclidean distance between the AR markers for each specific task is smaller than a pre-defined threshold which will ensure that the gripper is correctly placed to pick-up the bottle or that it is above the crate, ready for the bottle insertion. To avoid overshooting, the incremental displacements decrease as the Euclidean distance between the markers decreases. Once the desired position is reached, the algorithm updates the position of the bottle or the crate based on a final reading of the attached marker and then starts the Pick-Up or Peg-in-Hole action sequence.

3.2 Action Parametrization

In order to simulate and analyze the two tasks, the actions are first formalized and defined by a set of parameters for which to investigate the search space and find the best solution.

Pick-Up Action. In our context, the Pick-Up action is defined by three parameters (Fig. 3) and is performed in four steps (Fig. 4). These parameters are the ones that define the search space of the solution and they are defined as follows:
H_{pu}: The perpendicular distance from the horizontal plane defined by the bottle's neck.
X_{pu}: The distance with which the gripper will be moved towards the bottle, on the plane defined by the bottle's neck.
Φ_{pu}: The angle to which the gripper will be moved in an upward motion, relative to the bottle's vertical axis.

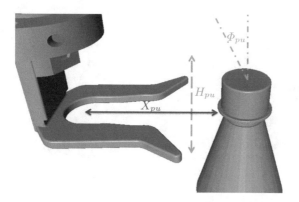

Fig. 3. The Pick-Up action parameters.

The four steps that compose the ideal Pick-Up action (i.e. not affected by uncertainties) are described as follows:

Initial Position: In the first step the robot moves the gripper from the position where it was left by the visual servoing algorithm towards the bottle, as shown in Fig. 4a. If we consider the horizontal plane defined by the bottle's neck, at the end of this motion, the gripper's TCP is at a horizontal distance of 0.05 m from the bottle's neck and at a vertical distance of H_{pu}.

First Linear Motion: In the second step, the gripper is moved linearly towards the bottle, as shown in Fig. 4b, with a distance equal to X_{pu}. At this point, the bottle is placed in-between the fingers of the gripper.

Circular Motion: After the first linear movement, the gripper is moved upward to a distance of 0.2 m and rotated to an angle of Φ_{pu}, relative to the vertical axis defined by the initial placement of the bottle, as depicted in Fig. 4c.

Second Linear Motion: In the final step, the gripper is again moved linearly upward for 0.1 m, while maintaining the Φ_{pu} angle defined in the previous step (see Fig. 4d). This step is done to reduce the risk of dropping the bottle. Due to it's weight and the motion of the arm, the bottle will slide backwards, in-between the gripper's fingers, thus leading to a better grasp.

The Pick-Up action is evaluated at the end of the second linear motion. For a successful action, the bottle has to first and foremost be placed in the gripper. However, the grasp itself can also be evaluated by checking the position and the rotation of the bottle frame with respect to the gripper's TCP frame. The better the grasp, the more alignment will there be between the two frames. In case the bottle is not in the gripper at the end of the sequence, the action is labeled as a failure.

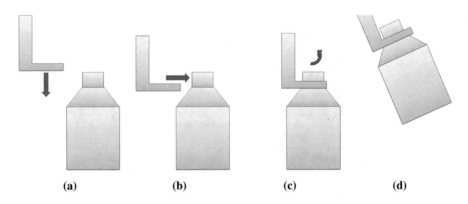

(a)   (b)   (c)   (d)

Fig. 4. The Pick-up action movements: (a) Initial position. (b) First linear motion. (c) Circular motion. (d) Final position.

Peg-in-Hole Action. Similar to the Pick-Up action, the Peg-in-Hole can be defined by two parameters (Fig. 5) as follows:

X_{pih}: The perpendicular distance between the bottom of the bottle and the horizontal plane defined by the hole.

Φ_{pih}: The angle of the bottle in its initial position relative to the vertical axis defined by the hole.

When considering an ideal PiH bottle insertion action, the four steps that compose such an action can be described as follows:

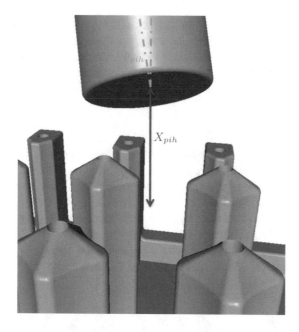

Fig. 5. The PiH action parameters.

Initial Position: The first step is to move the arm with the bottle in the gripper from the position where it was left by the visual servoing algorithm (above the crate), towards the specified crate hole. After this motion, the bottle is at a perpendicular distance of 0.05 m from the hole and tilted with an angle Φ_{pih}.

First Linear Motion: In this second step, the bottle is linearly moved towards the crate's hole until the bottom of the bottle is at a X_{pih} perpendicular distance from the hole, as shown in Fig. 6a.

Circular Motion: After the first linear movement, the bottle is rotated until $\Phi_{pih} = 0$, as depicted in Fig. 6b.

Second Linear Motion: In the final step, the bottle is moved linearly down the hole, while maintaining the alignment between their vertical axes, as shown in Fig. 6c. In our proposed solution, when the bottom of the bottle is inside the hole after the second linear motion, the robot just drops the bottle down the hole.

After the second linear motion, the PiH action is evaluated. For a successful action, the bottle has to be partially inside the crate hole, such that a simple tilt of the gripper would cause the bottle to slide from in-between the fingers into the hole (see Fig. 6d). Reaching this correct position can be confirmed by checking the alignment between the bottle frame and the hole frame. In the case in which the bottle is stuck by hitting one of the edges of the hole, the action is labeled as a failure.

Using the described steps, the relative path between the end effector and the bottle or between the bottle and the crate hole is computed for each parameter set and then executed in the simulated environment and tested on the real platform. The evaluation and labeling of the actions (i.e. success of failure) is done automatically in simulation and manually for the real world platform.

3.3   Action Simulation

Using the RobWorkSim [20] framework, together with the ODE [21] physics engine, the two actions are simulated using different values for their defining parameters. The ranges of the parameters are set beforehand and they define the search space. As such, for the Pick-Up action, the parameters have the following ranges: $H_{pu} \in [-0.02; 0.02]$ m, $X_{pu} \in [-0.06; 0]$ m and $\Phi_{pu} \in [-45; 45]°$, with step sizes of 0.0025 m, 0.0025 m and 2.5° respectively.

For the PiH action, the parameter ranges are set to: $X_{pih} \in [-0.03; 0.03]$ m and $\Phi_{pih} \in [-45; 45]°$, with the step sizes of 0.005 m and 2°.

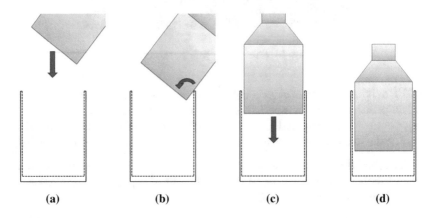

(a)   (b)   (c)   (d)

Fig. 6. The PiH action movements: (a) First linear motion. (b) Circular motion. (c) Second linear motion. (d) Final position.

It is worth mentioning that the parameter ranges are defined using the positive directions of the XYZ-axes of the gripper TCP frame (Fig. 7a) for the Pick-Up action and the crate hole frame (Fig. 7b) for the PiH action.

To investigate to what extent the system can deal with uncertainties, noise is added to the gripper, the bottle and the crate hole frames. This is done by randomly choosing a direction from the XYZ-axes defining the frame and multiplying it with a randomly selected value from a normal distribution. In the case of the Pick-Up action, for the gripper and the bottle frames, the noise normal distribution has a mean value of zero and a standard deviation of 0.002 m for the positional part. The rotation part of both the frames is not effected, because a pre-defined rotation is used for the action, in order to ensure that the gripper is correctly approaching the bottle. The range of the uncertainties is based on preliminary ALVAR marker tracking data.

For the PiH action, the noise normal distribution for the crate has a mean value of zero and a standard deviation of 0.0035 m for the positional part and 3° for the rotational part. In the case of the bottle, the standard deviation is 0.001 m and 0.5° for the positional and the rotational parts respectively. These ranges are based on both preliminary ARUCO marker tracking data and simulations of the action. Initial simulation tests have shown that due to the tight fit of the bottle and the crate hole, a positional error larger than 0.0025 m on one of the XY-axes or a rotational error larger than 0.8° on one of the PY-Euler angles will lead to failure or to the appearance of forces between the bottle and the crate.

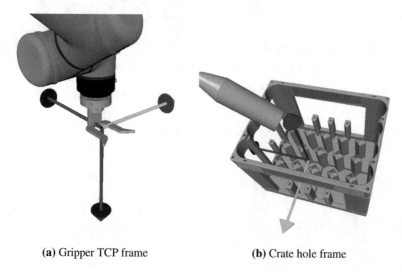

(a) Gripper TCP frame   **(b)** Crate hole frame

Fig. 7. Visualization of the two frames used to define the parameter ranges for (a) the Pick-Up action and (b) the PiH action.

3.4   Action Learning

To learn the parameter values that will lead to a successful task, both of the actions are simulated multiple times for each parameter set (10 repetitions for Pick-Up and 5 repetitions for PiH), where each repetition is affected by randomly selected perturbations, as explained in Subsect. 3.3. This leads to a total number of 157250 individual simulations for the Pick-Up action and 2925 simulations for PiH.

Because the search space for the Pick-Up action is 3-dimensional, the results are visualized as 2D success probability maps in the $H_{pu}\Phi_{pu}$-plane for fixed X_{pu} parameter values, as depicted in Fig. 8. To further narrow down the solution parameter space, each successful grasp is evaluated and the information is retained and used for the Kernel Density Estimation step, discussed in Subsect. 4.1.

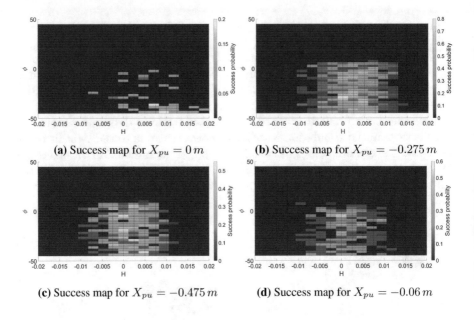

(a) Success map for $X_{pu} = 0\,m$

(b) Success map for $X_{pu} = -0.275\,m$

(c) Success map for $X_{pu} = -0.475\,m$

(d) Success map for $X_{pu} = -0.06\,m$

Fig. 8. The success probability maps for the Pick-Up action, for fixed X_{pu} values, before applying KDE.

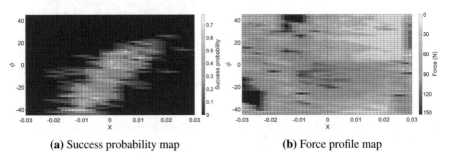

(a) Success probability map

(b) Force profile map

Fig. 9. The success probability map and the force profile map for the PiH action before applying KDE (Source: [16]).

For the PiH action, the results are also visualized using success probability maps in the $X_{pih}\Phi_{pih}$-plane as shown in Fig. 9a. To further improve the search for the optimal parameter set, the force profile of each simulated action is investigated. For each parameter set, the average of the highest contact forces that characterized each individual repetition is computed. The force is measured only between the bottle and the edges of the selected hole, where the maximum allowed force is limited to $150\,N$. The forces distribution can be seen in Fig. 9b.

4   Results

By applying KDE on the simulation results from Subsect. 3.4, the algorithm is able to determine the optimal parameter sets that would lead to a successful Pick-Up and PiH task. The best found parameter sets are tested on the Care-O-Bot platform. In order to further compare the simulation results with the real-world ones, for the PiH action, an additional subset of parameters is tested on the robot.

4.1   Optimal Action for Pick-Up

To get a good overview of the structure of the data and a good approximation of the best parameter sets that define the Pick-Up action, Kernel Density Estimation is used for the success probability maps shown in Fig. 8.

The Matlab implementation of the Multivariate Kernel Density Estimation method is used [22], which uses a multivariate Gaussian kernel. Each independent data sample is defined as a 5D vector $V = [X_{pu}; H_{pu}; \Phi_{pu}; S; G]$, where X_{pu}, H_{pu} and Φ_{pu} are the three Pick-Up action parameters, S is the success rate for that specific parameter set and G is the grasp evaluation score which is set to 0, in case of a failed action. The values of S and G are the weights that determine the smoothing contribution of each specific $[X_{pu}, H_{pu}, \Phi_{pu}]$ parameter set in the search space. In this Matlab implementation, the bandwidth is in the form of a d-by-d diagonal matrix, where d equals the dimensions of the search space. Additionally, the multivariate kernel is a product of all the one-dimensional kernel smoothing functions used for each dimension. Because for the Pick-Up action the search space is 3-dimensional, the bandwidth matrix has a 3-by-3 dimension. The values of the elements on the main diagonal are empirically chosen to be equal to the step size of each parameter range. The result of applying KDE on the simulated data can be seen in Fig. 10.

Finding the best parameter set for the Pick-Up action means determining the maximum of the estimated density surface. For complex cases, this can be done by applying optimization algorithms, such as gradient ascent. However, due to the nature of the data, in this case the best parameter set can be found just by searching for the maximum value in the density surface. As a result, the algorithm finds the best $[X_{pu}, H_{pu}, \Phi_{pu}]$ parameter set to be $[-0.0275\,\text{m}, 0.0025\,\text{m}, -7.5°]$.

4.2   Pick-Up Action Verification on the Real System

The best parameter set - as found by the simulation - is tested with ten trials on the Care-O-Bot platform, where the experiments showed a success rate of 90% in picking up the bottle, thus affirming the validity of the action learning technique. It is worth mentioning that the failed attempt was not due to the bottle tipping over, but rather because it slipped through the fingers while lifting it. This is an error from which the system can recover simply by re-performing the Pick-Up action.

4.3   Optimal Action for Peg-in-Hole

In the case of the PiH action, the Kernel Density Estimation via diffusion Matlab technique [23] is applied for both the success probability (Fig. 9a) and the force profile map (Fig. 9b), where the data is structured in a list of 3D vectors of type $V = [X_{pih}; \Phi_{pih}; D]$. X_{pih} and Φ_{pih} are the two PiH action parameters, while D is (depending on the case) either the success rate or the force value for that specific parameter set. As such, the smoothing factor of each specific $[X_{pih}, \Phi_{pih}]$ parameter set is determined according to the value of D.

The advantage of this KDE implementation, which also makes use of the multivariate Gaussian kernel, is that it utilizes a plug-in bandwidth selection method which implicitly computes the optimal bandwidth of the density estimator, without requiring any numerical optimization. Once this step is performed for both the success probability map and that of the forces distribution, the follow up step is to multiply the obtained KDEs. The result can be seen in Fig. 11c. This resulting KDE representation has the advantage of leading the search towards the set of parameters for which the PiH action is not only successful, but also performed with minimal contact forces between the bottle and the crate hole.

Due to the fact that the density surface is represented by a matrix where its size is defined by the X_{pih} and Φ_{pih} parameters and the individual density values depend on D, the best parameter set can be found by computing the matrix element with the maximum value. The best $[X_{pih}, \Phi_{pih}, D]$ parameter set is found to be $[0.0035\,\text{m}, 5°]$. Visualizing the KDE data representation, we can

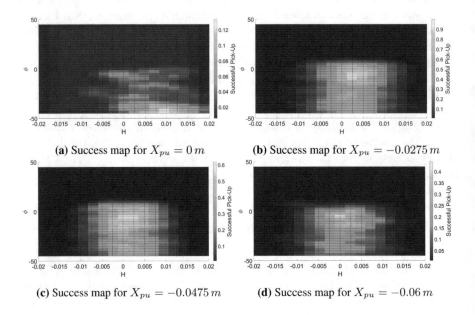

(a) Success map for $X_{pu} = 0\,m$ (b) Success map for $X_{pu} = -0.0275\,m$

(c) Success map for $X_{pu} = -0.0475\,m$ (d) Success map for $X_{pu} = -0.06\,m$

Fig. 10. The success probability maps for the Pick-Up action, for fixed X_{pu} values, after applying KDE.

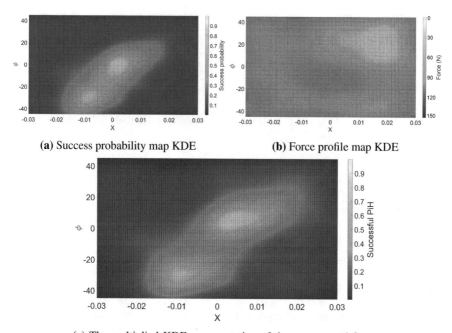

(a) Success probability map KDE

(b) Force profile map KDE

(c) The multiplied KDE representation of the success and force profile.

Fig. 11. The success probability map and the force profile map for the PiH action after applying KDE and the multiplied representation of the two (Source: [16]).

observe that a second peak is found around the $[-0.01\,\mathrm{m}, -30°]$ parameter set. By investigating the simulations, it can be determined that the success of the PiH action drops in-between these two peaks due to the fact that the bottle gets stuck on the edge of the hole and applies force on the crate.

4.4 PiH Action Verification on the Real System

Based on the results obtained in the previous subsection, a set of real world experiments are performed for the PiH action. As described in Sect. 3.1, the experiment begins with the visual servoing part, in which the Care-O-Bot is slowly bringing the gripper with the bottle towards the crate. Once the desired position is reached, the PiH sequence takes over, where specified $[X_{pih}, \Phi_{pih}]$ parameter sets are given as input. We investigate 10 different parameter sets manually selected from both promising and non-promising areas found in the simulation phase. Each of the sets is tested 10 times, thus a total number of 100 experiments are conducted. Each experiment is manually evaluated, which means that the D value for each of the individual samples is assigned a value from 1 to 5, after visually observing the experiment, according to weather or not the PiH action was successful and the intensity of the observed contact force.

The experiments show that the best parameter set - found to be $[0.0035\,\mathrm{m}, 5°]$ by the simulator - had a success rate of 100% on the real platform. The parameter

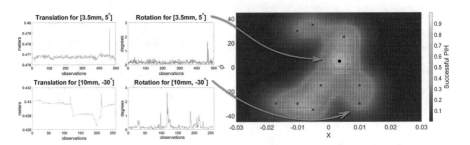

Fig. 12. Real world experiment (Source: [16]).

set $[-0.01\,\text{m}, -30°]$ corresponding to the second peak had a success rate of 70%, thus confirming the validity of the simulated result data. The $[-0.0025\,\text{m}, -15°]$ and $[0.005\,\text{m}, 25°]$ parameter sets also gave good results due to the compliance and curved shape of the bottle, which makes the insertion possible even if the edge of the hole is touched. Additionally, KDE is also applied on the outcome of the real world experiments (see Fig. 12). We can observe that the pattern of the density curve is similar to the one suggested by the simulation results in Fig. 11c.

In order to verify the assumption about the forces and because the Care-O-Bot is not equipped with a force-torque sensor, a third AR marker is placed on the side of the crate and used to track its movement during the PiH experiments. The underlying idea is that when the PiH action generates large contact forces, the crate is moved and thus the positional and rotational movement of the marker can be detected. Moreover, we verify if the scale of the movement is proportional to the strength of the contact force.

Figure 12 shows that the spikes in the translation and rotation of the AR marker appear at the moment any kind of contact force is applied. For the $[0.0035\,\text{m}, 5°]$ parameter set, the only registered spike appears around the $460th$ observation. The movement is generated by the bottle falling into the crate hole when it is released from the gripper. For the rest of the experiment, however, no major force is applied to the crate. On the other hand, for the $[0.01\,\text{m}, -30°]$ parameter set we can observe that when the bottle hits the edge of the crate, the tracker registers a movement of more than $0.003\,\text{m}$ in translation and $2.5°$ degrees in rotation. This suggests that a bigger contact force is applied, which aligns with the force estimation results in Fig. 11b and c.

To further test the validity of the simulation results, a second series of real experiments is performed, where extra positional uncertainty is added to the crate before performing the PiH sequence. The results of these experiments where presented in [16].

5   Conclusions

This paper proposes the transfer of industrial solutions for Peg-in-Hole (PiH) operations into the domain of service robotics. The actual PiH action is applied to the problem of handling bottles which are picked up and inserted into a crate for bottle depositing. Both the pick-up and the insertion operations are parameterized and explored in dynamic simulation with noise added to the sensor readings to identify actions which are not only successful but also lead to e.g. low contact forces. The optimal actions are subsequently determined using KDE and evaluated on a real service robot using the Care-O-Bot platform. The action execution on the actual robot is not relying on a precise hand-eye calibration or scene model, but rather makes use of visual servoing to instantiate the actions and to address inaccuracies.

The determined optimal actions for both the Pick-Up and the PiH operations have been tested on the real robot, with success rates of 90% and 100%. Besides the success or failure of the PiH action, the displacement of the deposit crate during the experiment has been recorded, showing that the action that was suggested to lead to the lowest contact forces also leads to the lower displacements of the crate in the real world. Since the crate can move freely, only limited by friction, it provides compliance to the system and the displacements of the crate relate to contact forces observed in simulation. Thereby, this paper demonstrates the successful adaptation of solutions developed for the industrial domain to solutions viable for service robotics. By including the uncertainties implied by the unstructured environment in the learning phase, actions that are robust to sensor noise have been determined.

Acknowledgements. This work was supported by the project Health-CAT, funded by the European Fund for regional development.

References

1. Schwarz, M., Schulz, H., Behnke, S.: RGB-D object recognition and pose estimation based on pre-trained convolutional neural network features. In: 2015 IEEE International Conference on Robotics and Automation (ICRA), pp. 1329–1335 (2015)
2. Kashiri, N., Laffranchi, M., Tsagarakis, N.G., Margan, A., Caldwell, D.G.: Physical interaction detection and control of compliant manipulators equipped with friction clutches. In: 2014 IEEE International Conference on Robotics and Automation (ICRA), pp. 1066–1071 (2014)
3. Mainprice, J., Hayne, R., Berenson, D.: Predicting human reaching motion in collaborative tasks using inverse optimal control and iterative re-planning. In: 2015 IEEE International Conference on Robotics and Automation (ICRA), pp. 885–892 (2015)
4. Dietrich, F., Buchholz, D., Wobbe, F., Sowinski, F., Raatz, A., Schumacher, W., Wahl, F.M.: On contact models for assembly tasks: experimental investigation beyond the peg-in-hole problem on the example of force-torque maps. In: 2010 IEEE/RSJ International Conference on Intelligent Robots and Systems, pp. 2313–2318 (2010)
5. Lin, L.L., Yang, Y., Song, Y.T., Nemec, B., Ude, A., Rytz, J.A., Buch, A.G., Krüger, N., Savarimuthu, T.R.: Peg-in-hole assembly under uncertain pose estimation. In: Proceeding of the 11th World Congress on Intelligent Control and Automation, pp. 2842–2847 (2014)

6. Sørensen, L.C., Buch, J.P., Petersen, H.G., Kraft, D.: Online action learning using kernel density estimation for quick discovery of good parameters for peg-in-hole insertion. In: 13th International Conference on Informatics in Control, Automation and Robotics (2016)
7. Gams, A., Petric, T., Nemec, B., Ude, A.: Learning and adaptation of periodic motion primitives based on force feedback and human coaching interaction. In: 2014 IEEE-RAS International Conference on Humanoid Robots, pp. 166–171 (2014)
8. Berio, D., Calinon, S., Leymarie, F.F.: Learning dynamic graffiti strokes with a compliant robot. In: 2016 IEEE/RSJ International Conference on Intelligent Robots and Systems (IROS), pp. 3981–3986 (2016)
9. Gams, A., Petrič, T.: On-line modifications of robotic trajectories: learning, coaching and force vs. position feedback. In: International Conference on Robotics in Alpe-Adria Danube Region, pp. 20–28. Springer (2016)
10. Jørgensen, J.A., Petersen, H.G.: Usage of simulations to plan stable grasping of unknown objects with a 3-fingered Schunk hand. In: IEEE International Conference on Intelligent RObots and Systems (IROS), Workshop - Robot Simulators: Available Software, Scientific Applications and Future Trends (2008)
11. Detry, R., Kraft, D., Kroemer, O., Bodenhagen, L., Peters, J., Krüger, N., Piater, J.: Learning grasp affordance densities. Paladyn J. Behav. Robot. **2**, 1–17 (2011)
12. Cherubini, A., Chaumette, F., Oriolo, G.: A position-based visual servoing scheme for following paths with nonholonomic mobile robots. In: 2008 IEEE/RSJ International Conference on Intelligent Robots and Systems, pp. 1648–1654 (2008)
13. Kyrki, V., Kragic, D., Christensen, H.I.: New shortest-path approaches to visual servoing. In: 2004 IEEE/RSJ International Conference on Intelligent Robots and Systems, (IROS 2004), Proceedings, vol. 1, pp. 349–354. IEEE (2004)
14. Gratal, X., Romero, J., Kragic, D.: Virtual visual servoing for real-time robot pose estimation. In: IFAC Proceedings, vol. 44, pp. 9017–9022 (2011)
15. Graf, B., Reiser, U., Hägele, M., Mauz, K., Klein, P.: Robotic home assistant Care-O-bot®3 - product vision and innovation platform. In: IEEE Workshop on Advanced Robotics and its Social Impacts (2009)
16. Suvei, S. D., Bodenhagen, L., Thulesen, T. N., Jami, M., Krüger, N.: Applying peg-in-hole actions with a service robot. In: 7th International Conference on Simulation and Modeling Methodologies, Technologies and Applications (2017)
17. Garrido-Jurado, S., noz Salinas, R.M., Madrid-Cuevas, F., Marín-Jiménez, M.: Automatic generation and detection of highly reliable fiducial markers under occlusion. Pattern Recognit. **47**, 2280–2292 (2014)
18. VTT: (Alvar 2.0)
19. Zuiderveld, K.: Contrast limited adaptive histogram equalization. In: Graphics Gems IV, pp. 474–485. Academic Press Professional, Inc. (1994)
20. Jørgensen, J.A., Ellekilde, L.P., Petersen, H.G.: Robworksim - an open simulator for sensor based grasping - conference papers - vde publishing house. In: ISR/ROBOTIK 2010 - ISR 2010 (41st International Symposium on Robotics) and ROBOTIK 2010 (6th German Conference on Robotics) (2010)
21. Smith, R., et al.: Open dynamics engine. IEEE Trans. Robot. (2005)
22. Scott, D.W.: Multivariate Density Estimation: Theory, Practice, and Visualization. Wiley (2015)
23. Botev, Z.I., Grotowski, J.F., Kroese, D.P.: Kernel density estimation via diffusion. Ann. Stat. **38**, 2916–2957 (2010)

Applying Deep Learning for Surrogate Construction of Simulation Systems

Zong-De Jian[1], Hung-Jui Chang[1,2], Tsan-sheng Hsu[1], and Da-Wei Wang[1(✉)]

[1] Institute of Information Science, Academia Sinica, Taipei, Taiwan
[2] Department of Computer Science and Information Engineering,
National Taiwan University, Taipei, Taiwan
{zdjian1988,chj,tshsu,wdw}@iis.sinica.edu.tw

Abstract. The deep learning approach has been successfully applied to various disciplines. When using optimization algorithms, there is a need to evaluate the performance of solutions found so far. The simulation system usually serve as the evaluator. However, to speedup the process, an approximation function, called surrogate, can replace the time consuming simulator. We propose to use deep learning to construct the surrogate function in epidemiology. The simulator is an agent-based stochastic model for influenza and the optimization problem is to find vaccination strategy to minimize the number of infected cases or economical impact. The optimizer is a genetic algorithm and the fitness function is the simulation program. An attempt to use the surrogate function with table lookup and interpolation was reported before. The results show that the surrogate constructed by deep learning approach outperforms the interpolation based one for both total case and economical impact. The average of the absolute value of relative error is less than 0.27%, which is quite close to the intrinsic limitation of the stochastic variation of the simulation software 0.2%, and the rank coefficients are all above 0.999. The vaccination strategy recommended is still to vaccine the school age children first which is consistent with the previous studies for minimizing total infected cases. As to minimize economical impact, the priority goes to the middle schoolers then to young working adults The results are encouraging and it should be a worthy effort to use machine learning approach to explore the vast parameter space of simulation models in epidemiology.

Keywords: Deep learning · Surrogate · Disease simulator

1 Introduction

Simulation models are forms of abstraction of the real world. They are especially useful to carry out experimentations which are difficult or impossible to do in the real world. With the recent success of machine learning techniques, especially the

---

An earlier extended abstract of this paper appears in [1].

© Springer Nature Switzerland AG 2019
M. S. Obaidat et al. (Eds.): SIMULTECH 2017, AISC 873, pp. 335–350, 2019.
https://doi.org/10.1007/978-3-030-01470-4_18

deep learning approach, how to apply them to learn from the simulated models is thus an interesting question. We explore the feasibility of learning a surrogate function from data generated by the simulation model.

Agent-based stochastic simulations have been applied widely for the study of infectious diseases [2,3]. One of the strength of the simulation model is its flexibility to incorporate detailed disease control strategies compared to mathematical models. However, it still consumes a significant amount of computing resources. Instead of searching larger parameter space, epidemiologists usually have to carefully craft the scenarios to demonstrate their points. The vaccine can mitigate the impact of the pandemic flu, however, the outcome depends on the vaccination strategy. We use the case of searching for appropriate vaccination strategy to demonstrate the effectiveness of our deep learning approach. Instead of comparing a few strategies selected by domain experts, we formulate it as an optimization problem with multiple objective functions. The search space can potentially contain many dimensions. In this paper, we focus on the dimension of vaccination allocation.

We usually has limited supply of the vaccine, thus the disease control agency has to decide the amount of vaccine allocated to various groups. It is a consensus that the health care professionals should get the highest priority. The next question is how much to distribute to different age groups. How to determine the number of doses to each age group is an important issue. The objective can also be complicated, for example, one can search for a strategy to minimize economic impact, or to minimize the total number of infected cases. In this paper, we study both objectives. To compute the economical impact, we follow the formulation by Meltzer [4].

When we employ the genetic algorithm to tackle the optimization problem, the gene encodes the vaccine distribution among age groups and the fitness function is the total number of infected cases or the economic impact. The input of the fitness evaluation function is the outcome of running the simulation module. In our case, each simulation run takes about 3 min, the fitness evaluation thus becomes the bottleneck of the process. Using a faster approximation in place of the true fitness function is called surrogate-assisted evolutionary computation [5]. The idea was first proposed in the mid-1980s [6]. A surrogate function combines table lookups and linear interpolation was suggested in [7]. In this paper, we construct surrogate functions with deep neural networks and compare them with interpolation ones.

We treat the simulation results as ground truth and measure the performance of the surrogate function with *relative error, the performance of genetic algorithm and rank correlation coefficient*. The relative error is the difference between the output of surrogate function and the simulation divided by the output of the simulation. The average of the absolute value of the relative errors of the surrogate functions constructed by the deep neural network approach range from 0.12% to 0.27% for total case surrogates and 0.07% to 0.21% for economical impact surrogate. The deep neural network surrogates performs better than interpolation surrogates. percent. The rank correlation coefficients of the surrogates are better than 99.9%.

2   Material and Method

Our approach to search for the optimal vaccination strategy is a case of the simulation-based optimization [8]. The discrete time agent-based simulation software developed in house is briefly introduced below [9], followed by a description of the optimization procedure which is a genetic algorithm coupled with a simulated annealing procedure called hybrid simulated annealing genetic algorithm (HSAGA).

In the software system, there are people, their contacts and parameters of natural history of the disease. The mock population is constructed based on the national demographics from Taiwan Census 2000 (http://eng.stat.gov.tw/). There are about 1.72 million *preschool children* (0–5 years old), about 2.36 million *elementary school children* (6–12 years old), about 0.99 million *middle school children* (13–15 years old), about 0.97 million *high school children* (16–18 years old), about 3.86 million *young adults* (19–29 years old), about 10.28 million *adults* (30–64 years old), and about 1.94 million *elders* (65+ years old).

The software implemented an SEIR system, which means that an individual can be in one of the following four states, susceptible(S), exposed(E), infectious(I), and recovered(R). When an individual in the susceptible state is infected by the virus, he entered the exposed state, after some time the individual became infectious, and finally he entered recovered state. An important virus-dependent parameter is the transmission probability, denoted by p_{trans}, which is the probability that an effective contact results in an infection. The probability of effective contact between two individuals depends on the contact groups they belong to. According to the disease natural history, an exposed individual becomes infectious and an infectious individual becomes recovered, in our setting the average incubation period is 1.9 days and the average infectious period is 4.1 days [2]. A contact group models a daily close interaction of individuals, where every member has the same contact probability with all other members in the same group. There are eleven contact groups which are community, neighborhood, household cluster, household, workgroup, high school, middle school, elementary school, daycare center, kindergarten, and playgroup [10]. The population size of Taiwan is about 22.12 million.

Each individual can belong to several contact groups simultaneously at any time. The duration of a simulation run is set at 365 days. Each day has two 12-hour periods, daytime and nighttime respectively. Behaviors of individuals are depicted in Fig. 1. During daytime, contact occurs in all contact group. School aged children go to schools. Preschool children go to a daycare centers, kindergartens or playgroups. Young adults and adults go to work thus join workgroups. In the nighttime, contact occurs only in communities, neighborhoods, household clusters, and households.

The model parameters are similar to ones in a study by [2], with modifications to fit Taiwan situation by a contact diary study [11]. We note that the stochastic variation of this simulation system is around 0.2% [9].

In this paper, our setting is similar to our previous study about surrogates functions and the optimization algorithm HSAGA is the same as one presented

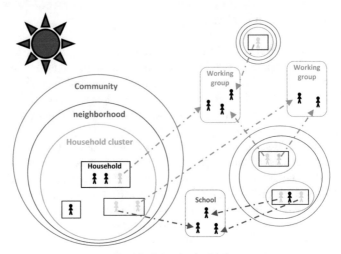

(a) Contact behavior in daytime.

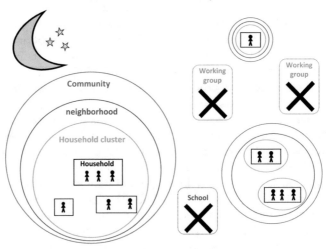

(b) Contact behavior in nighttime.

Fig. 1. Contact behavior [1].

in our previous paper [7]. We briefly recapy here. The $p_t rans$ is set at 0.1, the vaccine is available 30 days after the index case occurred, total 2.5 million of doses are applied to different age groups according to the priority. However, we only focus on the case that the vaccine efficacy, VEi and VEs, are fixed at 0.9 [12]. The vaccine is available 30 days after the index case occurred, total available vaccine are all administered to individuals.

A vaccination priority is denoted by $p = (x_1, x_2, ..., x_7)$, where x_i is the amount of vaccine for the age group i and it is less than the population of age group i. We use p_α to denote vectors with only those value of x_α are nonzero and α is a set with numberings of each age group. For example $p_{2,3}$ denote the set of vaccination priority with non-zero entries for age group 2 and age group 3 and $p_{i,j,k}$ to denote the set of vectors with 3 non-zero entries. Let $p = \mathbf{0}$ denote the baseline case with no vaccination. We use $Sim(p)$ to denote the number of infected cases reported by the simulation program with vaccination priority p. In HSAGA the gene population consists of ten vaccine allocations represented in prefix sum format. That is $p = (20, 50, 50, 70, 20, 30, 10)$ can be rewritten as $\hat{p} = (20, 70, 120, 190, 210, 240, 250)$, since the total amount of vaccine is always 2.5 millions the last coordinate can be dropped. Each iteration begins with simulated annealing step to perturb each candidate, followed by the selection, crossover, and mutation steps [7].

The flow of HSAGA is shown in Fig. 2. When the simulator serves as the fitness function, for a given allocation, we carry out 5 simulation runs and take the average. The best solution of the previous generation and the first nine solutions for this generation become the candidates of next generation. Each iteration begins with a simulated annealing step for each candidate. It is a temperature controlled mutation, i.e., we mutate each candidate according to the temperature (that is the number of iterations up to the point). The process stops at 200 iterations with an early stop condition when five consecutive iterations consist of the same candidates.

To mutate, we randomly pick index i and randomly generate a number x, replace x_i with x and sort the resultant sequence.

Deep learning models are built with multilayer perceptrons (MLP). Each perceptron has a transformation function to produce output to next perceptrons with the inputs from connected perceptrons at the previous stage. During the training phase, the prediction error is determined by the loss function, and the error triggering a weight adjustment procedure, and backpropagation is the most commonly applied method. We use deep neuron networks (DNN) as the model

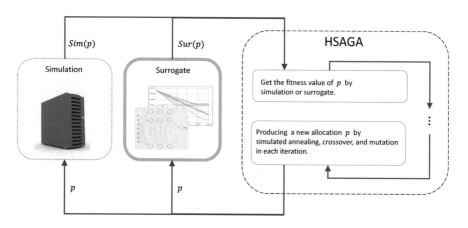

Fig. 2. Process of HSAGA [1].

of surrogate functions. In this study, we use Keras on Theano running on Nvidia GeForce GTX 1080 Graphics Card [13].

The surrogate function takes the vaccine allocation of seven age groups, p, as the input and the output is the total number of infected cases or economical impact.

We compare the surrogates learnt by neural networks with the interpolation based ones we constructed before. We first give a brief description of the previous work, and the details can be found in [7].

To apply interpolation, we need a set of reference points serve as the entries of the lookup table, denoted by P_t, the values of these points are the simulation results. Given p, if $p \in P_t$ the result, $Sim(p)$, is the total number of infected cases by simulation. If $p \notin P_t$, then $Sim(p)$ is the estimated total number of infected cases by interpolation. We first compute 26 reference points for each age group, there are 182 points in total. The 26 points are evenly spaced up to 2.5 million doses of vaccine.

We build a lookup table with entries in the form of $p_{i,j}$ to capture the interactions between age groups. For each age group i, five evenly spaced values are determined. Given two age groups, we take one value from each group to form a vaccination strategy and carry out the simulation. There are total $21 \times 5 \times 5 = 525$ such points.

For a point $p = (x_1, x_2, ..., x_7)$, we slightly abuse the notation and use $p_{j,k}$ denote the point with the same value as p in j^{th} and k^{th} dimensions and all other dimensions are zero. The adjustment term for the interaction between dimension j and k, denoted by $\delta(p_{j,k})$, is defined as following:

$$Sur_2^I(p) = Sim(0) + \sum_{i=1}^{7} \Delta(p_i) + \sum_{j=1}^{6} \sum_{k=j+1}^{7} \delta(p_{j,k}) \tag{1}$$

3   Results

We reuse data sets collected in previous studies to evaluate the surrogate functions developed. There are total 3120 points. The first set contains the data points sampled while developing interpolation based surrogate. The second set contains the data points evaluated during the execution of HSAGA with simulator as the fitness evaluator. And the third set are points with only 3 non-zero parameters. Each points contains seven features which are the amount of vaccine for age groups and two labels which are the total number of infected cases and the total economic cost. We use 10-fold cross-validation to test the trained surrogate. We use S_i denote the ith training set as well as the result of that set when there is no confusion. For each training process the epoch size is 100 thousands and batch size is 32.

In Fig. 3, we show that the performances of the naive linear interpolation and its deep learning counter part are much worse than surrogates incorporated the interaction between age groups. From now on we focus only on $Sur_2^D(p)$ and $Sur_2^I(p)$.

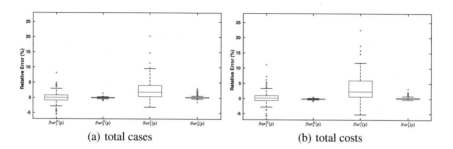

(a) total cases   (b) total costs

Fig. 3. 10-fold cross-validation: relative error for data sets S_0.

In Fig. 4, the relationship between the accuracy of the surrogate and the number of epochs is demonstrated by only one training incident since all of them are similar. It is apparent that 100 thousand epoch size performs better. However, due to time limit as well as the potential of overfitting we do not try larger epochs.

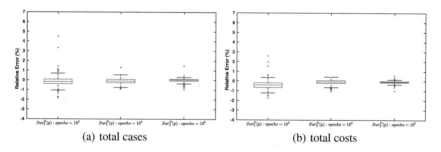

(a) total cases   (b) total costs

Fig. 4. 10-fold cross-validation: relative error for data sets S_0 with different training epochs.

In Figs. 5 and 6 are the testing results for total case surrogates and total cost surrogates. The results show that the surrogates constructed with deep neural learning outperforms interpolation based surrogates in terms of the average absolute error, the max absolute error, the average, the standard deviation and the interquartile range for both the total case and economical cost. Details are in Tables 1 and 2.

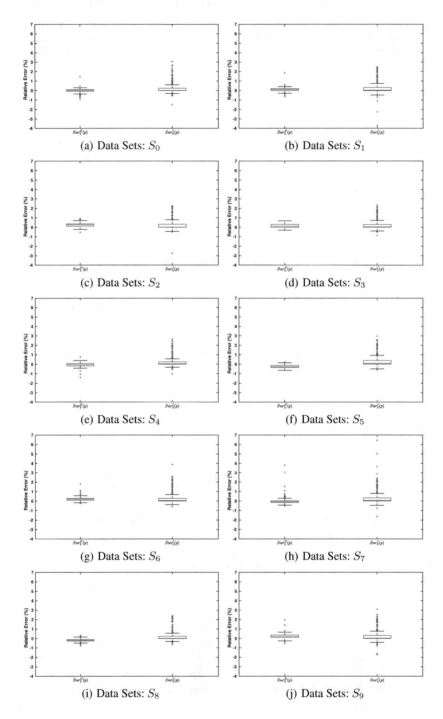

Fig. 5. 10-fold cross-validation: relative error of total cases for each testing set.

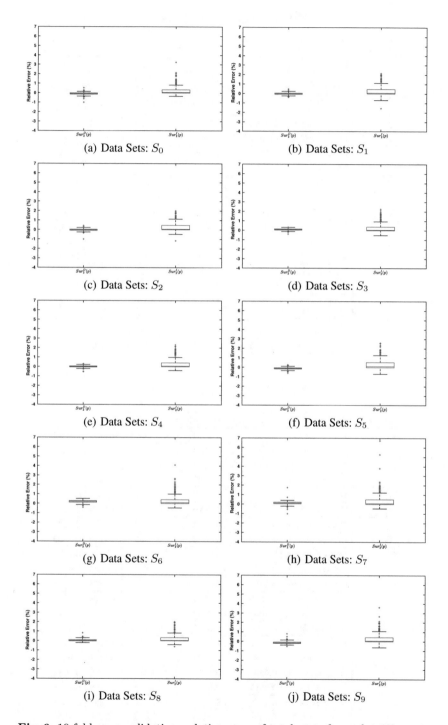

Fig. 6. 10-fold cross-validation: relative error of total costs for each testing set.

Table 1. Detail data of Fig. 5.

Data sets	Surrogate	Avg of abs	Max of abs	Avg	SD	IQR
S_0	Deep learning	0.1283	1.4675	−0.0086	0.1942	0.1848
	Interpolation	0.2838	3.0487	0.2199	0.5245	0.2428
S_1	Deep learning	0.1587	1.8535	0.1068	0.1899	0.2092
	Interpolation	0.3691	3.7037	0.2543	0.6637	0.3336
S_2	Deep learning	0.2585	0.9186	0.2380	0.1834	0.2491
	Interpolation	0.3062	2.7450	0.2419	0.5506	0.3232
S_3	Deep learning	0.1904	0.6789	0.1438	0.2030	0.3123
	Interpolation	0.3073	2.3511	0.2489	0.5355	0.2882
S_4	Deep learning	0.1539	1.3942	−0.0390	0.2047	0.2569
	Interpolation	0.2909	2.6368	0.2267	0.5217	0.2428
S_5	Deep learning	0.2573	0.6537	−0.2464	0.1652	0.2261
	Interpolation	0.3586	2.9493	0.3153	0.5936	0.3881
S_6	Deep learning	0.216	1.8165	0.1986	0.1982	0.1938
	Interpolation	0.3474	3.8968	0.3019	0.6297	0.2832
S_7	Deep learning	0.1728	3.8063	−0.0162	0.3499	0.1806
	Interpolation	0.3725	6.4301	0.3072	0.7419	0.3236
S_8	Deep learning	0.2146	0.7917	−0.1911	0.1588	0.1800
	Interpolation	0.2641	2.3873	0.1914	0.5067	0.2457
S_9	Deep learning	0.2627	1.9694	0.2298	0.2336	0.2691
	Interpolation	0.3446	3.0851	0.2533	0.6102	0.3208

Table 2. Detail data of Fig. 6.

Data Sets	Surrogate	Avg of abs	Max of abs	Avg	SD	IQR
S_0	Deep learning	0.1218	0.9984	−0.0876	0.1351	0.1417
	Interpolation	0.3036	3.2391	0.2663	0.4968	0.3382
S_1	Deep learning	0.0837	0.4704	−0.0098	0.1115	0.1318
	Interpolation	0.3714	2.1507	0.3009	0.5538	0.4702
S_2	Deep learning	0.0814	1.0057	−0.0266	0.1141	0.1268
	Interpolation	0.3265	1.9925	0.2788	0.4894	0.4629
S_3	Deep learning	0.1047	0.3884	0.0914	0.0916	0.1171
	Interpolation	0.3278	2.2572	0.2787	0.5129	0.3835
S_4	Deep learning	0.0748	0.5379	0.0066	0.1021	0.1105
	Interpolation	0.3211	2.3035	0.2793	0.4808	0.4130
S_5	Deep learning	0.1194	0.6113	−0.1044	0.1058	0.1192
	Interpolation	0.3919	2.5911	0.3560	0.5578	0.5400
S_6	Deep learning	0.2067	0.5163	0.1952	0.1299	0.1697
	Interpolation	0.3706	4.0913	0.3363	0.5911	0.4094
S_7	Deep learning	0.1560	1.7467	0.1151	0.1785	0.1703
	Interpolation	0.3934	6.7198	0.3437	0.7201	0.5016
S_8	Deep learning	0.086	0.8254	0.0289	0.1149	0.1431
	Interpolation	0.2887	2.0148	0.2358	0.4669	0.3368
S_9	Deep learning	0.1610	0.8104	−0.1375	0.1291	0.1501
	Interpolation	0.3495	3.6186	0.3043	0.5549	0.4362

Table 3. The best allocation of $HSAGA$ with each fitness function.

(a) total cases

S	C	I	N	p ($\times 10^4$ doses)			
				ES	MS	HS	YA
Interpolation	4.990	69	904	97	78	75	0
Deep Learning S_0	4.987	80	1072	99	81	70	0
S_1	4.998	74	1003	97	83	70	0
S_2	5.012	83	1100	93	83	75	0
S_3	5.011	87	1141	97	82	72	0
S_4	5.003	86	1120	97	82	72	0
S_5	4.980	82	1102	99	81	70	0
S_6	5.002	67	893	97	80	73	0
S_7	4.993	83	1083	98	81	71	0
S_8	4.993	73	939	100	77	73	0
S_9	5.000	65	856	98	80	72	0

(b) total costs

S	E	I	N	p ($\times 10^4$ doses)			
				ES	MS	HS	YA
Interpolation	9.745	79	1057	0	78	73	99
Deep Learning S_0	9.641	85	1121	0	83	67	100
S_1	9.647	91	1193	0	85	68	97
S_2	9.655	72	965	0	85	67	98
S_3	9.666	84	1084	0	86	64	100
S_4	9.657	103	1352	0	86	65	99
S_5	9.647	98	1288	0	85	67	98
S_6	9.681	97	1270	0	87	66	97
S_7	9.673	91	1179	0	85	70	95
S_8	9.654	94	1237	0	86	66	98
S_9	9.654	84	1142	0	85	67	98

' S ': data sets

' C ': total cases ($\times 10^6$)

' E ': total costs ($\times 10^9$ \$US)

' I ': total iterations

' N ': total allocations

' ES ': *elementary school children*

' MS ': *middle school children*

' HS ': *high school children*

' YA ': *young adults*

We next apply both surrogates by deep learning $(Sur_2^D(p))$ and interpolation $(Sur_2^I(p))$ to HSAGA algorithm to compare the quality of recommended vaccination strategy. The results are shown in Table 3. If the goal is to minimizing total infected cases, the appropriate strategy is to vaccinate middle school students first and then vaccinate elementary school students. As for minimizing economical impact, the first priority is still vaccinating middle school students,

Table 4. The gray level of Table 3(a).

	data sets	the best allocations of each iteration
Interpolation		
Deep Learning	S_0	
	S_1	
	S_2	
	S_3	
	S_4	
	S_5	
	S_6	
	S_7	
	S_8	
	S_9	

Table 5. The gray level of Table 3(b).

data sets	the best allocations of each iteration
Interpolationss	
Deep Learning S_0	
S_1	
S_2	
S_3	
S_4	
S_5	
S_6	
S_7	
S_8	
S_9	

Table 6. Rank correlation coefficient.

(a) total cases

	Data Sets	r_s for All Samples
Interpolation		0.99787
Deep Learning	S_0	0.99991
	S_1	0.99988
	S_2	0.99989
	S_3	0.99988
	S_4	0.99991
	S_5	0.99991
	S_6	0.99989
	S_7	0.99987
	S_8	0.99989
	S_9	0.99990

(b) total costs

	Data Sets	r_s for All Samples
Interpolation		0.99481
Deep Learning	S_0	0.99977
	S_1	0.99976
	S_2	0.99979
	S_3	0.99981
	S_4	0.99979
	S_5	0.99979
	S_6	0.99974
	S_7	0.99980
	S_8	0.99979
	S_9	0.99978

yet the rest of the vaccine is allocated to young people from 19 to 29 years old. And the explanation is that the economic loss of young working adults is much higher than elementary school students. We also use grey level picture developed before to help visualize the evolution process which is shown in Tables 4 and 5 [1].

As for the execution time, it takes about 80000 s to carry out 60 rounds when the fitness function is the simulator [7], it takes 44 s when surrogate is replaced by $Sur_2^I(p)$, the total execution time is 1068 s with $Sur_2^D(p)$, however, most of the time was overhead of GPU since it takes only several micro-seconds to finish one evaluation.

For genetic algorithms, the rank preserving surrogates are preferred. One metric to measure the fidelity of surrogates is rank correlation coefficient (r_s) [14]:

$$r_s = 1 - \frac{6 \times \sum_{i=1}^{N} (R_A[i] - R_B[i])^2}{N(N^2 - 1)} \tag{2}$$

For rank correlation coefficient, both the surrogates from deep learning and interpolation performs well 99%, we do note that deep learning has slightly edge and the explanation is that the neural network might capture some interactions between age groups better. The detail data is present in Table 6.

4 Conclusion and Discussion

This study explores the feasibility of using machine learning approach to constructing surrogates as the cost function for optimization schemes. Two different cost functions are studies, one is counting the total number of infected cases and the other one is to calculate the total economic loss due to epidemics. Our results show that surrogates learnt by deep neural networks usually outperform surrogates using anchor points and interpolation approach designed by human. Although interpolation surrogates run faster, we note that the machine learning approach can serve as a generic and automatic surrogate construction process. It does not need human to design a sensible surrogate function by past experience.

Our results confirm the finding of previous studies that school children should be vaccinated with high priority when the objective is to reduce the total number of infected cases. To reduce the economical impact, the advise is to vaccinate middle schoolers first then young working adults. One obvious future direction is to use machine learning to explore the vast landscape of scenarios with various objective functions and constraints. For example, the infectiousness of the virus strand might vary, the available date of vaccine may not be known in advance. The only variables are the amounts of vaccine allocated to different age groups now, more parameters will be included as input in the future. We also plan to try other neural network architecture such as the convolutional neural network and the recurrent neural network. We hope not only we can construct accurate surrogates with more parameters, but also can gain insight about the delicate interaction between model parameters and outcome by studying the neural networks.

Finally, we envision that an autonomous software searches through the huge scenario space with the help of surrogate function and adaptively executes simulation program to revise the surrogate function to produce higher fidelity surrogate and better search results by applying reinforcement learning methods.

Acknowledgements. We thank anonymous reviewers for their suggestions. The research is partially funded by the grant of "MOST104-2221-E-001-021-MY3", and "Multidisciplinary Health Cloud Research Program: Technology Development and Application of Big Health Data. Academia Sinica, Taipei, Taiwan".

References

1. Jian, Z.-D., Chang, H.-J., Hsu, T.-s., Wang, D.-W.: Learning from simulated world - surrogates construction with deep neural network. In: the 7th International Conference on Simulation and Modeling Methodologies, Technologies and Applications, pp. 83–92 (2017)
2. Germann, T.C., Kadau, K., Longini, I.M., Macken, C.A.: Mitigation strategies for pandemic influenza in the United States. Proc. Natl. Acad. Sci. **103**(15), 5935–5940 (2006)
3. Chao, D.L., Halloran, M.E., Obenchain, V.J., Longini, I.M.: FluTE, a publicly available stochastic influenza epidemic simulation model. PLOS Comput. Biol. **6**(1), 1–8 (2010)
4. Meltzer, M.I., Cox, N.J., Fukuda, K.: The economic impact of pandemic influenza in the United States: priorities for intervention. Emerg. Infect. Dis. **5**(5), 659–671 (1999)
5. Jin, Y.: Surrogate-assisted evolutionary computation: recent advances and future challenges. Swarm Evol. Comput. **2**(1), 61–70 (2011)
6. Grefenstette, J.J., Fitzpatrick, J.M.: Genetic search with approximate fitness evaluations. In: International Conference on Genetic Algorithms and Their Applications, pp. 112–120 (1985)
7. Jian, Z.-D., Hsu, T.-s., Wang, D.-W.: Searching vaccination strategy with surrogate-assisted evolutionary computing. In: the 6th International Conference on Simulation and Modeling Methodologies, Technologies and Applications, pp. 56–63 (2016)
8. Gosavi, A.: Simulation-based optimization. Parametric optimization techniques and reinforcement learning (2015)
9. Tsai, M.-T., Chern, T.-C., Chuang, J.-H., Hsueh, C.-W., Kuo, H.-S., Liau, C.-J., Riley, S., Shen, B.-J., Shen, C.-H.: Wang, D.-W., Hsu, T.-s.: Efficient simulation of the spatial transmission dynamics of influenza. PLoS ONE **5**(11), 1–8 (2010)
10. Chang, H.-J., Chuang, J.-H., Fu, Y.-C., Hsu, T.-s., Hsueh, C.-W., Tsai, S.-C., Wang, D.-W.: The impact of household structures on pandemic influenza vaccination priority. In: The 5th International Conference on Simulation and Modeling Methodologies, Technologies and Applications, pp. 482–487 (2015)
11. Fu, Y.-C., Wang, D.-W., Chuang, J.-H.: Representative contact diaries for modeling the spread of infectious diseases in Taiwan. PLoS ONE **7**(10), 1–7 (2012)
12. Basta, N.E., Halloran, M.E., Matrajt, L., Longini, I.M.: Estimating influenza vaccine efficacy from challenge and community-based study data. Am. J. Epidemiol. **168**(12), 1343–1352 (2008)
13. Keras Documentation. https://keras.io/
14. Loshchilov, I., Schoenauer, M., Sebag, M.: Comparison-Based Optimizers Need Comparison-Based Surrogates, pp. 364–373 (2010)

A General Method to Compare Different Co-simulation Interfaces: Demonstration on a Case Study

Georg Engel[1,2(✉)], Ajay S. Chakkaravarthy[2], and Gerald Schweiger[1,2]

[1] Institute for Software Technology, University of Technology Graz, Infeldgasse 16b, 8010 Graz, Austria
[2] AEE - Institute for Sustainable Technologies, Feldgasse 19, 8200 Gleisdorf, Austria
g.engel@aee.at
http://www.aee-intec.at

Abstract. A method is presented to compare different co-simulation interfaces. The comparison assesses user-friendliness and flexibility, computational costs and accuracy. Interfaces corresponding to different versions of loose and strong coupling are discussed. The specific implementations include the Functional Mockup Interface (FMI), the Building Controls Virtual Test Bed (BCVTB) and a Component Object Model (COM). A case study is introduced to present the method in a pedagogical way. The case study includes a compact thermal energy storage modelled in Trnsys and a heat sink modelled in Simulink. Generalizations of the method to realistic full-scale co-simulations are proposed.

Keywords: Co-simulation · FMI · BCVTB · Trnsys · Simulink
Compact thermal energy storage

1 Introduction

1.1 Motivation and Background

Co-simulation can be considered as a simulation method involving a collaboration of various tools or solvers. The motivation for co-simulation is manifold. From a theoretical and often also practical perspective, complex systems are composed by subsystems. Typically, specific tools and methods exist to discuss the different subsystems. These specialized tools are in general not capable of modelling the entire system. This holds in particular for cross-domain systems, which are getting more and more attraction in recent years and decades. A completely different kind of motivation for co-simulation is the fact that a variety of different tools and methods are used by different teams. Therefore, collaboration of teams with expertise in different tools can be aided dramatically by co-simulation techniques. Similarly, it potentially supports rapid-prototyping, IP protection, parallelization, X-in-the-loop and much more [1].

© Springer Nature Switzerland AG 2019
M. S. Obaidat et al. (Eds.): SIMULTECH 2017, AISC 873, pp. 351–365, 2019.
https://doi.org/10.1007/978-3-030-01470-4_19

Prominent applications of co-simulation can be found, e.g., in energy systems, the automotive sector and cyber-physical systems. In the field of energy systems, the need for co-simulation arises from the goal of the energy transition. As the share of fluctuating energy sources such as wind and solar energy increases, other parts of the energy systems must become more flexible to match the available energy from renewable resources with the demand in terms of location, time and quantity. Most options for increasing the flexibility of energy systems imply sector-coupling and thus combining different energy domains, but also increasing supply and demand flexibility or integrating energy storage technologies [2,3]. In the automotive sector, the transition to E-mobility poses a variety of challenges. For instance, the lack of waste heat and the narrow temperature window required by the battery require a smart thermal management of vehicles [4], possibly including thermal storage solutions [5]. This leads to new requirements for simulation approaches and tools. To increase the efficiency of existing systems, detailed models of all subsystems that capture all important dynamics are required.

There are specific tools for each individual domain of district-scale energy systems, but no single tool can cover all domains and aspects in order to simulate the entire system [6]. Co-simulation approaches allow for the combination and reuse of existing tools and methods that are robust and well-suited for their particular domain. A main drawback of co-simulating is that numerical stability problems may arise [7], code optimizations within a particular tool may be lost [8] and some co-simulation frameworks have inconvenient application programming interfaces so that such methods are inappropriate for engineering applications. An overview of co-simulation approaches and tools, research challenges, and research opportunities are presented in [1,9–11]. Co-simulation approaches can be divided into three categories: discrete event, continuous time and hybrid co-simulations [1]. Prominent standards are FMI for continuous time co-simulations and High Level Architecture for discrete event ones, while hybrid co-simulation is yet an open challenge for research. [12] present an error estimation for co-simulations based on classical Richardson extrapolation, and a modified algorithm for a reliable communication step size control based on an extension of the step size control of classical time integration. They conclude that the numerical efficiency of co-simulation algorithms may be improved by higher-order approximations of subsystem inputs.

The present work presents a comparison of different co-simulation interfaces for a simple case study, and suggests also a methodology how the comparison can be performed for realistic cases. Part of the work has been presented in [13–15]. The main contributions of this paper are:

– A general methodology to compare different co-simulation interfaces.

– An explicit comparison of different co-simulation interfaces between Trnsys and Simulink for a case study in thermal engineering.

– The comparison serves as padagogical discussion of different types of interfaces.

1.2 Tools and Interfaces

FMI [16] is a tool independent standard that has been developed in the ITEA2 European Advancement project MODELISAR. FMI supports both model exchange and co-simulation of dynamic models, providing a zip of an executable and xml-files describing metadata for the subsystem. FMI is currently supported by 95 tools and is used by various industries and universities.

Trnsys is a simulation environment for the transient simulation of thermal systems, originally written in the Fortran programming language [17]. Type155 is available in the Trnsys standard library and implements a direct link to Matlab. The connection uses the Matlab engine, which is launched as a separate process. The Fortran routine communicates with the Matlab engine through a COM interface. Type 155 can be used in different calling modes (standard component called in each iteration or real-time controller called only after convergence).

BCVTB is a software environment developed at Lawrence Berkeley National Laboratory [18]. It allows connecting different simulation tools to exchange data during the time integration. BCVTB is based on Ptolemy II, an open-source software framework supporting experimentation with actor-oriented design. BCVTB has interfaces to EnergyPlus, Dymola, Functionl Mock-up Units (FMU), Matlab and Simulink, Radiance, ESP-r, Trnsys and BACnet.

The coupling between the different tools can be done by either loose or strong coupling [9]. In loose coupling, the data exchange between simulators is realized only at certain points in time, without iteration between the coupled simulators. Loose coupling can be realized by sequential or parallel execution of the individual simulators. Strong coupling methods iterate the values needed from other partial systems in every time step. Generally, the strong coupling shows higher accuracy and higher stability at the costs of a higher computational time consumption [19]. More accurately, strong coupling co-simulation show the same accuracy as monolithic simulations, while loose coupling introduces additional integration error terms through the finite step size of the communication between the solvers.

2 Method

2.1 Comparison with a Reference Simulation

The comparison between the different interfaces considers user-friendliness, flexibility, accuracy and computational costs, following the discussion presented in [13]. The user-friendliness and the flexibility are judged only on a qualitative basis.

In order to estimate the accuracy, the results are compared to a reference, e.g. simulation results of a more accurate simulation. The corresponding deviation is often referred to as "global error", which potentially accumulates over simulation time. For the presented case study, we define as "reference simulation" the model implemented entirely in Trnsys, and employed with improved solver parameters (see Table 1) to ensure high accuracy results. The variables communicated via the co-simulation interface (inlet and outlet temperature of the heat transfer fluid) as well as the temperatures of the heat storage and the body are compared to the corresponding time-series results obtained in the reference simulation. The maximum deviation is considered as measure for the accuracy.

To discuss the computational costs, a simple batch-script is used to measure the overall simulation time. This includes overhead like starting Matlab etc., but this is in most cases the relevant timing for the user. Replica simulations serve to estimate the confidence interval.

2.2 Case Study

As case study to present the method and to compare different co-simulation interfaces, we introduce a toy example where a sorption-based compact thermal energy storage (modelled in Trnsys) is coupled thermally to a simple heat sink (modelled in Simulink), which was also presented in [14]. The corresponding system design is shown in Fig. 1. We discuss continuous time co-simulation only, which is why discrete events like control switches are avoided. Only discharging of the storage is considered, where the sorption process releases heat, increasing the temperature of the storage. The heat is extracted via a heat transfer fluid to the heat sink, which is represented by a simple body with one thermal node. The temperatures of the heat transfer fluid are communicated via the co-simulation interface.

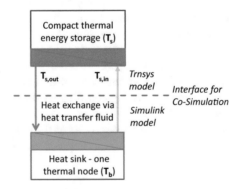

Fig. 1. Schematics of the case study employed to present the method and to discuss the different co-simulation interfaces [13]. A compact thermal energy storage is connected to a heat sink with one thermal node via a heat transfer fluid. The storage is modelled in Trnsys, while the heat sink is modelled in Simulink. The temperatures of the heat transfer fluid are communicated via the co-simulation interface between the two simulation tools.

3   Model

As shown in Fig. 1, the system is decomposed into two subsystems: a compact thermal energy storage modelled in Trnsys, and a thermal sink modelled in Simulink. This section briefly introduced the physical model for these subsystems.

3.1   Heat Storage Modelled in Trnsys

The compact thermal energy storage is modelled in Trnsys as depicted in Fig. 2. A more detailed description of the model is found in [20], related results have also been presented in [21].

The following set of ordinary differential equations models the inner states of the sorption store, which are the store temperature T_s (energy balance, Eq. (1)) and the water load of the sorption material x_s (mass balance, Eq. (2)) [20]:

$$C_{\text{tot}} \frac{dT_s}{dt} = \dot{Q}_{\text{HX}} + \dot{Q}_{\text{vap,in}} + \dot{Q}_{\text{ads}} + \dot{Q}_{\text{ambient}} \tag{1}$$

$$\frac{dx_s}{dt} = k_{\text{LDF}} \left[x_{s,\text{equ}}(p_{\text{vap}}, T_s) - x_s \right], \tag{2}$$

where t denotes time, C_{tot} the total (sensible) heat capacity of the sorption store, and k_{LDF} the linear driving force parameter for adsorption and desorption, respectively [22]. $x_{s,\text{equ}} = x_{s,\text{equ}}(p_{\text{vap}}, T_s)$ is the equilibrium water load of the sorption material, calculated for the current store temperature T_s and vapor

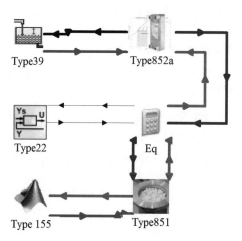

Fig. 2. Model of the compact thermal energy storage in Trnsys shown examplarily for the interface based on Type155 [13]. Type851 represents the sorption reactor and Type 852a the evaporator/condenser coupled to the water reservoir Type39. Type22 and the equation block serve to calculate the vapour pressure between the reactor and the evaporator/condenser.

pressure p_{vap} , e.g. by the Dubinin approach [23]. The various terms on the right hand side of Eq. 1 represent the heat flows in and out of the sorption store. The heat flow via the heat exchanger (subscript "HX"), using the one-node approximation, i.e., constant temperature $T_s = \text{const.}$, is calculated by

$$\dot{Q}_{\text{HX}} = U A_{\text{HX}} \Delta T_{\log}(T_s, T_{s,\text{in}}) \tag{3}$$

$$= \left[1 - e^{\frac{-U A_{s,\text{HX}}}{\dot{m}_{\text{HTF}} c_{\text{p,HTF}}}} \right] \dot{m}_{\text{HTF}} \, c_{\text{p,HTF}} \, [T_{s,\text{in}} - T_s] \tag{4}$$

$\dot{m}_{\text{HTF}}$ denotes the mass flow of the heat transfer fluid, and $c_{\text{p,HTF}}$ its heat capacity. $T_{s,\text{in}}$ (and $T_{s,\text{out}}$) are the inlet (and outlet) temperatures of the sorption reactor fixed bed heat exchanger. The vapour mass flow is given by $\dot{m}_{\text{vap}} = m_0 \frac{dx_s}{dt}$. The outlet temperature of the heat transfer fluid is the output variable communicated to Simulink and given by

$$T_{s,\text{out}} = T_{s,\text{in}} - \left[1 - e^{\frac{-U A_{S,\text{HX}}}{\dot{m}_{\text{HTF}} c_{\text{p,HTF}}}} \right] [T_{s,\text{in}} - T_s] \,. \tag{5}$$

The evaporation/condensation rate is modelled linearly in the driving pressure difference, resulting in a vapor mass flow according to

$$\dot{m}_{\text{vap}}(T_2) = \frac{\beta A \left(p_{\text{vap}} - p_{\text{sat}}(T_2) \right)}{R_{\text{vap}} T_2}, \tag{6}$$

where R_{vap} denotes the gas constant of water vapor, βA is the mass transfer coefficient characterizing the linearized kinetics, and $p_{\text{sat}}(T_2)$ is the saturation vapor pressure for a given temperature T_2. For consistency, the vapour mass flow must be matched between the sorption store and the evaporator/condenser, for which an additional iterative solver is employed (Type22).

3.2   Heat Sink Modelled in Simulink

As heat sink, a sensible heat capacity with one thermal node and a counter-flow heat exchanger is modelled by

$$\frac{dT_b}{dt} = \frac{\dot{m}_{\text{HTF}} \, c_{\text{p,HTF}}}{m_b c_{\text{p,b}}} \left[1 - e^{\frac{-U A_{\text{HX}}}{\dot{m}_{\text{HTF}} c_{\text{p,HTF}}}} \right] [T_{s,\text{out}} - T_b] \tag{7}$$

$$T_{s,\text{in}} = T_{s,\text{out}} - \left[1 - e^{\frac{-U A_{\text{HX}}}{\dot{m}_{\text{HTF}} c_{\text{p,HTF}}}} \right] [T_{s,\text{out}} - T_b] \,. \tag{8}$$

T_b denotes the temperature of the heat sink, m_b its mass and $c_{p,b}$ its heat capacity.

3.3   Co-simulation

The interface of the co-simulation is situated physically in the circuit of the heat transfer fluid. Correspondingly, the inlet and outlet temperatures $T_{s,\text{in}}$ and $T_{s,\text{out}}$ of the sorption reactor heat exchanger are the variables communicated via the interface between Trnsys and Simulink.

4   Settings

The various solver parameters used in this study are given in Table 1.

4.1   Type155

Type155 establishes a communication between Trnsys and Matlab. On the Matlab side, a script is executed, where input and output variables and also all Trnsys-specific solver informations ("info array") are communicated. This communication interface has been exploited to construct co-simulation schemes corresponding to strong coupling and sequential loose coupling with different input extrapolation functions.

In order to build a strong coupling between Trnsys and Simulink, a Matlab-script was developed to start and stop Simulink simulations at each iteration, as detailed in [15]. In this case, the Simulink's simulation start and end time match the current and the next time step of the Trnsys simulation, respectively. In order to build a sequential loose coupling scheme, the script is modified such that the Simulink simulation is executed only once for each time step of the Trnsys Simulation. This was implemented such that Simulink gets input only after convergence of the Trnsys subsystem.

4.2   BCVTB and FMI

BCVTB allows to integrate simulation tools like Trnsys and Simulink directly as "simulator", or alternatively as FMU. To make use of FMI, an actual implementation for FMU export and import is required. An FMU was exported from the Trnsys subsystemusing an open-source tool [24], and imported in Simulink using the tool provided by Modelon [25]. As an example, a scheme for BCVTB is shown in Fig. 3. BCVTB provides a loose coupling co-simulation, where all input variables are extrapolated as constants from one communication point to the next one. FMI is in general a flexible approach, but the specific implementation discussed here allows for loose coupling only.

5   Results

We present results for the system behaviour of the case study, generated by the monolithic Trnsys simulation, and then focus on the accuracy of the various co-simulation interfaces considered, including different settings. Details concerning maximum deviation and computational costs are found in Table 1, some of the results have been presented also in [13–15].

5.1   System Behaviour

The results for the system behaviour produced by the reference simulation are shown in Fig. 4. The enthalpy released by the reaction increases the temperature of the heat storage up to roughly 39°C, which is in the further progress reduced again by the cooling through the thermal coupling to the heat sink, until the various temperatures eventually converge.

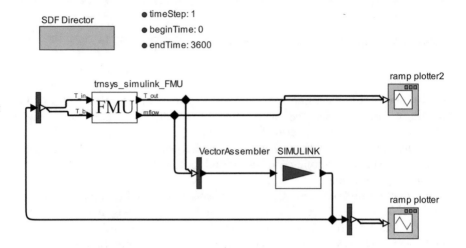

Fig. 3. Simulation scheme of BCVTB shown examparily for the particular case of Trnsys integrated as FMU and Simulink integrated as simulator [13].

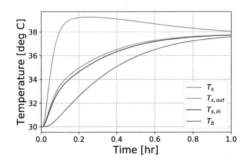

Fig. 4. Results for the system behaviour for the temperatures of the heat sink T_b, the heat storage T_s, the outlet of the heat storage $T_{s,out}$ and the inlet of the heat storage $T_{s,in}$. Generated by the reference monolithic simulation (modified from [13]).

5.2   Monolithic Simulation

To be able to evaluate the accuracies of the various co-simulation setups, we first discuss the accuracy of the monolithic simulation with default parameters. The deviation of the latter compared to the reference simulation (monolithic simulation with improved parameters, see Fig. 4) is shown in Fig. 5, left hand side. The initial peak in the deviation relates to the strong dynamics of the system in the early phase of discharging, where the state of charge of the storage is still high and hence the kinetics is sizeable. Parameters are given in Table 1.

5.3   Strong Coupling

In "callingMode = iterative component", Type155 calls a Matlab-script in each iteration. A custom script has been implemented such that Simulink is executed

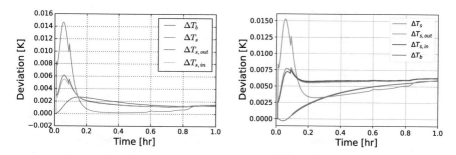

Fig. 5. Left: Deviation of the different temperatures of the monolithic Trnsys simulation with default parameters compared to the reference simulation. Right: Deviation of the different temperatures of the strong coupling scheme with default parameters compared to the reference simulation. The accuracy is of the same magnitude as the monolithic simulation, as expected for a strong coupling scheme.

in each iteration until convergence of the co-simulation in the respective time step is encountered. Such a procedure has integration errors under control and can be called a strong coupling co-simulation.

Figure 5, right hand side, shows the deviation of the results of the co-simulation based on Type155 in strong coupling scheme when compared to the results of the reference simulation. As expected, the deviation is of the same order as the one of the monolithic simulation itself. This verifies that the solver tolerances of the individual simulation tools are respected also by the co-simulation in this strong coupling scheme. Possible improvement of the accuracy in a strong coupling scheme has been investigated and was found to be possible to a high degree, but is not presented here in detail.

5.4   Loose Coupling: Sequential

Loose coupling implies that addional integration errors are introduced in the co-simulation compared to the monolithic one. The size of these errors depends on the scheme (parallel/sequential) and the input extrapolation function. We show results for the errors of sequential loose coupling co-simulation using a constant respectively linear input extrapolation function in Fig. 6. Both co-simulations have been generated using the Type155 interface. The constant input extrapolation function can be easily implemented using Type155 in the mode "callingMode = real-time controller". This way, the computational demand is significantly reduced (see Table 1).

The implementation of the linear extrapolation scheme requires "callingMode = real-time controller" and some more coding, which is briefly sketched in the following. In a sequential loose coupling scheme, only one side of the communication requires extrapolation. This is represented by $T_{s,in}$, communicated from Simulink to Trnsys, in the present case. Since Type155 in "callingMode = real-time controller" calls the Matlab-script only after convergence of the Trnsys solver, it cannot be used to modify input values before convergence. Hence,

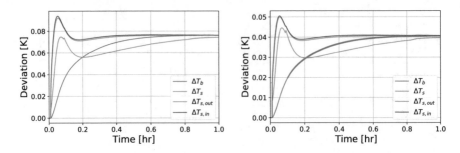

Fig. 6. Left: Deviation of the different temperatures of a sequential loose coupling scheme with default parameters, using a constant input extrapolation function. The scheme is implemented using Type155 in "callingMode = real-time controller". Right: Like left hand side, but applying a linear extrapolation for the input values. The deviation is reduced by almost 50% compared to the constant extrapolation.

the Matlab script is modified to mimic a "callingMode = real-time controller" behaviour, i.e., executing Simulink only after convergence even if called in each iteration, i.e., for Type155 in the setting "callingMode = iterative component". As a second step, values need to be stored to estimate the derivative and the extrapolation scheme has to be implemented. Finally, discontinuities like events should be accounted for to avoid sizeable extrapolation discrepancies. It is found that the deviation is reduced almost by a factor of two compared to the standard loose coupling scheme including a constant extrapolation of the input values. At the same time, the computational demand is increased by a factor of two (see Table 1), which, however, mainly originates in the mimicking of the "callingMode = real-time controller" behaviour (called but idle during the iterations) rather than the linear extrapolation itself.

5.5   Loose Coupling: Parallel

For the parallel loose coupling scheme we consider three different implementations. (i) Using BCVTB [18] as co-simulation master algorithm and integrating Trnsys and Simulink as actors according to Ptolemy theory. (ii) Like the first implementation, but integrating Trnsys as FMU. These results are shown in 7, left hand side. (iii) Integrating Trnsys as FMU directly in Simulink, without an addtional master algorithm. The Trnsys-FMU is generated using an open source tool [24]. These procedures implement a loose coupling scheme of parallel ("Jacobi") type, which allows a faster computation, however at the costs of a yet poorer accuracy as seen in Fig. 7, right hand side. Parallelization is straightforward, but not addressed in the present work.

A comparison including evaluation of different co-simulation interfaces is difficult if one of them is more accurate at higher computational costs. For a somewhat fair comparison, we suggest to tune the (solver/communication) parameters of one of the interfaces to match their accuracy (or optionally the computational demand). For that goal, we relax the solver parameters of the

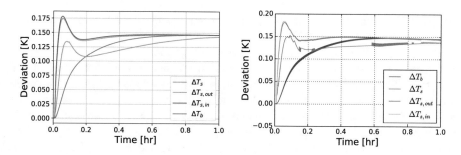

Fig. 7. Left: Deviation of the different temperatures of parallel loose coupling scheme with default parameters. Right: Results for the sequential loose coupling scheme with constant input extrapolation function (compare Fig. 6, left hand side) with parameters relaxed to match the accuracy of the parallel loose coupling.

sequential loose coupling (Type155) to match the accuracy of the parallel loose coupling (BCVTB/FMI) with default parameters. For the present use case, the time step can be increased by a factor of two, while the tolerance can be relaxed by a factor of a thousand. After this matching of the accuracy, the computational demand of the sequential loose coupling (Type155) co-simulation is still larger compared to the parallel one (BCVTB/FMI). The corresponding results for the accuracy are shown in Fig. 7, right hand side.

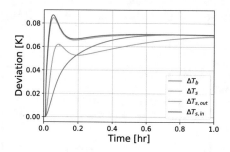

Fig. 8. Deviation of the different temperatures of the loose coupling scheme compared to the results of the strong coupling scheme (both in default setup). The curves differ only slightly from the ones shown in Fig. 6, left hand side, indicating that the strong coupling scheme can be used to estimate the accuracy of the loose coupling scheme.

6 General Method for Accuracy Estimation

In actual applications using co-simulation interfaces, there is typically no mono-lithic simulation available to be used as a reference to estimate the additional errors introduced by the communication of the co-simulation. While the errors of the strong coupling co-simulation respects the user-defined solver tolerances, the errors of loose coupling co-simulation are unknown in the first place. This is a major drawback of loose coupling co-simulations, which are otherwise often preferred over strong coupling because of their lower computational demands, simplicity and better availability.

We suggest to make use of the strong coupling co-simulation to estimate the inaccuracy of the loose coupling one. In this respect, a high flexibility of the interface (like Type155) is very valuable, as it allows to easily switch between strong and loose coupling. The deviation between the results of the loose coupling and the strong coupling co-simulation is shown in Fig. 8. Comparison to Fig. 6, left hand side, verifies that the errors are estimated to satisfactory precision. If required, the estimation of the error could be refined by choosing improved solver parameters for the strong coupling co-simulation, at the costs of a higher computational demand.

Further options exist for accuracy estimation in realistic cases without reference simultion [1]. Most of them rely on a series expansion involving the communication time step, or different versions of input extrapolation functions. Problems are known to potentially arise for specific cases, e.g. if algebraic loops occur [12]. Therefore, such information has to be provided as metadata in addition to the model of the subsystems.

7   Discussion

The accuracy and the computational demand of the various interfaces investigated are shown in Fig. 9. Obviously, increasing accuracy implies an increase of computational demand. Table 1 shows the settings of the various setups and detailed results.

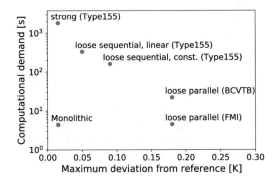

Fig. 9. An overview of different co-simulation interfaces, confronting accuracy and computational demand. "const." and "linear" refer to the respective input extrapolation function for loose coupling.

Table 1. The parameter settings for the different simulations as well as performance indicators for the (co-)simulations for accuracy and computational demands are given. "Ref." denotes the reference simulation, "Monolithic" a Trnsys simulation without co-simulation, "default" the default parameter setting ($t_{\text{step}} = 1\,\text{s}$ and tol. $= 10^{-6}$), "sequential" the sequential loose coupling scheme based on Type155, "parallel" the parallel loose coupling scheme based on BCVTB/FMI, "lin. extrapol." a linear extrapolation of the input variables. "relaxed" the relaxed solver parameters (to match the accuracy of the parallel loose coupling scheme). t_{step} denotes the Trnsys time step; "call.Mode" abbreviates "callingMode", which is the parameter of the Type155 governing the communication pattern; and "tol." gives the relative tolerance. ΔT_b, ΔT_s, $\Delta T_{s,\text{out}}$ and $\Delta T_{s,\text{in}}$ refer to the respective maximum deviation of these variables in units of Kelvin, and "comp." denotes the computational demand of the respective simulation. The simulations were executed on an Intel Xeon E5 6C/12T 2.2 GHz with 96 GB RAM.

	t_{step} [s]	call.Mode	tol.	ΔT_b	ΔT_s	$\Delta T_{s,\text{out}}$	$\Delta T_{s,\text{in}}$	comp. [s]
Ref.	0.1	-	10^{-7}	-	-	-	-	31
Monolithic (default)	1	-	10^{-6}	0.003	0.015	0.006	0.006	4
Strong (default)	1	0	10^{-6}	0.005	0.015	0.008	0.007	1800
Sequential (default)	1	10	10^{-6}	0.07	0.07	0.08	0.09	160
Sequential (lin. extrapol.)	1	10	10^{-6}	0.04	0.04	0.05	0.05	330
Sequential (relaxed)	2	10	10^{-3}	0.15	0.15	0.18	0.18	72
Parallel	1	-	10^{-6}	0.15	0.14	0.17	0.18	22

8   Conclusions

A method is presented to discuss and compare different co-simulation interfaces. Strong coupling and loose coupling in sequential scheme was implemented using the Type155 provided in the Trnsys standard library and a custom Matlab-script. Loose coupling in parallel scheme was implemented in three different version using BCVTB and FMI. The interfaces are discussed using a simple case study relating to recent research of thermal engineering, where a compact heat storage is modelled in Trnsys and a simple body acting as heat sink is modelled in Simulink.

Considering the handling of the interface, the Type155-based interface offers a lot of flexibility to the user, where also customized settings can be realized. BCVTB, on the other hand, shows limited flexibility, but offers out-of-the-box models and is convenient for rapid prototyping. FMI is a becoming a widely supported standard, and offers a lot of functionality. However, the actual functionality depends on the specific tool and implemented interface at hand. For the present case study, this functionality is limited to parallel loose coupling.

As expected, strong coupling requires higher computational resources while yielding higher accuracy. The accuracy of the strong coupling co-simulation results is comparable to the classical simulation, which verifies that the tolerance settings of the individual solvers can be trusted also for the co-simulation in this setup. In a fair comparison with tuned parameters to match the accuracy, the parallel loose coupling (FMI/BCVTB) was faster than the sequential one (Type155). The choice between the various interfaces is not obvious, but depends on the particular requirements on flexibilty, user friendliness, accuracy and computational resources.

Acknowledgements. The research leading to these results has received funding from the Austrian FFG Programme Energieforschung under grant agreement no. 845020, and the Research Studio Austria no. 844732. The authors acknowledge valuable discussions with W. Glatzl, H. Schranzhofer, G. Lechner, I. Hafner and E. Widl.

References

1. Gomes, C., Thule, C., Broman, D., Larsen, P.G., Vangheluwe, H.: Co-simulation: State of the art, CoRR, vol. abs/1702.0, February 2017
2. Lund, P.D., Lindgren, J., Mikkola, J., Salpakari, J.: Review of energy system flexibility measures to enable high levels of variable renewable electricity. Renew. Sustain. Energy Rev. **45**, 785–807 (2015)
3. Schweiger, G., Rantzer, J., Ericsson, K., Lauenburg, P.: The potential of power-to-heat in Swedish district heating systems. Energy (2017 in press). https://doi.org/10.1016/j.energy.2017.02.075
4. Bandhauer, T.M.: A Critical review of thermal issues in Lithium-ion batteries. J. Electrochem. Soc. **158**(3), R1 (2011)
5. Engel, G.: Sorption cold storage for thermal management of the battery of a hybrid vehicle. In: 12th International Renewable Energy Storage Conference (2018)
6. Allegrini, J., Orehounig, K., Mavromatidis, G., Ruesch, F., Dorer, V., Evins, R.: A review of modelling approaches and tools for the simulation of district-scale energy systems. Renew. Sustain. Energy Rev. **52**, 1391–1404 (2015)
7. Trcka, M., Hensen, J.L.M., Wetter, M.: Co-simulation of innovative integrated HVAC systems in buildings. J. Build. Perform. Simul. **2**(3), 209–230 (2009)
8. Wetter, M., Fuchs, M., Nouidui, T.S.: Design choices for thermofluid flow components and systems that are exported as Functional Mockup Units. In: 11th International Modelica Conference, no. iv, pp. 31–41 (2015)
9. Trcka, M.: Co-simulation for performance prediction of innovative integrated mechanical energy systems in buildings. Ph.d. thesis (2008)
10. Atam, E.: Current software barriers to advanced model-based control design for energy-efficient buildings. Renew. Sustain. Energy Rev. **73**, 1031–1040 (2017)
11. Mathias, O., Gerrit, W., Leon, U.: Life cycle simulation for a process plant based on a two-dimensional co-simulation approach. In: Computer Aided Chemical Engineering, vol. 37 (2015)
12. Arnold, M., Clauss, C., Schierz, T.: Error analysis and error estimates for Co-Simulation in FMI for model exhange and Co-Simulation V2.0. Arch. Mech. Eng. **LX**, 75–94 (2013)
13. Engel, G., Chakkaravarthy, A., Schweiger, G.: A methodology to compare different co-simulation interfaces: a thermal engineering case study. In: SimulTech 2017 - Proceedings of the 7th International Conference on Simulation and Modeling Methodologies, Technologies and Applications (2017)
14. Engel, G., Schweiger, G.: A comparison of co-simulation interfaces between Trnsys and Simulink: a thermal engineering case study. In: 9th Vienna International Conference on Mathematical Modelling (2018)

15. Engel, G., Chakkaravarthy, A., Schweiger, G.: Co-simulation between Trnsys and Simulink based on type155. In: Lecture Notes in Computer Science (including subseries Lecture Notes in Artificial Intelligence and Lecture Notes in Bioinformatics), vol. 10729 (2018)

16. Blochwitz, T., Otter, M., Arnold, M., Bausch, C., Clauß, C., Elmqvist, H., Junghanns, A., Mauss, J., Monteiro, M., Neidhold, T., Neumerkel, D., Olsson, H., Peetz, J.V., Wolf, S.: The functional mockup interface for tool independent exchange of simulation models. In: 8th International Modelica Conference 2011, pp. 173–184 (2009

17. Klein, S.A., Beckman, W.A., Duffie, J.A.: TRNYSYS - a transient simulation program (1976)

18. Wetter, M.: Co-simulation of building energy and control systems with the Building Controls Virtual Test Bed. J. Build. Perform. Simul. 4(3), 185–203 (2011)

19. Hafner, I., Heinzl, B., Roessler, M.: An investigation on loose coupling cosimulation with the BCVTB. SNE Simul. Notes Eur. 23, 45–50 (2013)

20. Engel, G., Asenbeck, S., Koell, R., Kerskes, H., Wagner, W., van Helden, W., Kerskes, H.: Simulation of a seasonal, solar-driven sorption storage heating system. J. Energy Storage 13, 40–47 (2017)

21. Köll, R., van Helden, W., Engel, G., Wagner, W., Dang, B., Jänchen, J., Kerskes, H., Badenhop, T., Herzog, T.: An experimental investigation of a realistic-scale seasonal solar adsorption storage system for buildings. Solar Energy 155, 388–397 (2017)

22. Glueckauf, E.: Theory of chromatography. Part 10 - Formulae for diffusion into spheres and their application to chromatography. Trans. Faraday Soc. 51, 1540–1551 (1955)

23. Dubinin, M.M.: Adsorption in micropores. J. Colloid Interface Sci. 23(4), 487–499 (1967)

24. Widl, E.: TRNSYS FMU Export Utility (2015). https://sourceforge.net/projects/trnsys-fmu/

25. Modelon: FMI toolbox for Matlab/Simulink (2017)

Author Index

© Springer Nature Switzerland AG 2019
M. S. Obaidat et al. (Eds.): SIMULTECH 2017, AISC 873, pp. 367–368, 2019.
https://doi.org/10.1007/978-3-030-01470-4

Printed in the United States
By Bookmasters